Introduction to Singularities

Shihoko Ishii

Introduction to Singularities

Second Edition

Springer

Shihoko Ishii
The University of Tokyo
Tokyo, Japan

ISBN 978-4-431-56871-1 ISBN 978-4-431-56837-7 (eBook)
https://doi.org/10.1007/978-4-431-56837-7

Original Japanese edition published by Springer-Verlag Tokyo, Tokyo, 1997

Softcover re-print of the Hardcover 2nd edition 2018

Printed on acid-free paper

This Springer imprint is published by the registered company Springer Japan KK part of Springer Nature
The registered company address is: Shiroyama Trust Tower, 4-3-1 Toranomon, Minato-ku, Tokyo 105-6005, Japan

Preface to the Second Edition

In the preface to the first edition, the author wrote that there were three remarkable developments in singularity theory after 1997, the publication year of the Japanese version of this book. The three are:

(1) applications of jet schemes to singularities;
(2) big progress in the Minimal Model Problem in higher-dimensional varieties;
(3) the positive characteristic method.

In the first edition of the English version published in 2014, the author reflected the results of (2), but not (1) or (3). In this new version, the author introduces (1) and (3) very briefly but enough to show the basic ideas which would help the further studies of the reader. This new version also fixes some typographic and language problems of the first edition of the English version. The author would like to express her hearty thanks to Kohsuke Shibata and Nguyen Duc Tam for helping her in this process.

Tokyo, Japan
February 2018

Shihoko Ishii

Preface to the First Edition

"When I started working on mathematics, singularities were very fancy for me" said Lê Dung Tráng, a Vietnamese mathematician, to the audience in his talk at Trieste in 1991. He had been a leading researcher of singularities for more than 20 years at that time. Hearing this, I understood that his encounter with singularities was a happy event.

On the contrary, my encounter with singularities was not happy. At that time, I was working on the moduli problem of compact varieties, and singularities were just "troublemakers". Even only one singularity disturbs my knowledge of the global structure of a variety; it disturbs the duality of the cohomologies, the vanishing of the cohomology groups, the stability of the projective variety and so on. But, it is no use to complain about such things. Sun Tzu said in *The Art of War*, "If you know the enemy and your army well, any of your fights would not be risky". You would not need to worry about the existence of singularities if you knew the singularities well and could distinguish bad singularities and not-so-bad singularities. I thought so and started to work on singularities. In this work I became fascinated by singularities. Gradually, singularities were becoming not an enemy but fancy things for me. In those days, M. Reid posed the Minimal Model Conjecture stating that there exists a minimal model when one admits the existence of mild singularities on the models, and then singularities got people's attention in a different way. Then, S. Mori, Y. Kawamata, M. Reid, V. Shokurov, J. Kollár and others' contributions solved the Minimal Model Conjecture for 3-folds. In these works, an important point was to classify mild singularities which are admitted by the minimal models.

Now the Minimal Model Conjecture is also going to be solved for higher dimensions. By this, we can expect development of research of singularities in different ways.

The goal of this book is based on the "classical" classification theory of 2-dimensional singularities, and shows a classification of singularities by regular differential forms. The title is *Introduction to Singularities*, but this book does not cover all topics on singularities; the topics were selected by the author's "biased" viewpoint.

I expect the reader to have some but not much knowledge on algebraic geometry and to wish to know what singularities of algebraic varieties are. In accordance with the editors' request, I introduced sheaves and also varieties briefly. But I could not show everything necessary in this book, so I quoted appropriate references at the necessary points.

The structure of this book is as follows:

In Chap. 2, we give preliminary knowledge on sheaves, algebraic varieties, analytic spaces and cohomologies. We start to work on singularities in Chap. 4. The reader can start to read from Chap. 4 and refer to Chaps. 2 and 3 as necessary.

In Chap. 4, we introduce the concept of singularities and resolution of singularities. We also show the concrete construction of resolution of the singularities of a non-degenerate hypersurface. In Chap. 5, we prepare the canonical sheaf and obtain an adjunction formula. In Chap. 6, we introduce a classification of singularities based on canonical sheaves. In Chap. 7, we study 2-dimensional singularities, looking at singularities from the viewpoint of the canonical sheaf. In Chap. 8, we study higher-dimensional singularities, in particular 3-dimensional singularities. In Chap. 9, we study deformations of singularities. We present tables that show how properties of singularities change under a deformation.

This book is the translation of a Japanese textbook by the author published in 1997. Since the publication of the Japanese version, we have had big developments in many unexpected directions in algebraic geometry related to singularities:

(1) applications of jet schemes to singularities;
(2) big progress in the Minimal Model Problem in higher-dimensional varieties;
(3) the positive characteristic method.

Since there are good expository works for each direction, the author did not explore these new developments deeply. The reader interested in these directions can refer to [EM] or [I14] for (1), [BCHM] for (2) and [ST] for (3).

The author thanks Ph.D. student Ngyuen Duc Tam for reading the manuscript through and giving valuable comments. The author also thanks Osamu Fujino for providing the author with the information on the recent development of the Minimal Model Problem.

Tokyo, Japan Shihoko Ishii
June 2014

Contents

Chapter 1
Warming Up

All great discoveries are made by mistakes.
(Young [Mur])

What do you imagine by the word "singularity"? Let us consider the following curves defined by the equations

(1) $y = x^3$, (2) $y^2 = x^3$, (3) $y^2 = x^3 + x^2$

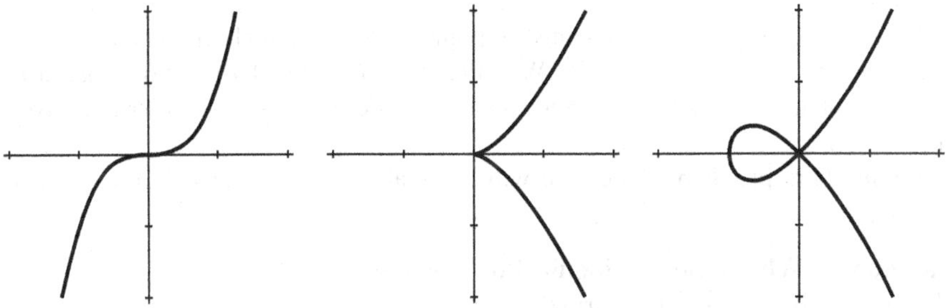

Where are the singular points in (1)? The origin may be singular, since it is an inflexion point. How about (2)? Everybody (even a child at elementary school who does not know the definition of an inflexion point) answers that the origin is singular. In (3), everybody would agree with that the origin is singular. Comparing with the "singularity" of (1), the singularity of the origin in (2) or (3) is much clearer. We call the latter ones singular points. Then, what are the differences between these points and the other non-singular points? You may notice that all points except for the origin in (2) and (3) have a small neighborhood which is a segment of a line. But the origin in (2) and (3) does not have such a neighborhood; any small neighborhood of (2) is a curve with a cusp and any small neighborhood of (3) consists of two lines crossing at the point. These cannot be regarded as a segment of a line. We can explain with an equation what it means that a neighborhood is regarded as a

S. Ishii, *Introduction to Singularities*,
https://doi.org/10.1007/978-4-431-56837-7_1

segment of line as follows: Denote the equation of a curve by $f(x, y) = 0$. If a partial differential $\frac{\partial f}{\partial y}(a, b)$ at a point (a, b) is not 0, then, by the Implicit Function Theorem, any point (a', b') on the curve in a neighborhood can be represented by $(a', h(a'))$ by using its x-coordinate, therefore the curve is regarded to be the same as the x-axis. The same holds true when we exchange the roles of x and y. Therefore, if $\operatorname{rank}\left(\frac{\partial f}{\partial x}(a, b), \frac{\partial f}{\partial y}(a, b)\right) = 1$, then the curve is regarded as a line in a neighborhood of (a, b). The converse of this statement also holds.

Indeed, as a line on an x, y plane is defined by $mx + ny = \ell$ ($m \neq 0$ or $n \neq 0$), we have $\operatorname{rank}\left(\frac{\partial f}{\partial x}(a, b), \frac{\partial f}{\partial y}(a, b)\right) = \operatorname{rank}(m, n) = 1$.

This equivalence holds for more variables and for even complex variables. Actually the Implicit Function Theorem for a complex analytic function of n variables shows the equivalence. Let X be the set of zeros of m analytic functions $f_1(x_1, x_2, \cdots, x_n) = f_2(x_1, x_2, \cdots, x_n) = \cdots = f_m(x_1, x_2, \cdots, x_n) = 0$ in $\mathbb{C}^n$ of dimension r. Then the following equivalence holds:

1. $\operatorname{rank}\begin{pmatrix} \frac{\partial f_1}{\partial x_1}(P) \cdots \frac{\partial f_1}{\partial x_n}(P) \\ \\ \frac{\partial f_m}{\partial x_1}(P) \cdots \frac{\partial f_m}{\partial x_n}(P) \end{pmatrix} = n - r.$
2. In a neighborhood of P, X is regarded as an open subset of r-dimensional complex space $\mathbb{C}^r$.

Now we call a point P a non-singular point if it has a neighborhood analytically isomorphic to an open subset of $\mathbb{C}^r$. We call a point P a singular point or singularity if it does not have such a neighborhood. (The exact definition of a singularity is given in Chap. 4.)

We list here problems which we will think about in this book. Some of them overlap.

Question 1. What kinds of singularities are possible?
More precisely, how are singularities made?

Question 2. How do properties of singularities change under any procedure?

Question 3. What kinds of singularities are considered "mild"?
Some kinds of singularities behave like a non-singular point. We think that these singularities are mild. We would like to recognize if it is a mild singularity or not, when we are given a singularity.

Question 4. Is there an algorithm/criterion to find a singularity with a certain property?
For example, an algorithm for finding a mild singularity.

Question 5. Is there a simple classification of singularities, in which each class has a distinguishable property?

Talking of Question 1, there is a big diversity in the making of singularities. In many cases, we can reduce it to typical constructions of singularities. Here, we show some typical ways to construct singularities. We have ways to get a new singularity by cutting, dividing, lifting, and contracting. We will see how Questions 2–5 appear in each case.

1.1 By Cutting — Hypersurface Singularity, Hypersurface Cut of a Singularity

Let x be a point of an analytic space U. Take a holomorphic function f such that $f(x) = 0$ and let $X = \{y \in U \mid f(y) = 0\}$, then X is again an analytic space containing x. In this way, constructing (X, x) from (U, x) is called "construction by cutting", and (X, x) is called a hypersurface cut of (U, x). In particular, if U is an open subset of $\mathbb{C}^n$, we call (X, x) a hypersurface singularity. Now, consider Question 2 in this case.

Question 2. How have the properties of (X, x) changed from the properties of (U, x)?
It is also possible to ask the following question:

Question 2′. Let (X, x) be a hypersurface singularity defined by the equation $f = 0$. Regard the coefficients of f as parameters and deform them. Then, what kind of singularities do we obtain?
This question is considered in Chap. 9.

Question 3. What kinds of singularities are mild? Of course, it depends on viewpoints. But in many cases we have a common group of mild singularities in various viewpoints. In the 2-dimensional case, the hypersurfaces in $\mathbb{C}^3$ defined by the following equations are shown as the mildest singularities:

$$\begin{array}{lll} A_n \textit{ type} & x^2 + y^2 + z^{n+1} = 0, & (n \geq 1) \\ D_n \textit{ type} & x^2 + y^2 z + z^{n-1} = 0, & (n \geq 4) \\ E_6 \textit{ type} & x^2 + y^3 + z^4 = 0, & \\ E_7 \textit{ type} & x^2 + y^3 + yz^3 = 0, & \\ E_8 \textit{ type} & x^2 + y^3 + z^5 = 0. & \end{array}$$

These are called Du Val singularities.

Question 4. How can we find a Du Val singularity among hypersurface singularities?
This question is considered in Sects. 7.3 and 7.5.

Question 5. How we can classify hypersurface singularities?
According to the classification introduced in Chap. 6, hypersurface singularities are classified into three classes.

1.2 By Dividing — Quotient Singularity, Quotient of a Singularity

Let $\mathrm{Aut}(X)$ be the automorphism group of an analytic space X.

Let G be a finite subgroup of $\mathrm{Aut}(X)$ and let X/G be the set of cosets by the following equivalence relation $\approx$:

$$x \approx y \iff y = \sigma x, \quad \exists \sigma \in G.$$

Then, we have the canonical analytic structure on X/G.

The construction X/G from X in this way is said to be a construction by dividing by G, and we call X/G the quotient space of X by G. In particular, if X is an open subset of $\mathbb{C}^n$, we call the singularities on X/G quotient singularities.

Question 2. How are properties of the singularities on X/G different from properties of the singularities on X?
Refer to Theorem 6.2.9 regarding this question.

Question 2′. How do the properties of a quotient singularity change under a deformation?
This question is treated in Chap. 8.

Question 3. What kinds of singularities are mild among quotient singularities?
Actually we can see that all quotient singularities are quite mild in Theorem 7.4.9 and Corollary 7.4.10.

Question 4. Is there a method to distinguish quotient singularities from other singularities?
Yes, for the 2-dimensional case, there is a method to distinguish it clearly (Theorem 7.4.19).

Question 5. To which class do quotient singularities belong?
Actually all quotient singularities belong to the class of mildest singularities according to the classification in Chap. 6 (Proposition 6.3.12, also Theorem 7.4.9).

1.3 By Lifting Up — Covering Singularity

The inverse of the above procedure is called "lifting up".

I.e., if we are given an analytic space Y, we construct X and finite group G appropriately such that $G \subset \mathrm{Aut}(x)$, $Y = X/G$.

Actually, properties of the singularities on X are not so different from those of the singularities on Y (Theorem 6.2.9), and sometimes better than those on Y. Therefore, we often study the singularities on X in order to study the singularities on Y. This is the case of the canonical cover of $\mathbb{Q}$-Gorenstein singularities (Definition 6.2.2).

1.4 By Contracting

For an analytic space Y and a sub-analytic space E on X, consider a morphism $f : Y \to X$ such that the restriction morphism $f\mid_{Y\setminus E} : Y \setminus E \to X \setminus f(E)$ is an isomorphism. If $\dim f(E) < \dim E$, we call X an analytic space contracting E in Y.

Question 2. How do properties of E reflect properties of X?
We will treat this question in many places in this book, for example Proposition 8.1.13, and Corollary 8.1.14.

Questions 3, 4. Can we characterize "mild" singularities in terms of properties of E?
For the 2-dimensional case, it goes quite well (Sects. 7.3 and 7.5).

Question 5. Does the classification of singularities relate to the classification of compact spaces E?
In particular, if E is irreducible and reduced, the classification of E by the Kodaira dimension corresponds to the classification in Chap. 6 of the singularity obtained by contracting E (Example 6.3.13).

As you see in this warming up, this book does not answer all questions we pose, but will give you motivation to work on the unsolved questions. The author will be happy if you use this book as a springboard and challenge new problems in this field.

Chapter 2
Sheaves, Algebraic Varieties and Analytic Spaces

Efforts and results are sometimes proportional.

In this chapter we introduce the concept of sheaves and introduce briefly algebraic varieties and analytic spaces which are in our interest. Readers who know these concepts well can skip this chapter.

2.1 Preliminaries on Category

First, we should clarify in which "world" we are working. A "world" is a category. We begin with the definition of a category.

Definition 2.1.1. *A category* $\mathscr{C}$ consists of a class $\mathrm{Ob}(\mathscr{C})$ of *objects* and sets $\mathrm{Hom}_{\mathscr{C}}(A, B)$ of *morphisms* for any two objects $A, B \in \mathrm{Ob}(\mathscr{C})$ satisfying the following:

(1) If $A \neq A'$ or $B \neq B'$, then $\mathrm{Hom}_{\mathscr{C}}(A, B) \cap \mathrm{Hom}_{\mathscr{C}}(A', B') = \emptyset$.
(2) For $A, B, C \in \mathrm{Ob}(\mathscr{C})$ there is a map

$$\begin{array}{ccc} \mathrm{Hom}_{\mathscr{C}}(A, B) \times \mathrm{Hom}_{\mathscr{C}}(B, C) & \longrightarrow & \mathrm{Hom}_{\mathscr{C}}(A, C) \\ (u, v) & \longmapsto & v \circ u \end{array}$$

which gives the composite of morphisms such that:

(2.a) For each object $A \in \mathrm{Ob}(\mathscr{C})$ there exists an identity morphism $1_A \in \mathrm{Hom}_{\mathscr{C}}(A, A)$, such that $u \circ 1_A = u$, $1_A \circ v = v$ hold for every $u \in \mathrm{Hom}_{\mathscr{C}}(A, B)$, $v \in \mathrm{Hom}_{\mathscr{C}}(C, A)$.

S. Ishii, *Introduction to Singularities*,
https://doi.org/10.1007/978-4-431-56837-7_2

(2.b) For every $u \in \mathrm{Hom}_{\mathscr{C}}(A, B)$, $v \in \mathrm{Hom}_{\mathscr{C}}(B, C)$ and $w \in \mathrm{Hom}_{\mathscr{C}}(C, D)$, the associative law $w \circ (v \circ u) = (w \circ v) \circ u$ holds.

For simplicity we sometimes denote $A \in \mathrm{Ob}(\mathscr{C})$ by $A \in \mathscr{C}$ and $u \in \mathrm{Hom}_{\mathscr{C}}(A, B)$ by $u : A \to B$.

Example 2.1.2. The following are examples of categories:

$\mathscr{C} =$ (Sets): The objects are all sets. $\mathrm{Hom}_{\mathscr{C}}(A, B) := \{u \mid u : A \to B \text{ map}\}$.
$\mathscr{C} =$ (Ab): The objects are all abelian groups. $\mathrm{Hom}_{\mathscr{C}}(A, B) = \{u \mid u : A \to B$ group homomorphism$\}$.
$\mathscr{C} =$ (Rings): The objects are all commutative rings. $\mathrm{Hom}_{\mathscr{C}}(A, B) = \{u \mid u : A \to B$ ring homomorphism$\}$.
$\mathscr{C} = (\mathrm{Mod}_R)$: The objects are all R-modules for a fixed commutative ring R. $\mathrm{Hom}_{\mathscr{C}}(A, B) = \{u \mid u : A \to B$ R-module homomorphism$\}$.
$\mathscr{C} =$ (Top): The objects are all topological spaces. $\mathrm{Hom}_{\mathscr{C}}(A, B) = \{u \mid u : A \to B$ continuous map$\}$.
$\mathscr{C} = (\mathrm{Ouv}_X)$: The objects are all open subsets of a fixed topological space X. $\mathrm{Hom}_{\mathscr{C}}(A, B)$ is $\{u : A \to B$ inclusion$\}$ if $A \subset B$ and is the empty set if $A \not\subset B$.

Here, the composite of morphisms in the category sense is the composite of maps in the usual sense. An identity morphism in the category sense is the identity map in the usual sense.

Definition 2.1.3. The *dual category* $\mathscr{C}^0$ of a category $\mathscr{C}$ is as follows:

(1) $\mathrm{Ob}(\mathscr{C}^0) = \mathrm{Ob}(\mathscr{C})$.
(2) For every $A, B \in \mathscr{C}$ we define

$$\mathrm{Hom}_{\mathscr{C}^0}(A, B) := \mathrm{Hom}_{\mathscr{C}}(B, A).$$

(3) For every $A, B, C \in \mathscr{C}$, $u \in \mathrm{Hom}_{\mathscr{C}^0}(A, B)$, $v \in \mathrm{Hom}_{\mathscr{C}^0}(B, C)$ the composite $v \circ u$ in the category $\mathscr{C}^0$ is defined by $u \circ v$ in the category $\mathscr{C}$.

We have notions "injective" and "surjective" for a map of sets. We generalize these to the common concepts for any "world".

Definition 2.1.4. In a category $\mathscr{C}$ a morphism $u : A \to B$ is called a *monomorphism* if two morphisms $v_1, v_2 : C \to A$ coincide ($v_1 = v_2$) whenever they satisfy $u \circ v_1 = u \circ v_2$. In this case (A, u) is called a subobject of B.

A morphism $u : A \to B$ is called an *epimorphism* if two morphisms $w_1, w_2 : B \to C$ coincide whenever they satisfy $w_1 \circ u = w_2 \circ u$. In this case (u, B) is called a quotient object of A.

If a morphism $u : A \to B$ is monomorphism and epimorphism, then we call it a *bijection*.

The concepts monomorphism and epimorphism in the categories $\mathscr{C} =$ (Sets), (Ab), (Mod_R) coincide with usual injections and surjections.

Definition 2.1.5. Let $u : A \to B$ be a morphism in a category $\mathscr{C}$. If a morphism $v : B \to A$ satisfies $u \circ v = 1_B$, $v \circ u = 1_A$, we call v the *inverse morphism* of u and denote it by u^{-1} ("inverse morphism" is unique). A morphism $u : A \to B$ which has the inverse morphism is called an *isomorphism*. In this case we say that A is *isomorphic* to B.

Note 2.1.6. An isomorphism is a bijection. But the converse does not hold in general. For example in the category $\mathscr{C} =$ (Top), we have the following example. Let X be a non-trivial topological space and Y the trivial topological space on X. Here, trivial topological space means that the open subsets are only the total space and the empty set $\emptyset$. In this case the identity map $id_X : X \to Y$ is a continuous map, therefore a morphism in $\mathscr{C} =$ (Top). And it is also a bijection, but the inverse morphism does not exist, so id_X is not an isomorphism.

Examples in Example 2.1.2: (Sets), (Ab), (Mod_R). A bijection is always an isomorphism.

Definition 2.1.7. If for two monomorphisms $u : B \to A$, $u' : B' \to A$ there exist $v : B \to B'$, $v' : B' \to B$ such that $u = u' \circ v$, $u' = u \circ v'$ we say that u and u' are *equivalent* and denote it by $u \sim u'$ ($\sim$ is an equivalence relation). For two epimorphisms $v : A \to C$, $v' : A \to C'$ we can also define an equivalence similarly.

Definition 2.1.8. For two objects $A_1, A_2 \in \mathscr{C}$, a triple (A, p_1, p_2) satisfying the following conditions $(DP1)$, $(DP2)$ is called *the direct product* of A_1 and A_2, and is denoted by $A_1 \times A_2$.

(DP1) There exist morphisms
$$\begin{array}{ccc} A & \xrightarrow{p_1} & A_1 \\ {\scriptstyle p_2}\downarrow & & \\ A_2 & & \end{array}$$

(DP2) For every morphism
$$\begin{array}{ccc} B & \xrightarrow{q_1} & A_1 \\ {\scriptstyle q_2}\downarrow & & \\ A_2 & & \end{array}$$
there exists a unique morphism $\varphi : B \to A$ such that the following diagram is commutative:

$$\begin{array}{ccc} B & \xrightarrow{q_1} & A_1 \\ {\scriptstyle q_2}\downarrow & \searrow^{\varphi} & \uparrow{\scriptstyle p_1} \\ A_2 & \xleftarrow[p_2]{} & A \end{array}$$

This morphism φ is denoted by (q_1, q_2).

For two objects $A_1, A_2 \in \mathscr{C}$ let $u_i : A_i \to A$ $(i = 1, 2)$ be morphisms to an object A. Then, the triple (A, u_1, u_2) is called a *direct sum* of A_1 and A_2 if (A, u_1, u_2) is the direct product of A_1 and A_2 in the dual category $\mathscr{C}^0$. The direct sum is denoted by $A_1 \oplus A_2$.

By $(DP2)$, a direct product and a direct sum are unique up to isomorphisms if they exist.

Definition 2.1.9. If a category $\mathscr{C}$ satisfies the following conditions, we call it an *additive category*.

(1) The set of morphisms $\mathrm{Hom}_{\mathscr{C}}(A, B)$ is an abelian group for every pair of objects $A, B \in \mathscr{C}$ and satisfies $(v_1 + v_2) \circ u = v_1 \circ u + v_2 \circ u, v \circ (u_1 + u_2) = v \circ u_1 + v \circ u_2$.
(2) The direct product $A \times B$ and the direct sum $A \oplus B$ exist for every pair of objects $A, B \in \mathscr{C}$.
(3) There exists an object $A \in \mathscr{C}$ such that $1_A = 0$. Such an object is called a null object and is denoted by 0. (The null object is unique up to isomorphisms.)

Definition 2.1.10. For a morphism $u : A \to B$ in an additive category $\mathscr{C}$, a pair (A', i) is called the *kernel* of u if it satisfies the following conditions:

(1) $i : A' \to A$ is a monomorphism.
(2) $u \circ i = 0$.
(3) For every morphism $v : C \to A$ such that $u \circ v = 0$, there exists a morphism $v' : C \to A'$ such that $v = i \circ v'$. (v' is unique by injectivity of i.)

In this case we denote (A', i) or A' by $\mathrm{Ker}\, u$.

On the other hand, we call (p, B') the *cokernel* of u, if (B', p) is the kernel of u in the dual category $\mathscr{C}^0$. In this case we denote (p, B') or B' by $\mathrm{Coker}\, u$.

We denote Ker (Coker u) by $\mathrm{Im}\, u$ and call it the *image* of u. On the other hand, we denote Coker (Ker u) by $\mathrm{Coim}\, u$ and call it the *coimage* of u.

By the definition the kernel $\mathrm{Ker}\, u$ is a subobject of A and the cokernel $\mathrm{Coker}\, u$ is a quotient object of B, the image $\mathrm{Im}\, u$ is a subobject of B and the coimage Coim is a quotient object of A. These are unique if these exist.

Proposition 2.1.11. *Assume that for morphism $u : A \to B$ in an additive category $\mathscr{C}$, there are the image* $(\mathrm{Im}\, u, i)$ *and the coimage* $(p, \mathrm{Coim}\, u)$. *Then there is a unique morphism $\overline{u} : \mathrm{Coim}\, u \to \mathrm{Im}\, u$ such that $u = i \circ \overline{u} \circ p$.*

Proof. By the definition of the kernel $(\mathrm{Ker}\, u, j)$ we have $u \circ j = 0$. Therefore, by the definition of the coimage $\mathrm{Coim}\, u = \mathrm{Coker}\, j$, there is a unique morphism $u' : \mathrm{Coim}\, u \to B$ such that $u = u' \circ p$. By the definition of the cokernel $(\mathrm{Coker}\, u, q)$ we have $q \circ u = q \circ u' \circ p = 0$. Here, as p is an epimorphism, it follows that $q \circ u' = 0$. Now, by the definition of the image $\mathrm{Im}\, u$ there exists a unique morphism $\overline{u} : \mathrm{Coim}\, u \to \mathrm{Im}\, u$ such that $u' = i \circ \overline{u}$.

Definition 2.1.12. A category $\mathscr{C}$ is called an *abelian category* if the following conditions are satisfied:

(1) $\mathscr{C}$ is an additive category.
(2) The kernel $\mathrm{Ker}\, u$ and the cokernel $\mathrm{Coker}\, u$ exist for every morphism $u : A \to B$.

(3) For every morphism $u : A \to B$ the morphism $\overline{u} : \mathrm{Coim}\, u \to \mathrm{Im}\, u$ in Proposition 2.1.11 is an isomorphism.

Example 2.1.13. (Ab), (Rings), (Mod_R) are additive categories. In the categories (Ab), (Mod_R) for every morphism $u : A \to B$ the usual kernel $\mathrm{Ker}\, u = \{x \in A \mid u(x) = 0\}$ and the usual cokernel $\mathrm{Coker}\, u = B/u(A)$ are the kernel $\mathrm{Ker}\, u$ and the cokernel $\mathrm{Coker}\, u$ in the category, respectively. These categories are both abelian categories.

Definition 2.1.14. In an abelian category $\mathscr{C}$ a sequence of morphisms

$$A \xrightarrow{f} B \xrightarrow{g} C$$

is called *exact* if the equality $\mathrm{Im}\, f = \mathrm{Ker}\, g$ holds.

Next we consider correspondences between two categories.

Definition 2.1.15. Let $\mathscr{C}, \mathscr{C}'$ be two categories. $F : \mathscr{C} \to \mathscr{C}'$ is called a *covariant functor* if $F = (F_{\mathrm{Ob}}, \{F_{A,B}\}_{A,B \in \mathscr{C}})$ consists of two kinds of correspondence $F_{\mathrm{Ob}} : \mathrm{Ob}(\mathscr{C}) \to \mathrm{Ob}(\mathscr{C}')$ and $F_{A,B} : \mathrm{Hom}_{\mathscr{C}}(A, B) \to \mathrm{Hom}_{\mathscr{C}'}\big(F_{\mathrm{Ob}}(A), F_{\mathrm{Ob}}(B)\big)$ $(\forall A, B \in \mathscr{C})$ satisfying the following (for simplicity we denote both F_{Ob} and $F_{A,B}$ by F):

(1) For every object $A \in \mathscr{C}$ the equality $F(1_A) = 1_{F(A)}$ holds.
(2) For morphisms $u : A \to B$, $v : B \to C$ in $\mathscr{C}$ it follows that

$$F(v \circ u) = F(v) \circ F(u).$$

On the other hand, $F : \mathscr{C} \to \mathscr{C}'$ is called a *contravariant functor* if $F = (F_{\mathrm{Ob}}, \{F_{A,B}\}_{A,B \in \mathscr{C}})$ consists of two correspondences $F_{\mathrm{Ob}} : \mathrm{Ob}(\mathscr{C}) \to \mathrm{Ob}(\mathscr{C}')$, $F_{A,B} : \mathrm{Hom}_{\mathscr{C}}(A, B) \to \mathrm{Hom}_{\mathscr{C}'}\big(F_{\mathrm{Ob}}(B), F_{\mathrm{Ob}}(A)\big)$ $(\forall A, B \in \mathscr{C})$ satisfying (1) and the following:

(2′) For two morphisms $u : A \to B$, $v : B \to C$ in $\mathscr{C}$, we have

$$F(v \circ u) = F(u) \circ F(v).$$

Clearly a contravariant functor $F : \mathscr{C} \to \mathscr{C}'$ is the same as the covariant functor $F : \mathscr{C}^0 \to \mathscr{C}'$. Henceforth when we say just a functor, we mean a covariant functor.

Definition 2.1.16. Let $\mathscr{C}, \mathscr{C}'$ be additive categories. We call a functor $F : \mathscr{C} \to \mathscr{C}'$ an *additive functor* if every pair of morphisms $u, v \in \mathrm{Hom}_{\mathscr{C}}(A, B)$ satisfy $F(u + v) = F(u) + F(v)$.

Definition 2.1.17. Let $\mathscr{C}$, $\mathscr{C}'$ be abelian categories and $0 \to A \to B \to C \to 0$ an exact sequence in $\mathscr{C}$. A functor $F : \mathscr{C} \to \mathscr{C}'$ is called an *exact functor* if the induced sequence $0 \to F(A) \to F(B) \to F(C) \to 0$ is again an exact sequence. If $0 \to F(A) \to F(B) \to F(C)$ is an exact sequence, we call F a *left exact functor* and if $F(A) \to F(B) \to F(C) \to 0$ is exact, we call F a *right exact functor.*

Example 2.1.18. (i) Let $\mathscr{C} = (\mathrm{Mod}_R)$ and fix an object $M \in \mathscr{C}$. Define $F : \mathscr{C} \to \mathscr{C}$ by $F(N) = N \otimes_R M$ for an object $N \in \mathscr{C}$ and by $F(u) : N \otimes M \to N' \otimes M, n \otimes m \mapsto u(n) \otimes m$ for a morphism $u : N \to N'$. Then, F is a right exact functor. In particular, if M is a flat R-module, F is an exact functor.

(ii) Let $\mathscr{C} = (\mathrm{Mod}_R)$ and fix an object $M \in \mathscr{C}$. Define $F : \mathscr{C} \to \mathscr{C}$ by $F(N) := \mathrm{Hom}_{\mathscr{C}}(M, N)$ for an object $N \in \mathscr{C}$ and by $F(u) : \mathrm{Hom}_{\mathscr{C}}(M, N) \to \mathrm{Hom}_{\mathscr{C}}(M, N'), f \mapsto u \circ f$ for a morphism $u : N \to N'$. Then F is a left exact functor. In particular, if M is a free M-module, F is an exact functor.

Now, we will consider a correspondence between two functors.

Definition 2.1.19. For two functors $F, G : \mathscr{C} \to \mathscr{C}'$, a correspondence $u : F \to G$ is called a *morphism of functors* (we also call it a *natural transformation*), if $u = \{u_A\}_{A \in \mathscr{C}}$ satisfies the following:

(1) For every $A \in \mathscr{C}$, $u_A : F(A) \to G(A)$ is a morphism of the category $\mathscr{C}'$.
(2) For a morphism $f : A \to B$ the following diagram is commutative:

$$\begin{array}{ccc} F(A) & \xrightarrow{u_A} & G(A) \\ {\scriptstyle F(f)}\downarrow & & \downarrow{\scriptstyle G(f)} \\ F(B) & \xrightarrow[u_B]{} & G(B). \end{array}$$

Definition 2.1.20. Two functors $F, G : \mathscr{C} \to \mathscr{C}'$ are called *isomorphic* if there exist morphisms $u : F \to G, v : G \to F$ of functors such that for every $A \in \mathscr{C}$ the equalities $v_A \circ u_A = 1_{F(A)}, u_A \circ v_A = 1_{G(A)}$ hold. In this case we write $F \simeq G$.

A functor $F : \mathscr{C} \to \mathscr{C}'$ is called an *equivalent functor* if there exists a functor $G : \mathscr{C}' \to \mathscr{C}$ such that $F \circ G : \mathscr{C}' \to \mathscr{C}', G \circ F : \mathscr{C} \to \mathscr{C}$ are both isomorphic to the identity functors.

Now we will introduce the direct system and inductive limit, which will be useful in the discussion on sheaves.

Definition 2.1.21. If a set Λ has a relation $\succ$ and it satisfies the following, we call $(\Lambda, \succ)$ a *directed set*:

(1) $\alpha \succ \alpha$ holds for every $\alpha \in \Lambda$.
(2) For $\alpha \succ \beta, \beta \succ \gamma$, we have $\alpha \succ \gamma$.
(3) For every $\alpha, \beta \in \Lambda$, there exists $\gamma \in \Lambda$, such that $\gamma \succ \alpha, \gamma \succ \beta$.

Definition 2.1.22. Let $\mathscr{C}$ be a category and Λ a directed set. Assume that every $\alpha \in \Lambda$ corresponds to $A_\alpha \in \mathscr{C}$ and every pair $\alpha, \beta \in \Lambda$ with $\alpha \succ \beta$ corresponds to $f_\alpha^\beta : A_\beta \to A_\alpha$. If the family $\mathbb{A} = (A_\alpha, f_\alpha^\beta)_{\alpha,\beta\in\Lambda}$ satisfies the following, then we call $\mathbb{A}$ an *inductive system* in $\mathscr{C}$.

(1) For every $\alpha \in \Lambda$, we have $f_\alpha^\alpha = 1_{A_\alpha}$.
(2) For every triple $\alpha \succ \beta \succ \gamma$, we have $f_\alpha^\beta \circ f_\beta^\gamma = f_\alpha^\gamma$.

Definition 2.1.23. Let $\mathbb{A} = (A_\alpha, f_\alpha^\beta)_{\alpha,\beta\in\Lambda}$ be an inductive system in $\mathscr{C}$. A pair $(A, f_\alpha)_{\alpha\in\Lambda}$ consisting of objects $A \in \mathscr{C}$ and morphisms $f_\alpha : A_\alpha \to A$ with the following properties is called the *inductive limit* of $\mathbb{A}$ (sometimes just A is called the inductive limit of $\mathbb{A}$):

(1) For every $\alpha \succ \beta$, the equality $f_\beta = f_\alpha \circ f_\alpha^\beta$ holds.
(2) If a pair $(B, g_\alpha)_{\alpha\in\Lambda}$ consisting of $B \in \mathscr{C}$ and $g_\alpha : A_\alpha \to B$ satisfy the condition (1), there exists a unique morphism $g : A \to B$ such that $g_\alpha = g \circ f_\alpha$ $(\alpha \in \Lambda)$.

We denote this A by $\varinjlim_{\alpha\in\Lambda} A_\alpha$.

By condition (2), the inductive limit is unique up to isomorphisms, if the inductive limit exists. The inductive limit does not exist in general. But some categories have the inductive limit for every inductive system. In this case, we say that the category is closed under the inductive limits. We will check whether the previous categories are closed under the inductive limits.

Example 2.1.24. (Ouv_X) is closed under the inductive limits. Indeed, for an inductive system $(U_\alpha, f_\alpha^\beta)_{\alpha,\beta\in\Lambda}$ put $U = \bigcup_{\alpha\in\Lambda} U_\alpha$ and let $f_\alpha : U_\alpha \to U$ be the inclusion map. Then, we can see that (U, f_α) satisfies conditions (1), (2).

Example 2.1.25. The category (Sets) is closed under the inductive limits. Indeed, let $(A_\alpha, f_\alpha^\beta)_{\alpha,\beta\in\Lambda}$ be an inductive system. Let $A' = \coprod_{\alpha\in\Lambda} A_\alpha$, and define a relation $\sim$ in A' as follows:

For $a, b \in A'$ we define $a \sim b$ if for $a \in A_\alpha, b \in A_\beta$ there exists $\gamma \in \Lambda$ such that $\gamma \succ \alpha, \gamma \succ \beta$ and $f_\gamma^\alpha(a) = f_\gamma^\beta(b)$. This relation $\sim$ is an equivalent relation, therefore we obtain the set of equivalence classes $A = A'/\sim$. Here, let $f_\alpha : A_\alpha \to A$ be the canonical morphism, then (A, f_α) satisfies (1), (2).

Example 2.1.26. Categories (Ab), (Ring), (Mod_R) are closed under the inductive limits. Indeed, let $(A_\alpha, f_\alpha^\beta)_{\alpha,\beta\in\Lambda}$ be an inductive system. First let $(A, f_\alpha)_{\alpha\in\Lambda}$ be the inductive limit of the set and then define canonical operations in A as follows: For addition note that arbitrary elements $\overline{a}, \overline{b}$ in A are represented as $f_\alpha(a)$, $f_\beta(b)$ for suitable elements $a \in A_\alpha$, $b \in A_\beta$. Take $\gamma \in \Lambda$ such that $\gamma \succ \alpha$, $\gamma \succ \beta$, then define $\overline{a} + \overline{b}$ by $f_\gamma\big(f_\gamma^\alpha(a) + f_\gamma^\beta(b)\big)$. This definition is independent of the choice of γ by the definition of an inductive system. The multiplication of a ring and scalar multiplication of a module are defined in the same way.

Example 2.1.27. The category (Top) is closed under the inductive limits. This is proved similarly; first take the inductive limit of sets and introduce the quotient topology.

Proposition 2.1.28. *Let $\mathscr{C}$ be a category in Examples 2.1.24–2.1.27. The inductive limit $A = \varinjlim_{\alpha\in\Lambda} A_\alpha$ has the following properties:*

(i) For every $a \in A$, there exist $\alpha \in \Lambda$ and $a_\alpha \in A_\alpha$ such that $f_\alpha(a_\alpha) = a$.
(ii) If $a_\alpha \in A_\alpha$ and $a_\beta \in A_\beta$ satisfy $f_\alpha(a_\alpha) = f_\beta(a_\beta)$, then there exists $\gamma \in \Lambda$ with $\gamma \succ \alpha$, $\gamma \succ \beta$ such that $f_\gamma^\alpha(a_\alpha) = f_\gamma^\beta(a_\beta)$.
(iii) In particular, if $\mathscr{C} = (\mathrm{Ab}), (\mathrm{Rings}), (\mathrm{Mod}_R)$, for $a \in A_\alpha$ such that $f_\alpha(a) = 0$ there exists $\beta \succ \alpha$ satisfying $f_\beta^\alpha(a) = 0$.

Proof. The proof is obvious from the construction of the categories in Examples 2.1.24–2.1.27.

Next we introduce the dual concept of the inductive limit.

Definition 2.1.29. Let $\mathscr{C}$ be a category and Λ a directed set. Assume that for every $\alpha \in \Lambda$, an object $B_\alpha \in \mathscr{C}$ is fixed and a morphism $g_\beta^\alpha : B_\alpha \to B_\beta$ is defined for $\alpha, \beta \in \Lambda$ such that $\alpha \succ \beta$. We call $\mathbb{B} = (B_\alpha, g_\beta^\alpha)_{\alpha,\beta\in\Lambda}$ a *projective system* of $\mathscr{C}$, if it satisfies the following:

(1) For every $\alpha \in \Lambda$, it follows that $g_\alpha^\alpha = 1_{B_\alpha}$.
(2) For $\alpha \succ \beta \succ \gamma$, the equality $g_\gamma^\alpha = g_\gamma^\beta \circ g_\beta^\alpha$ holds.

Definition 2.1.30. Let $\mathbb{B}$ be a projective system of a category $\mathscr{C}$. A pair $(B, g_\alpha)_{\alpha\in\Lambda}$ consisting of an object $B \in \mathscr{C}$ and morphisms $g_\alpha : B \to B_\alpha (\forall\alpha \in \Lambda)$ is called the *projective limit* of $\mathbb{B}$:

(1) For every $\alpha \succ \beta$, the equality $g_\beta = g_\beta^\alpha \circ g_\alpha$ holds.
(2) If a pair $(C, h_\alpha)_{\alpha\in\Lambda}$ consisting of objects $C \in \mathscr{C}$ and morphisms $h_\alpha : C \to B_\alpha (\forall\alpha \in \Lambda)$ satisfies (1), then there exists a unique $g : C \to B$ satisfying $h_\alpha = g_\alpha \circ g (\forall\alpha \in \Lambda)$.

The projective limit B is written as $\varprojlim_{\alpha\in\Lambda} B_\alpha$.

The projective limit is also unique up to isomorphisms if it exists. The categories above, where we checked that these are closed under the inductive limits, are also closed under the projective limits. This is left to the reader as an exercise.

2.2 Sheaves on a Topological Space

Let X be a topological space and $\mathscr{C}$ a category. Let (Ouv_X) be the category of open subsets of X introduced in the previous section.

Definition 2.2.1. $\mathscr{F}$ is called a *$\mathscr{C}$-presheaf* on X if it is a contravariant functor $\mathscr{F} : (\mathrm{Ouv}_X) \to \mathscr{C}$, i.e., the following hold:

(1) For an open subset $U \subset X$ an object $\mathscr{F}(U) \in \mathscr{C}$ is determined.
(2) Let $V \subset U$ be open subsets of X. For the inclusion map i_U^V there exists a morphism $\mathscr{F}(i_U^V) : \mathscr{F}(U) \to \mathscr{F}(V)$ in the category $\mathscr{C}$ and it satisfies the following (henceforth, for simplicity we write ρ_V^U for $\mathscr{F}(i_U^V)$ and call it the *restriction morphism*):

(2-1) For every open subset $U \subset X$, the equality $\rho^U_U = 1_{\mathscr{F}(U)}$ holds;
(2-2) For open subsets $W \subset V \subset U \subset X$, the equality $\rho^U_W = \rho^V_W \circ \rho^U_V$ holds.

(3) In particular, if $\mathscr{C}$ is an additive category, then the equality $\mathscr{F}(\phi) = 0$ holds.

If the category $\mathscr{C}$ is (Sets), (Ab), (Rings), (Mod_R), we call $\mathscr{F}$ a *presheaf of sets*, a *presheaf of ableian groups* and so on.

Definition 2.2.2. Let $\mathscr{F}, \mathscr{G}$ be $\mathscr{C}$-presheaves on X. If $\varphi : \mathscr{F} \to \mathscr{G}$ is a morphism of functors, then we call it a *morphism of $\mathscr{C}$-presheaves*. By this, we obtain a category whose objects are $\mathscr{C}$-presheaves and the set of morphisms $\mathrm{Hom}(\mathscr{F}, \mathscr{G})$ is the set of morphisms of $\mathscr{C}$-presheaves. This category is denoted by $\mathrm{PFais}^{\mathscr{C}}_X$.

Next we introduce the notion of sheaves that have better properties than presheaves.

Definition 2.2.3. A $\mathscr{C}$-presheaf $\mathscr{F}$ on X is called a *$\mathscr{C}$-sheaf* if for every open subset $U \subset X$ and an open covering $U = \bigcup_{i \in I} U_i$ of U the following hold:

(1) If $s, s' \in \mathscr{F}(U)$ satisfy $\rho^U_{U_i}(s) = \rho^U_{U_i}(s')$ $(\forall i \in I)$, then the equality $s = s'$ holds.
(2) If $s_i \in \mathscr{F}(U_i)$ $(i \in I)$ satisfy $\rho^{U_i}_{U_i \cap U_j}(s_i) = \rho^{U_j}_{U_i \cap U_j}(s_j)$, then there exists $s \in \mathscr{F}(U)$ such that $\rho^U_{U_i}(s) = s_i$ $(\forall i \in I)$.

Henceforth, for a sheaf $\mathscr{F}$, we denote $\rho^U_V(s)$ by $s|_V$ and $\mathscr{F}(U)$ sometimes by $\Gamma(U, \mathscr{F})$. An element $s \in \Gamma(U, \mathscr{F})$ is called a *section* of $\mathscr{F}$ on U.

For two $\mathscr{C}$-sheaves $\mathscr{F}$, $\mathscr{G}$ a morphism $\mathscr{F} \to \mathscr{G}$ of $\mathscr{C}$-presheaves is called a *morphism of $\mathscr{C}$-sheaves* if $\mathscr{F}$ and $\mathscr{G}$ are both $\mathscr{C}$-presheaves.

By this we have the category $\mathscr{C}$-sheaves on X. This category is denoted by $\mathrm{Fais}^{\mathscr{C}}_X$.

For a topological space X, the set Ouv_X is a directed set by defining the relation $\succ$ by the inclusion $\subset$. A subset $\mathrm{Ouv}_{X,x} = \{U \mid U \text{is an open subset such that } x \in U\}$ of Ouv_X is also a directed set by the same $\succ$. By (2-1), (2-2) in Definition 2.2.1, the system $(\mathscr{F}(U), \rho^U_V)_{U,V \in \mathrm{Ouv}_{X,x}}$ is an inductive system.

If $\mathscr{C}$ is (Ab), (Rings), (Mod_R), by Example 2.1.26 for a $\mathscr{C}$-presheaf $\mathscr{F}$ the inductive limit $\varinjlim_{U \in \mathrm{Ouv}_{X,x}} \mathscr{F}(U)$ exists. This is denoted by $\mathscr{F}_x$ and is called the *stalk* of $\mathscr{F}$ at x.

For a section $s \in \mathscr{F}(U)$, the image of s in $\mathscr{F}_x$ is denoted by s_x. Here, we define the *support* $\mathrm{Supp}\,(s)$ of s as follows:

$$\mathrm{Supp}\,(s) = \{x \in U \mid s_x \neq 0\},$$

where 0 is the zero of the abelian group or the R-module $\mathscr{F}_x$.

We have the following property about the support. From now on, $\mathscr{C}$ is always one of (Ab), (Rings), (Mod_R).

Proposition 2.2.4. *Take $\mathscr{F} \in \mathrm{PFais}^{\mathscr{C}}_X$. For an open subset $U \subset X$, the support* $\mathrm{Supp}\,(s)$ *of a section $s \in \mathscr{F}(U)$ is a closed subset of U.*

Proof. If there is no $x \in U$ such that $s_x = 0$, then the support is $\mathrm{Supp}(s) = U$ and the assertion is clear. If there is a point $x \in U$ such that $s_x = 0$, then there is an open subset V such that $x \in V \subset U$ and $s|_V = 0$, because $\mathscr{F}_x$ is the inductive limit and therefore we can use Proposition 2.1.28 (iii). By this we have $s_y = 0$ $(\forall y \in V)$.

Here, we consider when two sections coincide.

Proposition 2.2.5. *Let* $\mathscr{F} \in \mathrm{Fais}_X^{\mathscr{C}}$. *For every open subset* $U \subset X$ *and* $s, s' \in \Gamma(U, \mathscr{F}) = \mathscr{F}(U)$, *the following are equivalent:*

(i) $s = s'$,
(ii) $s_x = s'_x$ $(\forall x \in U)$.

Proof. (1) $\Rightarrow$ (2) is obvious. If we assume (2), by Proposition 2.1.28 (ii) for every $x \in U$ there exists an open neighborhood $U_x \subset U$ of x such that $s|_{U_x} = s'|_{U_x}$. As $U = \bigcup_{x \in U} U_x$ and $\mathscr{F}$ is a sheaf, we obtain $s = s'$ by Definition 2.2.3 (1).

Proposition 2.2.6. *For* $\mathscr{F} \in \mathrm{PFais}_X^{\mathscr{C}}$ *there exists* $\overline{\mathscr{F}} \in \mathrm{Fais}_X^{\mathscr{C}}$ *satisfying the following conditions (this is called the sheafification of* $\mathscr{F}$ *and it is unique up to isomorphisms):*

1. *There exists a morphism* $\theta : \mathscr{F} \to \overline{\mathscr{F}}$ *of presheaves.*
2. *If* $\varphi : \mathscr{F} \to \mathscr{G}$ *is a morphism of* $\mathscr{C}$*-sheaves to a* $\mathscr{C}$*-sheaf* $\mathscr{G}$, *there exists a unique morphism* $\psi : \overline{\mathscr{F}} \to \mathscr{G}$ *such that* $\varphi = \psi \circ \theta$.

Proof. If there exists $\overline{\mathscr{F}}$ satisfying (1) and (2), it is unique up to isomorphisms by the uniqueness of ψ in (2). For an open subset $U \subset X$ we define:

$$\overline{\mathscr{F}}(U) = \{s = \{s(x)\}_{x \in U} \in \prod_{x \in U} \mathscr{F}_x \mid \text{for every } x \in U \text{ there exist}$$
$$\text{an open neighborhood } V \text{ of } x \text{ and } t \in \mathscr{F}(V) \text{ satisfying } s(y) = t_y \ (\forall y \in V)\}.$$

For open subsets $V \subset U$ let $\rho_V^U : \overline{\mathscr{F}}(U) \to \overline{\mathscr{F}}(V)$ be the canonical map induced from the projection $\prod_{x \in U} \mathscr{F}_x \to \prod_{x \in V} \mathscr{F}_x$.

Then, clearly $\overline{\mathscr{F}}$ is a $\mathscr{C}$-presheaf. It is easily proved that it is also a $\mathscr{C}$-sheaf. Let us check condition (1) of Definition 2.2.3 and leave (2) to the reader.

Assume $\rho_{U_i}^U(s) = \rho_{U_i}^U(s')$ $(\forall i)$ for $s, s' \in \overline{\mathscr{F}}(U)$. Then by the definition of $\overline{\mathscr{F}}$, it follows that $s(x) = s'(x)$ $(\forall x \in U)$, therefore we have $s = s'$.

Now we will check (1) and (2) in the proposition. The map $\theta : \mathscr{F} \to \overline{\mathscr{F}}$ in (1) is given by $\mathscr{F}(U) \to \overline{\mathscr{F}}(U)$; $s \mapsto \{s_x\}_{x \in U}$. In (2), for $\varphi_U : \mathscr{F}(U) \to \mathscr{G}(U)$ we define ψ by $\psi_U : \overline{\mathscr{F}}(U) \to \mathscr{G}(U)$, $\{s(x)\}_{x \in U} \mapsto \{\varphi_{U_i}(t_i)\}_i$, where $\{U_i\}$ is an open covering of U and $t_i \in \mathscr{F}(U_i)$ is a section satisfying $s(x) = t_{ix}$ $(\forall x \in U_i)$. Here, by Proposition 2.2.5 we have $\rho_{U_i \cap U_j}^{U_i}(\varphi_{U_i}(t_i)) = \rho_{U_i \cap U_j}^{U_j}(\varphi_{U_j}(t_j))$ $(\forall i, j)$. Hence, as $\mathscr{G}$ is a sheaf, we obtain an element of $\mathscr{G}(U)$ by $\{\varphi_{U_i}(t_i)\}_i$. It is clear that ψ is a morphism of presheaves satisfying $\varphi = \psi \circ \theta$. On the other hand, there is no other map ψ' satisfying $\varphi = \psi' \circ \theta$.

Let $\overline{\mathscr{F}}$ be the sheafification of $\mathscr{F}$ on X, then the equality $\overline{\mathscr{F}}_x = \mathscr{F}_x$ holds for every $x \in X$ by the construction.

Definition 2.2.7. Let $\mathscr{F}$ be an (Ab)-sheaf on X and $\mathscr{G}$ a $\mathscr{C}$-sheaf on X. $\mathscr{F}$ is called a *subsheaf* of $\mathscr{G}$ if there is a morphism $u : \mathscr{F} \to \mathscr{G}$ of (Ab)-sheaves such that $u_U : \mathscr{F}(U) \to \mathscr{G}(U)$ is the inclusion map of a subgroup for every open subset $U \subset X$. Let $\mathscr{F}$ be a subsheaf of $\mathscr{G}$. For an open subset $U \subset X$ define $\mathscr{H}(U) = \mathscr{G}(U)/\mathscr{F}(U)$ and for $V \subset U$ define $\mathscr{H}(U) \to \mathscr{H}(V)$ by the induced map from $\mathscr{G}(U) \to \mathscr{G}(V)$. Then $\mathscr{H}$ is an (Ab)-presheaf. The sheafification of this sheaf is denoted by $\mathscr{G}/\mathscr{F}$ and is called the *quotient sheaf* of $\mathscr{G}$ by $\mathscr{F}$.

Definition 2.2.8. Let $\varphi : \mathscr{F} \to \mathscr{G}$ be a morphism of $\mathscr{C}$-sheaves. We will define the *kernel* and *image* of the morphism φ. A subsheaf $\mathscr{K}er\,\varphi$ of $\mathscr{F}$ and a subsheaf $\mathscr{I}m\,\varphi$ of $\mathscr{G}$ are defined as follows:

For an open subset $U \subset X$

$$\begin{aligned}\mathscr{K}er\,\varphi(U) &:= \{s \in \mathscr{F}(U) \mid \varphi(s)_x = 0,\ \forall x \in U\},\\ \mathscr{I}m\,\varphi(U) &:= \{t \in \mathscr{G}(U) \mid \text{for every } x \in U \text{ there are } x \in V_x \subset U \text{ and}\\ &\qquad s \in \mathscr{F}(V_x) \text{ such that } t_x = \varphi_x(s_x)\}.\end{aligned}$$

For open subsets $V \subset U$ the restriction maps are defined as the canonical map induced from the restriction maps of sheaves $\mathscr{F}$ and $\mathscr{G}$, respectively.

Then, both $\mathscr{K}er\,\varphi$, $\mathscr{I}m\,\varphi$ are subsheaves. These are called the *kernel* of φ and the *image* of φ, respectively. The quotient sheaf $\mathscr{G}/\mathscr{I}m\,\varphi$ of $\mathscr{G}$ by $\mathscr{I}m\,\varphi$ is called the *cokernel* and denoted by Coker φ.

The kernel, image and cokernel above are the kernel, image and cokernel, respectively, in the category $\mathrm{Fais}_X^{(\mathrm{Ab})}$. Therefore, we can define exactness of a sequence of morphisms of (Ab)-sheaves (Definition 2.1.14). Here, if $\mathscr{C} = (\mathrm{Mod}_R)$, the kernel, image and cokernel are also $\mathscr{C}$-sheaves.

Proposition 2.2.9. *For morphisms $\varphi : \mathscr{F} \to \mathscr{G}$, $\psi : \mathscr{G} \to \mathscr{H}$ of (Ab)-sheaves, the following are equivalent:*

(i) $0 \longrightarrow \mathscr{F} \xrightarrow{\varphi} \mathscr{G} \xrightarrow{\psi} \mathscr{H} \longrightarrow 0$ *is an exact sequence.*

(ii) For every $x \in X$ the sequence $0 \longrightarrow \mathscr{F}_x \xrightarrow{\varphi_x} \mathscr{G}_x \xrightarrow{\psi_x} \mathscr{H}_x \longrightarrow 0$ is an exact sequence of abelian groups.

Proof. In general, for a subsheaf $\mathscr{I} \subset \mathscr{K}$, the equality $\mathscr{I} = \mathscr{K}$ is equivalent to the equalities $\mathscr{I}_x = \mathscr{K}_x$ for all $x \in X$. On the other hand, for a morphism $\alpha : \mathscr{A} \to \mathscr{B}$ we have $(\mathscr{K}er\,\alpha)_x = \mathrm{Ker}\,(\alpha_x)$, $(\mathscr{I}m\,\alpha)_x = \mathrm{Im}\,(\alpha_x)$, which yields the proposition.

Example 2.2.10. Let k be a field and X a topological space. For an open subset $U \subset X$ we define $\mathscr{F}_{X,k}(U) := \{f \mid f : U \to k \text{ a map}\}$. For open subsets $V \subset U$, define $\rho_V^U : \mathscr{F}_{X,k}(U) \to \mathscr{F}_{X,k}(V)$ as the restriction $f \mapsto f|_V$ of a map. Then $\mathscr{F}_{X,k}$ is a sheaf of commutative rings on X.

Example 2.2.11. Let $\mathbb{C}^n$ be the topological space with the usual metric and $W \subset \mathbb{C}^n$ an open subset. For an open subset $U \subset W$ define $\mathscr{O}_W^{\text{hol}}(U) := \{f \mid f$ is a holomorphic function on $U\}$. For open subsets $V \subset U$, define $\rho_V^U : \mathscr{O}_W^{\text{hol}}(U) \to \mathscr{O}_W^{\text{hol}}(V)$ as the restriction $f \mapsto f|_V$ of a map. Then $\mathscr{O}_W^{\text{hol}}$ is a sheaf of commutative rings on W. Moreover, it is a sheaf of $\mathbb{C}$-algebras and a subsheaf of $\mathscr{F}_{W,\mathbb{C}}$.

Example 2.2.12. Let $W \subset \mathbb{C}^n$ be an open subset and $X \subset W$ the zero set of a finite number of holomorphic functions $f_1, \ldots, f_r$ on W. Such a subset X is called an *analytic set*. For an open subset $U \subset W$, define $\mathscr{I}(U) := \{f \mid f$ is a holomorphic function on U such that $f|_{X \cap U} = 0\}$. Define ρ_V^U as the restriction of functions as before. Then $\mathscr{I}$ is a sheaf of abelian groups and it is a subsheaf of $\mathscr{O}_W^{\text{hol}}$. Here, as $\mathscr{I}(U)$ is an ideal of $\mathscr{O}_W^{\text{hol}}(U)$, the presheaf defined by $\mathscr{O}_W^{\text{hol}}(U)/\mathscr{I}(U)$ is a presheaf of commutative rings. Therefore the sheafification $\mathscr{O}_W^{\text{hol}}/\mathscr{I}$ is a sheaf of commutative rings and, moreover, a sheaf of $\mathbb{C}$-algebras. This is a sheaf on W, but it is also considered as a sheaf on X. Indeed a subset V of X is represented as $V = U \cap X$ by using an open subset U of W. Define $\mathscr{O}_X^{\text{hol}}(V) := \mathscr{O}_W^{\text{hol}}/\mathscr{I}(U)$. Then the right-hand side is independent of a choice of an open subset U, therefore $\mathscr{O}_X^{\text{hol}}$ is a sheaf of commutative rings on X and a subsheaf of $\mathscr{F}_{X,\mathbb{C}}$. Here, the pair $(X, \mathscr{O}_X^{\text{hol}})$ is called a *reduced analytic local model* and $\mathscr{O}_X^{\text{hol}}$ the structure sheaf. Actually $\mathscr{O}_X^{\text{hol}}(U)$ is a reduced ring, i.e., it does not have a nilpotent element. This is equivalent to $\sqrt{\mathscr{I}} = \mathscr{I}$ and it is proved by the Hilbert Nullstellensatz (see, for example [Mu, I, §2 Theorem 1]). For a subsheaf $\mathscr{J} \subset \mathscr{I}$ such that $\mathscr{J}(U)$ is a $\mathscr{O}_W^{\text{hol}}(U)$-subideal of $\mathscr{I}(U)$ and satisfies $\sqrt{\mathscr{J}(U)} = \mathscr{I}(U)$ for every open subset U, the pair $(X, \mathscr{O}_W^{\text{hol}}/\mathscr{J})$ is called a (not necessarily reduced) *analytic local model*, $\mathscr{O}_W^{\text{hol}}/\mathscr{J}$ the structure sheaf and $\mathscr{J}$ the defining ideal. Here we note that pairs with the different structure sheaves are considered as different analytic local models even if the topological space X is the same. We also note that $X = \{x \in W \mid f(x) = 0,\ \forall f \in \mathscr{J}(W)\}$ for $\mathscr{J}$ as above.

Example 2.2.13. For an algebraically closed field k an algebraic set in k^n is defined as the zero set of $f_1, \ldots, f_r \in k[x_1, \ldots, x_n]$. As the set of all algebraic sets satisfies the axiom of closed subsets, we can introduce a topology on k^n. This topology is called *Zariski topology* on k^n. For an open subset $U \subset k^n$ in this topology we define $\mathscr{O}_{k^n}(U) := \{f/g \mid f, g \in k[x_1, \ldots, x_n], g \neq 0 \text{ on } U\}$. Then we obtain the canonical homomorphism $\rho_V^U : \mathscr{O}_{k^n}(U) \to \mathscr{O}_{k^n}(V)$ for every pair of open subsets $V \subset U$. By these definitions, $\mathscr{O}_{k^n}$ is a sheaf of commutative rings on k^n and, moreover, a sheaf of k-algebras. When in particular $k = \mathbb{C}$, we denote this sheaf by $\mathscr{O}_{\mathbb{C}^n}^{\text{alg}}$ in order to distinguish it from $\mathscr{O}_{\mathbb{C}^n}^{\text{hol}}$.

Example 2.2.14. Let $X \subset k^n$ be an algebraic set. Introduce a topology on X induced from the Zariski topology on k^n. For $U \subset k^n$ we define $\mathscr{I}(U) := \{\varphi \in \mathscr{O}_{k^n}(U) \mid \varphi|_{U \cap X} = 0\}$, then $\mathscr{I}$ is a subsheaf of $\mathscr{O}_{k^n}$. In the same way as in Example 2.2.12 the quotient sheaf $\mathscr{O}_{k^n}/\mathscr{I}$ is a sheaf of commutative rings on X. A pair $(X, \mathscr{O}_{k^n}/\mathscr{I})$ is called a *reduced affine variety* and $\mathscr{O}_{k^n}/\mathscr{I}$ the structure sheaf. Actually in this case also the structure sheaf is reduced by the Hilbert Nullstellensatz. Let $\mathscr{J} \subset \mathscr{I}$ be a subsheaf such that for every open subset $U \subset k^n$ $\mathscr{J}(U)$ is an $\mathscr{O}_{k^n}(U)$-subideal of $\mathscr{I}(U)$ such that $\sqrt{\mathscr{J}(U)} = \mathscr{I}(U)$. In this case a pair $(X, \mathscr{O}_{k^n}/\mathscr{J})$ is called a (not necessarily reduced) *affine variety* over k, $\mathcal{O}_{k^n}/\mathcal{J}$ the structure sheaf and $\mathcal{J}$ defining ideal. In the same way as in Example 2.2.12, we can check that $X = \{x \in k^n \mid f(x) = 0, \forall f \in \mathcal{J}(k^n)\}$.

2.3 Analytic Spaces, Algebraic Varieties

In this section we introduce an analytic space and an algebraic variety.

Definition 2.3.1. Let X be a topological space, $\mathcal{O}_X$ a sheaf of commutative algebra on X. A pair $(X, \mathcal{O}_X)$ is called a *ringed space*. In particular, if the germ $\mathcal{O}_{X,x}$ is a local ring for every $x \in X$, we call $(X, \mathcal{O}_X)$ a *locally ringed space*. A pair $(\varphi, \varphi^*) : (X, \mathcal{O}_X) \to (Y, \mathcal{O}_Y)$ is called a *morphism of ringed spaces* if $\varphi : X \to Y$ is a continuous map and $\varphi^* : \mathcal{O}_Y \to \varphi_*\mathcal{O}_X$ is a morphism of sheaves of rings. Here, $\varphi_*\mathcal{O}_X$ is the sheaf of commutative rings defined by $\varphi_*\mathcal{O}_X(U) = \Gamma(\varphi^{-1}(U), \mathcal{O}_X)$ for open subset $U \subset Y$. We call $(\varphi, \varphi^*) : (X, \mathcal{O}_X) \to (Y, \mathcal{O}_Y)$ a *morphism of locally ringed spaces*, if it is a morphism of ringed spaces and the homomorphism $\mathcal{O}_{Y,f(x)} \xrightarrow{\varphi_x^*} \mathcal{O}_{X,x}$ induced from φ^* is a local homomorphism, i.e., $\varphi_x^{*-1}(\mathfrak{m}_{X,x}) = \mathfrak{m}_{Y,f(x)}$ holds.

By this definition we can define the category of ringed spaces and also locally ringed spaces. Examples 2.2.10–2.2.14 give examples of ringed spaces $(X, \mathcal{F}_{X,k})$, $(W, \mathcal{O}_W) \sim (X, \mathcal{O}_{k^n}/\mathcal{J})$. In particular, Examples 2.2.11–2.2.14 are examples of locally ringed spaces.

Definition 2.3.2. We call a locally ringed space $(X, \mathcal{O}_X)$ an *analytic set* if X is a Hausdorff space and there exist an open covering $X = \bigcup U_i$ and an analytic local model $(V_i, \mathcal{O}_{V_i})$ which is isomorphic to $(U_i, \mathcal{O}_X|_{U_i})$ for each i as locally ringed spaces. Here, $\mathcal{O}_X|_{U_i}$ is the sheaf of commutative rings on U_i defined by $\mathcal{O}_X|_{U_i}(U) := \mathcal{O}_X(U)$ for an open subset $U \subset U_i$. This is called the *restricted sheaf* of $\mathcal{O}_X$ onto U_i. Henceforth, for an analytic space $(X, \mathcal{O}_X)$, we sometimes call X an *analytic space* and $\mathcal{O}_X$ the *structure sheaf*.

Definition 2.3.3. A locally ringed space $(X, \mathcal{O}_X)$ is called an *algebraic prevariety* if there exist an open covering $X = \bigcup_{i=1}^r U_i$ and an affine variety $(V_i, \mathcal{O}_{V_i})$ which is isomorphic to $(U_i, \mathcal{O}_X|_{U_i})$ for each i. In this case, we also call X an *algebraic prevariety* and $\mathcal{O}_X$ the *structure sheaf*. In particular, when $(X, \mathcal{O}_X)$ is an affine variety, the ring $A(X) := \Gamma(X, \mathcal{O}_X)$ is called the *affine coordinate ring* or just the *affine ring* of $(X, \mathcal{O}_X)$. When $k = \mathbb{C}$, we denote an analytic space and an algebraic prevariety by $\mathcal{O}_X^{\text{hol}}$ and $\mathcal{O}_X^{\text{alg}}$, respectively, in order to distinguish them.

Definition 2.3.4. A *morphism* $f : (X, \mathcal{O}_X) \to (Y, \mathcal{O}_Y)$ of analytic spaces or algebraic prevarieties over k is defined as a morphism of ringed spaces such that $\varphi^* : \mathcal{O}_Y \to \varphi_*\mathcal{O}_X$ is a morphism of sheaves of $\mathbb{C}$-algebras or k-algebras. By this we have the category of analytic spaces and the category of prevarieties over k. If there is no risk of confusion, we denote the morphism only by $f : X \to Y$, omitting $\mathcal{O}$.

Definition 2.3.5. An analytic space or an algebraic prevariety $(X, \mathcal{O}_X)$ over k is called an *open subspace* or a *open subprevariety* of $(Y, \mathcal{O}_Y)$ if there is a morphism $(i, i^*) : (X, \mathcal{O}_X) \to (Y, \mathcal{O}_Y)$ such that $i : X \hookrightarrow Y$ is the inclusion map of the open subsets and i induces an isomorphism $i^* : (X, \mathcal{O}_X) \simeq (X, \mathcal{O}_Y|_X)$. We call $(X, \mathcal{O}_X)$ a *closed subspace* or a *closed subprevariety* of $(Y, \mathcal{O}_Y)$ if there is a morphism $(i, i^*) : (X, \mathcal{O}_X) \to (Y, \mathcal{O}_Y)$ such that $i : X \hookrightarrow Y$ is the inclusion map of the closed subsets and i induces an isomorphism $i^* : \mathcal{O}_Y/\mathcal{I} \simeq \mathcal{O}_X$ for a subsheaf $\mathcal{I}$ of $\mathcal{O}_Y$ such that $\mathcal{I}(U) \hookrightarrow \mathcal{O}_Y(U)$ is an ideal for every open subset U. In this case, $\mathcal{I}$ is called the *defining ideal of* X.

A morphism $(\varphi, \varphi^*) : (X, \mathcal{O}_X) \to (Y, \mathcal{O}_Y)$ is called an *open immersion* if there is an open subspace or an open subprevariety $(U, \mathcal{O}_U)$ of $(Y, \mathcal{O}_Y)$ (φ, φ^*) induces an isomorphism $(X, \mathcal{O}_X) \simeq (U, \mathcal{O}_U)$. A closed immersion is defined in the same way.

Definition 2.3.6. Let $(V, \mathcal{O}_V)$ be a closed subspace or a closed subprevariety of $(Y, \mathcal{O}_Y)$. For a morphism $f : (X, \mathcal{O}_X) \to (Y, \mathcal{O}_Y)$, the *inverse image* $(f^{-1}(V), \mathcal{O}_{f^{-1}(V)})$ of $(V, \mathcal{O}_V)$ is defined as follows:

The set $f^{-1}(V)$ is the inverse image of V as a set and $\mathcal{O}_{f^{-1}(V)}$ is defined by the quotient sheaf $\mathcal{O}_X/\mathcal{I}_V\mathcal{O}_X$, where $\mathcal{I}_V$ is the defining ideal of $(V, \mathcal{O}_V)$ in $(Y, \mathcal{O}_Y)$. Here, the sheaf $\mathcal{I}_V\mathcal{O}_X$ is the sheafification of the presheaf defined by

$$\mathcal{F}(U) := \Big(\varinjlim_{W \supset f(U)} \mathcal{I}_V(W) \Big) \mathcal{O}_X(U).$$

The following are important properties of an affine variety.

Theorem 2.3.7. *Defining* $A : \mathcal{C} = (category\ of\ affine\ varieties) \to \mathcal{D} =$ *(category of finitely generated k -algebras) as following (i) and (ii), then A is an equivalent contravariant functor:*

(i) To an object $(X, \mathcal{O}_X) \in \mathrm{Ob}(\mathcal{C})$ we associate the affine ring $A(X, \mathcal{O}_X) = A(X)$.
(ii) For a morphism $u = (\varphi, \varphi^*) : (X, \mathcal{O}_X) \to (Y, \mathcal{O}_Y)$ in $\mathcal{C}$ we define $A(u) := \varphi_Y^* : A(Y) \to A(X)$.

By this functor, a closed immersion $(X, \mathcal{O}_X) \to (Y, \mathcal{O}_Y)$ corresponds to a surjection $A(Y) \to A(X)$ of k-algebras.

(Sketch of proof) For a finitely generated k-algebra A, we check how to decide the affine variety $(X, \mathcal{O}_X)$ such that $A(X) = A$. As A is finitely generated over k, we can represent it as $A \cong k[x_1, \ldots, x_n]/I$, where $k[x_1, \ldots, x_n]$ is the polynomial ring and I is an ideal.

Define the subset X of k^n as the zero set of polynomials in I. On the other hand, let $\mathcal{I}_X$ be the subsheaf of $\mathcal{O}_{k^n}$ defined by $\mathcal{I}_X(U) = (\mathcal{O}_{k^n}(U)$-ideal generated by $I)$. Then $(X, \mathcal{O}_{k^n}/\mathcal{I}_X)$ is the required affine variety.

Proposition 2.3.8. *There is the direct product for every pair of the objects in the category of analytic spaces or of algebraic prevarieties.*

(Sketch of proof) See the proof for analytic spaces in [Fis], p. 25. For the case of algebraic prevarieties we can construct the direct product as follows: First, in the case where both $(X, \mathcal{O}_X)$, $(Y, \mathcal{O}_Y)$ are affine varieties with the affine coordinate rings $A(X)$, $A(Y)$, respectively, the affine variety $(Z, \mathcal{O}_Z)$ with the affine coordinate ring $A(X) \otimes_k A(Y)$ is the direct product of $(X, \mathcal{O}_X)$ and $(Y, \mathcal{O}_Y)$. Here, we should note that the topological space of Z is bijective to the product space $X \times Y$, but the topology of Z is not the product topology. Next, in the case where $(X, \mathcal{O}_X)$, $(Y, \mathcal{O}_Y)$ are general algebraic prevarieties over k, take affine open coverings $\{U_i\}_{i=1}^r$ and $\{V_j\}_{j=1}^s$ of X, Y, respectively and then patch the direct products $(U_i, \mathcal{O}_X|_{U_i}) \times (V_j, \mathcal{O}_Y|_{V_j})$ together, which yields the direct product $(X, \mathcal{O}_X) \times (Y, \mathcal{O}_Y)$. In this case also the topological space is bijective to the product space $X \times Y$, but the topology is not the product topology.

The direct product $(X, \mathcal{O}_X) \times (Y, \mathcal{O}_Y)$ is sometimes denoted by $(X \times Y, \mathcal{O}_{X \times Y})$ or by $X \times Y$ if there is no risk of confusion.

Definition 2.3.9. If an algebraic prevariety X over k satisfies the following property (S), we call it an *algebraic variety*:

(S) $\Delta := (1_X, 1_X) : X \to X \times X$ is a closed immersion.

Proposition 2.3.10. *An open subprevariety or a closed subprevariety of an algebraic variety over k is again an algebraic variety. We call this an open subvariety and closed subvariety, respectively.*

Proof. Let $(U, \mathcal{O}_U)$ be a subprevariety of $(X, \mathcal{O}_X)$, then we have $\Delta(U) = \Delta(X) \cap (U \times U)$. The condition (S) is equivalent to saying that $\Delta(X) \subset X \times X$ is a closed subset ([Ha2, II, Cor 4.2]), therefore we obtain that $\Delta(U) \subset U \times U$ is a closed subset, which yields (S) for U.

Example 2.3.11. An affine variety over k is an affine algebraic variety.

Indeed, for a diagram:

$$\begin{array}{ccc} A(X) & \xleftarrow{1_{A(x)}} & A(X) \\ {\scriptstyle 1_{A(X)}}\uparrow & & \\ A(X) & & \end{array}$$

consider the map $\Delta^* : A(X) \otimes_k A(X) \to A(X)$, $f \otimes g \mapsto f \cdot g$, then we have the following commutative diagram:

$$\begin{array}{ccccc} A(X) & \xleftarrow{1_{A(X)}} & A(X) & \ni & f \\ {\scriptstyle 1_{A(X)}}\uparrow & \nwarrow{\scriptstyle \Delta^*} & \downarrow & & \downarrow \\ A(X) & \longrightarrow & A(X) \otimes_k A(X) & \ni & f \otimes 1 \\ \Downarrow & & \Downarrow & & \\ g & \longmapsto & 1 \otimes g & & \end{array}$$

Therefore, by Theorem 2.3.7, the map Δ^* is a homomorphism of affine rings corresponding to Δ in the condition (S). Obviously Δ^* is surjective, the corresponding morphism Δ is a closed immersion by Theorem 2.3.7.

In particular, an affine variety $(k^n, \mathcal{O}_{k^n})$ is denoted by $\mathbb{A}^n_k$.

Example 2.3.12 (projective space $\mathbb{P}^n_k$). Let $U_i = \mathbb{A}^n_k$ $(i = 0, 1, \ldots, n)$. The algebraic prevariety obtained by patching these as follows becomes an algebraic variety. This variety is denoted by $\mathbb{P}^n_k$ and called the *projective space* of dimension n over k.

Represent the affine coordinate ring $A(U_i)$ of U_i by $k[y_{i0}, \ldots, \widehat{y_{ii}}, \ldots, y_{in}]$, where $\widehat{y_{ii}}$ means that y_{ii} does not appear. Let U_{ij} be the open subset determined by $y_{ij} \neq 0$ $(i \neq j)$ on U_i, then it is the affine variety with the affine coordinate ring $k[y_{i0}, \ldots, \widehat{y_{ii}}, \ldots, y_{in}]_{y_{ij}}$. Define a k-algebra homomorphism $k[y_{i0}, \ldots, \widehat{y_{ii}}, \ldots, y_{in}]_{y_{ij}} \to k[y_{j0}, \ldots, \widehat{y_{jj}}, \ldots, y_{jn}]_{y_{ji}}$ by the correspondence $y_{i\ell} \mapsto y_{j\ell}/y_{ji}$ $(\ell \neq i, j)$, $y_{ij} \mapsto 1/y_{ji}$. Then, as this map has the inverse map $y_{j\ell} \mapsto y_{i\ell}/y_{ij}$ $(\ell \neq i, j)$, $y_{ji} \mapsto 1/y_{ij}$, we obtain an isomorphism $\varphi_{ij} : A(U_{ij}) \xrightarrow{\sim} A(U_{ji})$. By Theorem 2.3.7 we have an isomorphism $U_{ij} \cong U_{ji}$.

By these isomorphisms, we patch $\{U_i\}$ together. We can see that the patching has no conflict on the overlapped spaces as $\varphi_{jk} \cdot \varphi_{ij} = \varphi_{ik}$ on $(U_i)_{y_{ij} y_{ik}}$.

See, for example, [Ha2, II, 4.9] for the proof that $\mathbb{P}^n_k$ is an algebraic variety over k.

Next look at $\mathbb{P}^n_k$ as the points set. Let $(x_0, \ldots, x_n)$ be a point of k^{n+1}. Define a relation $\sim$ on $k^{n+1} \setminus \{0\}$ as follows:

$$(x_0, \ldots, x_n) \sim (x'_0, \ldots, x'_n) \iff \exists \lambda \in k \setminus \{0\},\ x'_i = \lambda x_i\ (\forall i).$$

Then $\sim$ is an equivalence relation and the quotient set is denoted by $k^{n+1} \setminus \{0\}/\sim$. An element of this set is denoted by $\overline{x}$ $(x \in k^{n+1} \setminus \{0\})$. Then the map $U_i \to k^{n+1} \setminus \{0\}/\sim$; $(y_{i0}, \ldots, \widehat{y_{ii}}, \ldots, y_{in}) \mapsto \overline{(y_{i0}, \ldots, y_{ii-1}, 1, y_{ii+1}, \ldots, y_{in})}$ can be patched together and we obtain a bijection $\mathbb{P}^n_k \to k^{n+1} \setminus \{0\}/\sim$. We represent a point of $\mathbb{P}^n_k$ by $(x_0 : x_1 : \cdots : x_n)$, where $(x_0, \ldots, x_n)$ is a point of k^{n+1} in the equivalence class. This is called the *homogeneous coordinates* of the point.

Let $f(X_0, \ldots, X_n)$ be a homogeneous polynomial of degree d. Since, for every $\lambda \in k \setminus \{0\}$ the equality $f(x_0, \ldots, x_n) = 0 \Leftrightarrow f(\lambda x_0, \ldots, \lambda x_n) = \lambda^d f(x_0, \ldots, x_n) = 0$ holds, we can define the zero set of f independently of the choice of homogeneous coordinates. Restricting this zero set on U_i, it coincides with the zero set $f_i(Y_{i0}, \ldots, \widehat{Y_{ii}}, \ldots, Y_{in}) := f(Y_{i0}, \ldots, Y_{ii-1}, 1, Y_{ii+1}, \ldots, Y_{in})$ in U_i. Therefore, the zero set of f is a closed subset on $\mathbb{P}^n_k$. The polynomial f_i is called the *dehomoginization* of f and we call f the *homogenization* of f_i. Between the homogenization and the dehomoginization we have the relation $f(X_0, \ldots, X_n) = X_i^d f_i\left(\frac{X_0}{X_i}, \ldots, \widehat{\frac{X_i}{X_i}}, \ldots, \frac{X_n}{X_i}\right)$.

Next we show an example which is an algebraic prevariety and not an algebraic variety.

Example 2.3.13. Let $(X_i, \mathcal{O}_{X_i}) \cong \mathbb{A}_k^1$ for $i = 1, 2$. Denote the affine rings $A(X_i) = k[x_i]$ $(i = 1, 2)$. Let $U_i \subset X_i$ be the open subset defined by $x_i \neq 0$, and patch X_1 and X_2 by the isomorphism $\Gamma(U_1 \mathcal{O}_{X_1}|_{U_1}) \xrightarrow{\sim} \Gamma(U_2, \mathcal{O}_{X_2}|_{U_2})$, $x_1 \mapsto x_2$. Let X be the algebraic prevariety obtained by this patching and U be the open subset corresponding to U_i. Then X is not an algebraic variety. Indeed, we can prove that X does not satisfy the condition (S). The closure of $\Delta(U) = \{(x, x) \mid x \in U\}$ in $X \times X$ is $\overline{\Delta(U)} = \Delta(U) \cup \{(P_1, P_1), (P_1, P_2), (P_2, P_1), (P_2, P_2)\}$, where $\{P_1, P_2\} = X \setminus U$. On the other hand $\Delta(X)$ contains $\Delta(U)$ as an open dense subset and $\Delta(X) = \Delta(U) \cup \{(P_1, P_1), (P_2, P_2)\} \subsetneq \overline{\Delta(U)}$, therefore $\Delta(X)$ is not a closed subset in $X \times X$. Now we obtain that $\Delta : X \hookrightarrow X \times X$ is not a closed immersion.

Definition 2.3.14. An analytic space or an algebraic variety over k is called *reduced* if the structure sheaf $\mathcal{O}_X$ is reduced, i.e., for every point $x \in X$ the stalk $\mathcal{O}_{X,x}$ has no nilpotent element.

An algebraic variety (or an analytic space) $(X, \mathcal{O}_X)$ is called *irreducible* if there is no proper closed subset $X_i \subsetneq X$ $(i = 1, 2)$ such that $X = X_1 \cup X_2$.

If an algebraic variety $(X, \mathcal{O}_X)$ is irreducible and reduced, we call it an *integral variety*.

An algebraic variety $(X, \mathcal{O}_X)$ is integral if and only if X is connected and $\mathcal{O}_{X,x}$ is an integral domain for every $x \in X$. If an algebraic variety X is represented as the union $X = \bigcup_{i=1}^r X_i$ of proper closed irreducible subsets X_i $(X_i \not\subset X_j, i \neq j)$, each X_i is called an *irreducible component* of X.

A closed subvariety of a projective space is called a *projective variety*. An open subvariety of a projective variety is called a *quasi-projective variety*.

Definition 2.3.15. We define the *dimension* $\dim_x X$ at a point $x \in X$ of an algebraic variety or an analytic space $(X, \mathcal{O}_X)$ as follows:

$$\dim_x X = \sup \left\{ r \;\middle|\; \begin{array}{l} r \text{ is the length of a sequence of irreducible closed subvarieties} \\ x \in Z_0 \subsetneq Z_1 \cdots \subsetneq Z_k \subset U,\ U \text{ open neighborhood of } x \end{array} \right\}.$$

The *dimension* of X is defined as $\sup_{x \in X} \dim_x X$ and denoted by $\dim X$. The dimension $\dim_x X$ at a point $x \in X$ coincides with the Krull dimension of $\mathcal{O}_{X,x}$ (see [Ma2, p. 30] for Krull dimension). We can see also that $\dim \mathbb{A}_k^n = n$, $\dim \mathbb{P}_k^n = n$.

2.4 Coherent Sheaves

Definition 2.4.1. Let $(X, \mathcal{O}_X)$ be a ringed space. A sheaf $\mathcal{F}$ of abelian groups on X is called a *sheaf of $\mathcal{O}_X$-modules* (or just $\mathcal{O}_X$-Module) if for every open subset $U \subset X$, $\mathcal{F}(U)$ is an $\mathcal{O}_X(U)$-module such that for every pair of open subsets $V \subset U \subset X$ the diagram

$$\begin{array}{ccccc}
\mathcal{O}_X(U) \times \mathcal{F}(U) & \longrightarrow & \mathcal{F}(U), & (h, f) & \longmapsto h \cdot f \\
\downarrow & & \downarrow & \downarrow & \downarrow \\
\mathcal{O}_X(V) \times \mathcal{F}(V) & \longrightarrow & \mathcal{F}(V), & (h|_V, f|_V) & \longmapsto h \cdot f|_V
\end{array}$$

is commutative.

Let $\mathcal{F}$, $\mathcal{G}$ be sheaves of $\mathcal{O}_X$-modules. A morphism $\varphi : \mathcal{F} \to \mathcal{G}$ of sheaves of abelian groups is called a *morphism of $\mathcal{O}_X$-Modules* if it induces an $\mathcal{O}_X(U)$-modules homomorphism $\varphi_U : \mathcal{F}(U) \to \mathcal{G}(U)$ for every open subset $U \subset X$. By this we can define the category of $\mathcal{O}_X$-modules. We denote this category by $(\mathrm{Mod}_{\mathcal{O}_X})$. This category $(\mathrm{Mod}_{\mathcal{O}_X})$ is an abelian category.

In particular, if $\mathcal{F}(U)$ is an $\mathcal{O}_X(U)$-ideal for every open subset U, we call $\mathcal{F}$ an *$\mathcal{O}_X$-Ideal*. The defining ideal of an analytic space or an algebraic variety $(X, \mathcal{O}_X)$ is an $\mathcal{O}_X$-ideal.

We will show some constructions of new $\mathcal{O}_X$-Modules from a given $\mathcal{O}_X$-Module.

Definition 2.4.2. (1) For $\mathcal{F}, \mathcal{G} \in (\mathrm{Mod}_{\mathcal{O}_X})$, define

$$\mathrm{Hom}_{\mathcal{O}_X}(\mathcal{F}, \mathcal{G}) = \{\varphi : \mathcal{F} \to \mathcal{G} \mid \text{a morphism of } \mathcal{O}_X\text{-Modules}\}.$$

Then this is an $\mathcal{O}_X(X)$-module. For an open subset $U \subset X$ define $\mathcal{H}om_{\mathcal{O}_X}(\mathcal{F}, \mathcal{G})(U) := \mathrm{Hom}_{\mathcal{O}_{X|U}}(\mathcal{F}|_U, \mathcal{G}|_U)$. For every pair of open subsets $V \subset U$ define a map $\mathcal{H}om_{\mathcal{O}_X}(\mathcal{F}, \mathcal{G})(U) \to \mathcal{H}om_{\mathcal{O}_X}(\mathcal{F}, \mathcal{G})(V)$ by the canonical restriction map, then $\mathcal{H}om_{\mathcal{O}_X}(\mathcal{F}, \mathcal{G})$ is an $\mathcal{O}_X$-Module.

(2) For $\mathcal{F}, \mathcal{G} \in (\mathrm{Mod}_{\mathcal{O}_X})$, the sheafification of the presheaf $U \mapsto \mathcal{F}(U) \otimes_{\mathcal{O}_X(U)} \mathcal{G}(U)$ (cf. Proposition 2.2.6) is denoted by $\mathcal{F} \otimes_{\mathcal{O}_X} \mathcal{G}$ and called the *tensor product* of $\mathcal{F}$ and $\mathcal{G}$ over $\mathcal{O}_X$. This is also an $\mathcal{O}_X$-Module.

(3) Let $f : (X, \mathcal{O}_X) \to (Y, \mathcal{O}_Y)$ be a morphism of ringed spaces. For an object $\mathcal{F} \in (\mathrm{Mod}_{\mathcal{O}_X})$, a presheaf $U \mapsto \mathcal{F}(f^{-1}(U))$ $(U \subset Y)$ is an $\mathcal{O}_Y$-Module. This is denoted by $f_*\mathcal{F}$ and is called the *direct image sheaf* of $\mathcal{F}$ by f.

(4) Let $f : (X, \mathcal{O}_X) \to (Y, \mathcal{O}_Y)$ be a morphism of ringed spaces. For $\mathcal{G} \in (\mathrm{Mod}_{\mathcal{O}_Y})$, the sheafification of the presheaf $V \mapsto \varinjlim_{U \supset f(V)} \mathcal{G}(U)$ $(V \subset X)$ is denoted by $f^{-1}\mathcal{G}$. Then $f^{-1}\mathcal{G}$ is an $f^{-1}\mathcal{O}_Y$-Module. On the other hand, there is a canonical homomorphism $f^{-1}\mathcal{O}_Y \to \mathcal{O}_X$ of sheaves of rings, therefore $\mathcal{O}_X$ is also an $f^{-1}\mathcal{O}_Y$-Module.

(5) Let $f, \mathcal{G}$ be as above. The tensor product $f^{-1}\mathcal{G} \otimes_{f^{-1}\mathcal{O}_Y} \mathcal{O}_X$ is denoted by $f^*\mathcal{G}$ and called the *inverse image* of $\mathcal{G}$ by f. This is also an $\mathcal{O}_X$-Module. Also by the definition we have $f^*\mathcal{O}_Y = \mathcal{O}_X$.

Some important sheaves on a variety are "coherent sheaves" which are introduced below. First we will prepare some notions for the introduction of coherent sheaves.

Definition 2.4.3. (1) An $\mathscr{O}_X$-Module $\mathscr{F}$ is called of *finite type* if for every $x \in X$ there exists an open neighborhood U of x such that there is a surjection $\mathscr{O}_X|_U^r \to \mathscr{F}|_U$ of $\mathscr{O}_X|_U$-Modules.
Here, $\mathscr{O}_X|_U^r$ means r product of $\mathscr{O}_X|_U$. A product $\mathscr{G} \times \mathscr{H}$ of $\mathscr{G}$ and $\mathscr{H}$ is a sheaf defined by $(\mathscr{G} \times \mathscr{H})(U) = \mathscr{G}(U) \times \mathscr{H}(U)$ $(U \subset X)$.

(2) An $\mathscr{O}_X$-Module $\mathscr{F}$ is called of *finite representation* if for every $x \in X$ there exists an open neighborhood U of x such that there is an exact sequence $\mathscr{O}_X|_U^m \to \mathscr{O}_X|_U^n \to \mathscr{F}|_U \to 0$ of $\mathscr{O}_X|_U$-Modules on U.

Definition 2.4.4. An $\mathscr{O}_X$-Module $\mathscr{F}$ is called *coherent* if the following two conditions are satisfied:

(1) $\mathscr{F}$ is of finite type.
(2) For every open subset $U \subset X$ and every morphism $\alpha : \mathscr{O}_X|_U^r \to \mathscr{F}|_U$, the kernel $\operatorname{Ker} \alpha$ is of finite type.

Obviously, coherence implies of finite representation, but the converse does not hold in general. Before introducing examples of coherent modules, we introduce some properties of coherent sheaves.

The following gives basic properties on inheritance of coherence.

Proposition 2.4.5. *Let $(X, \mathscr{O}_X)$ be a ringed space, then the following hold for $\mathscr{O}_X$-Modules $\mathscr{F}, \mathscr{G}, \mathscr{F}', \mathscr{F}''$:*

(i) Let $\mathscr{G}$ be a subsheaf of $\mathscr{F}$. Assume that $\mathscr{F}$ is coherent, then $\mathscr{G}$ is coherent if and only if $\mathscr{G}$ is of finite type.
(ii) Assume that $0 \to \mathscr{F}' \to \mathscr{F} \to \mathscr{F}'' \to 0$ is an exact sequence. If two objects in the sequence is coherent, then the rest is also coherent.
(iii) Let $\alpha : \mathscr{F} \to \mathscr{G}$ be a morphism of $\mathscr{O}_X$-Modules. If $\mathscr{F}, \mathscr{G}$ are coherent, then the kernel $\operatorname{Ker} \alpha$ *and the cokernel* $\operatorname{Coker} \alpha$ *are also coherent.*
(iv) If $\mathscr{F}, \mathscr{G}$ are both coherent, then $\mathscr{F} \times \mathscr{G}$ and $\mathscr{H}om_{\mathscr{O}_X}(\mathscr{F}, \mathscr{G})$ are coherent.

For the proof, see [Se1, I, §2].

Proposition 2.4.6. *For a ringed space $(X, \mathscr{O}_X)$, if $\mathscr{O}_X$ is coherent, then the following are equivalent for every $\mathscr{O}_X$-Module $\mathscr{F}$:*

(i) $\mathscr{F}$ is coherent;
(ii) $\mathscr{F}$ is of finite representation.

Proof. (1) $\Rightarrow$ (2) is obvious. If we assume (2), then there exists an open covering $\{U_i\}$ of X such that $\mathscr{O}_X|_{U_i}^r \xrightarrow{\varphi_i} \mathscr{O}_X|_{U_i}^s \xrightarrow{\psi_i} \mathscr{F}|_{U_i} \longrightarrow 0$ is an exact sequence. For the coherence of $\mathscr{F}$, it is sufficient to prove that for every U_i $\mathscr{F}|_{U_i}$ is coherent.

By the assumption and Proposition 2.4.5, (iv), it follows that $\mathscr{O}_X|_{U_i}^r$ and $\mathscr{O}_X|_{U_i}^s$ are coherent. By Proposition 2.4.5, (iii), $\mathscr{F}|_{U_i} \cong \operatorname{Coker} \varphi_i$ is also coherent.

Theorem 2.4.7. *The structure sheaf of an analytic space or an algebraic variety is itself coherent.*

Proof. For an analytic space, first we can prove that $\mathcal{O}^{\text{hol}}_{\mathbb{C}^n}$ is coherent by Oka's coherence theorem (see, for example [Nar]). Next, Cartan proved that the defining ideal $\mathcal{I}^{\text{hol}}_W$ of an analytic set W is coherent (see, for example, [Ab]), then Proposition 2.4.5 (ii) yields our theorem for the analytic case. Serre proved the theorem for an algebraic variety ([Se1, II, §2]).

Proposition 2.4.8 (Mu, III, §1, 3 & §2, Def.1). *Let A be the affine coordinate ring of an affine variety $(X, \mathcal{O}_X)$. For an $\mathcal{O}_X$-Module $\mathcal{F}$ the following are equivalent:*

(i) *$\mathcal{F}$ is coherent;*
(ii) *$\Gamma(X, \mathcal{F})$ is a finitely generated A-module and for every $f \in A$ and $\Gamma(X_f, \mathcal{F}) = \Gamma(X, \mathcal{F}) \otimes_A A_f$ holds, where $X_f := \{x \in X \mid f(x) \neq 0\}$.*

Let us see some important coherent sheaves.

Example 2.4.9. Let $(X, \mathcal{O}_X)$ be an analytic space or an algebraic variety over k and $\mathcal{I}$ the defining ideal of the closed immersion $\Delta : (X, \mathcal{O}_X) \to (X \times X, \mathcal{O}_{X\times X})$. Then, $\mathcal{I}/\mathcal{I}^2$ is a sheaf on X. Denote it by Ω_X and call it the *sheaf of differentials on* X. This is a coherent $\mathcal{O}_X$-Module. For each $x \in X$, $\Omega_{X,x}$ is generated by $df := f \otimes 1 - 1 \otimes f$ $(f \in \mathcal{O}_{X,x})$ over $\mathcal{O}_{X,x}$. The sections $\Gamma(U, \Omega_X)$ are also generated by the elements of the same type for an affine open subset U.

Proposition 2.4.10. *The following hold for Ω_X:*

((i) *if $f \in k$, then $df = 0$;*
(ii) *if $a \in k$, $f, g \in \mathcal{O}_X$, then $d(f+g) = df + dg$, $d(fg) = fdg + gdf$, $d(af) = adf$.*

Proof. As $\Omega_{X,x} \cong \mathcal{I}_{(x,x)}/\mathcal{I}^2_{(x,x)}$, by properties of $\mathcal{O}_{X,x}$-differential module [Ma2, 9, 25] the properties (i), (ii) hold on $\Omega_{X,x}$, therefore hold also on Ω_X.

Next we state a theorem on the topology of an analytic space, which is important and used often.

Theorem 2.4.11. *Let X be an analytic space and Y a closed analytic subspace of X. If X is embedded into a non-singular analytic space M (the definition of non-singular is given in Chap. 4), then there is an open neighborhood U of Y in X, such that Y is a deformation retract of U and also of $\overline{U}$.*

(Sketch of proof) By a theorem of Whitney [Wh], M is embedded into a real Euclidean space $\mathbb{R}^n$. Then X is a real analytic subspace of $\mathbb{R}^n$ and Y is a closed analytic subspace.

Therefore there exist simplicial subdivisions K and L of X and Y, respectively, such that L is a subcomplex of K. Now take the star-shaped neighborhood $St(L) = \bigcup_{\substack{\sigma\cap|L|\neq\emptyset \\ \sigma\in K}} \sigma$ of L in K, then $Y = |L|$ is a deformation retract in $St(L)$ by barycentric subdivision of K. Take $St(L)$ as $\overline{U}$ and let $U = \overline{U}^{\circ}$.

Chapter 3
Homological Algebra and Duality

An ordinary person forgets examples
or is drowned in examples.

In this chapter, we introduce cohomology groups by means of injective resolutions and show the duality theorem and spectral sequences. Readers who already know these concepts can skip this chapter.

3.1 Injective Resolution

In this section a category $\mathscr{C}$ is one of (Ab), (Mod_R), $(\mathrm{Mod}_{\mathscr{O}_X})$, where $(X, \mathscr{O}_X)$ is a ringed space. In order to define cohomology groups, we need injective resolutions. First we introduce an injective object.

Proposition 3.1.1. *Fix an object M of $\mathscr{C}$ and define $h_M : \mathscr{C} \to$ (Ab) as follows: For an object $A \in \mathscr{C}$ define $h_M(A) := \mathrm{Hom}_{\mathscr{C}}(A, M)$. For a morphism $f : A \to B$ of $\mathscr{C}$ define $h_M(f) : \mathrm{Hom}_{\mathscr{C}}(B, M) \to \mathrm{Hom}_{\mathscr{C}}(A, M)$ by $\varphi \mapsto \varphi \circ f$. Then h_M is a contravariant functor. If $A \xrightarrow{f} B \xrightarrow{g} C \longrightarrow 0$ is an exact sequence of $\mathscr{C}$, then the induced sequence $0 \longrightarrow h_M(C) \xrightarrow{h_M(g)} h_M(B) \xrightarrow{h_M(f)} h_M(A)$ is an exact sequence of (Ab).*

Proof. For an element $\varphi \in h_M(B)$ assume $h_M(f)(\varphi) = \varphi \circ f = 0$. This implies $\mathrm{Im}\, f \subset \mathrm{Ker}\, \varphi$. Here, by $\mathrm{Im}\, f = \mathrm{Ker}\, g$, the morphism $\varphi : B \to M$ factors through $g : B \to C$. Because of this, it follows that $\mathrm{Ker}\, h_M(f) \subset \mathrm{Im}\, h_M(g)$. The rest of the proof will be simple.

S. Ishii, *Introduction to Singularities*,
https://doi.org/10.1007/978-4-431-56837-7_3

Definition 3.1.2. An object $M \in \mathscr{C}$ is called an *injective object* if h_M is an exact functor, i.e., the sequence $0 \to h_M(C) \to h_M(B) \to h_M(A) \to 0$ in (Ab) induced from an exact sequence $0 \to A \to B \to C \to 0$ in $\mathscr{C}$ is again exact.

The following is easily proved.

Proposition 3.1.3. *The following are equivalent:*

(i) $M \in \mathscr{C}$ *is an injective object.*

(ii) For a monomorphism $A \overset{f}{\hookrightarrow} B$ *in* $\mathscr{C}$*, the induced map* $\mathrm{Hom}_{\mathscr{C}}(B, M) \to \mathrm{Hom}_{\mathscr{C}}(A, M)$, $\psi \mapsto \psi \circ f$ *is surjective.*

(iii) For a monomorphism $A \overset{f}{\hookrightarrow} B$ *in* $\mathscr{C}$ *and for a morphism* $\varphi : A \to M$*, there exists* $\psi : B \to M$ *such that* $\varphi = \psi \circ f$.

Proposition 3.1.4. *For an arbitrary object* $M \in \mathscr{C}$*, there exist an injective object* $L \in \mathscr{C}$ *and a monomorphism* $M \hookrightarrow L$.

Proof. For the categories (Ab), (Mod_R) the proof is left to the reader. In the case of the category ($\mathrm{Mod}_{\mathscr{O}_X}$), for an object $\mathscr{M} \in (\mathrm{Mod}_{\mathscr{O}_X})$ define $\mathscr{L}$ as follows:

For each $x \in X$, take an injective $\mathscr{O}_{X,x}$-Module $L(x)$ such that $\mathscr{M}_x \hookrightarrow L(x)$ and then define $\mathscr{L}(U) := \prod_{x \in U} L(x)$ for open subset $U \subset X$ and define $\rho_V^U : \mathscr{L}(U) \to \mathscr{L}(V)$ by $(s_x)_{x \in U} \mapsto (s_x)_{x \in V}$ for open subsets $V \subset U$. Then, the sheaf $\mathscr{L}$ is a required object.

By this proposition, we obtain the following:

Proposition 3.1.5. *The following are equivalent:*

(i) M *is an injective object.*

(ii) An arbitrary exact sequence $0 \longrightarrow M \overset{\alpha}{\longrightarrow} L \overset{\beta}{\longrightarrow} N \longrightarrow 0$ *splits, i.e., there exists an isomorphism* $\rho : L \overset{\sim}{\to} M \oplus N$ *such that the following diagram is commutative:*

$$\begin{array}{ccccccccc} 0 & \longrightarrow & M & \longrightarrow & L & \longrightarrow & N & \longrightarrow & 0 \\ & & \wr\big\downarrow 1_M & & \wr\big\downarrow \rho & & \wr\big\downarrow 1_N & & \\ 0 & \longrightarrow & M & \longrightarrow & M \oplus N & \longrightarrow & N & \longrightarrow & 0 \end{array}$$

Proof. If M is an injective object, for a monomorphism $M \overset{\alpha}{\hookrightarrow} L$ and the identity map $M \overset{1_M}{\longrightarrow} M$ there exists $\psi : L \to M$ such that $1_M = \psi \circ \alpha$ by Proposition 3.1.3, (iii). It is easy to induce the splitting. Conversely, assume (2). Let L be the injective object obtained in Proposition 3.1.4 for M. Consider the exact sequence $0 \to M \to L \to L/M \to 0$. By (2) we have an isomorphism $L \overset{\sim}{\to} M \oplus L/M$. As a direct summand of an injective object is also an injective object, we have assertion (1).

Definition 3.1.6. (1) We call $A^\bullet = (A^i, d^i)_{i\in\mathbb{Z}}$ a *cochain complex* in a category $\mathscr{C}$ or just a complex in $\mathscr{C}$, if $A^i \in \mathscr{C}$ and $d^i : A^i \to A^{i+1}$ is a morphism of $\mathscr{C}$ for each $i \in \mathbb{Z}$ such that $d^i \circ d^{i-1} = 0$.
Here we call d^i a *boundary operator* of $A^\bullet$. The symbol i in d^i and also the symbol d^i in $(A^i, d^i)_{i\in\mathbb{Z}}$ are often dropped.

(2) Let $A^\bullet$, $B^\bullet$ be cochain complexes. We call $u = (u^i)_{i\in\mathbb{Z}}$ a *morphism* $A^\bullet \to B^\bullet$ of cochain complexes, if $u^i : A^i \to B^i$ is a morphism of $\mathscr{C}$ for each i and the following diagram

$$\begin{array}{ccccccccc} \longrightarrow & A^{i-1} & \xrightarrow{d^{i-1}} & A^i & \xrightarrow{d^i} & A^{i+1} & \longrightarrow \\ & \downarrow{\scriptstyle u^{i-1}} & & \downarrow{\scriptstyle u^i} & & \downarrow{\scriptstyle u^{i+1}} & \\ \longrightarrow & B^{i-1} & \xrightarrow{d^{i-1}} & B^i & \xrightarrow{d^i} & B^{i+1} & \longrightarrow \end{array}$$

is commutative.

By (1) and (2) we define the category of cochain complexes in $\mathscr{C}$, which is denoted by $C_o(\mathscr{C})$.

Proposition 3.1.7. *The category $C_o(\mathscr{C})$ is an abelian category.*

(Sketch of proof) For a morphism $u : A^\bullet = (A^i, d^i_A) \to B^\bullet = (B^i, d^i_B)$, we obtain $\operatorname{Ker} u = (\operatorname{Ker} u^i, d^i_A|_{\operatorname{Ker} u^i})$ and $\operatorname{Coker} u = (\operatorname{Coker} u^i, \overline{d}^i_B)$, where $\overline{d}^i_B : \operatorname{Coker} u^i \to \operatorname{Coker} u^{i+1}$ is the morphism induced from d^i_B.

Unless otherwise stated, we treat only a complex $A^\bullet$ with $A^i = 0$ $(i < 0)$. This is denoted by $A^\bullet = (A^i)_{i\geq 0}$.

Now we are going to define cohomologies in $\mathscr{C}$.

Definition 3.1.8. For a cochain complex $C^\bullet \in C_o(\mathscr{C})$ we define

$$Z^i(C^\bullet) := \operatorname{Ker}(C^i \xrightarrow{d^i} C^{i+1}) \subset C^i$$
$$B^i(C^\bullet) := \operatorname{Im}(C^{i-1} \xrightarrow{d^{i-1}} C^i) \subset C^i,$$

then, by $d^i \circ d^{i-1} = 0$, we have $B^i(C^\bullet) \subset Z^i(C^\bullet)$. Here, we define $H^i(C^\bullet) := Z^i(C^\bullet)/B^i(C^\bullet)$ and call it the *ith cohomology*.

Note 3.1.9. A morphism $u : K^\bullet \to M^\bullet$ in $C_o(\mathscr{C})$ induces a morphism $H^i(u) : H^i(K^\bullet) \to H^i(M^\bullet)$ in $\mathscr{C}$ canonically as follows: By the commutativity of

$$\begin{array}{ccccccccc} \longrightarrow & K^{i-1} & \xrightarrow{d^{i-1}} & K^i & \xrightarrow{d^i} & K^{i+1} & \longrightarrow \\ & \downarrow{\scriptstyle u^{i-1}} & \curvearrowright & \downarrow{\scriptstyle u^i} & \curvearrowright & \downarrow{\scriptstyle u^{i+1}} & \\ \longrightarrow & M^{i-1} & \xrightarrow[d^{i-1}]{} & M^i & \xrightarrow[d^i]{} & M^{i+1} & \longrightarrow \end{array}$$

induces the inclusions $u^i(B^i(K^\bullet)) \subset B^i(M^\bullet)$ and $u^i\left(Z^i(K^\bullet)\right) \subset Z^i(M^\bullet)$, which yield the map $H^i(u) : Z^i(K^\bullet)/B^i(K^\bullet) \to Z^i(M^\bullet)/B^i(M^\bullet)$. By this $H^i : K(\mathscr{C}) \to \mathscr{C}$, $K^\bullet \mapsto H^i(K^\bullet)$ is a covariant functor.

By a short exact sequence of complexes we obtain a long exact sequence of cohomologies.

Theorem 3.1.10. *(i) For an exact sequence in $C_o(\mathscr{C})$:*

$$0 \longrightarrow K^\bullet \xrightarrow{u} L^\bullet \xrightarrow{v} M^\bullet \longrightarrow 0$$

there exists a morphism $H^i(M^\bullet) \xrightarrow{\partial^i} H^{i+1}(K^\bullet)$ for each $i \geq 0$ such that

$$\begin{array}{l} \cdots\cdots\cdots\cdots\cdots\cdots \\ \xrightarrow{\partial^{i-1}} H^i(K^\bullet) \xrightarrow{H^i(u)} H^i(L^\bullet) \xrightarrow{H^i(v)} H^i(M^\bullet) \\ \xrightarrow{\partial^i} H^{i+1}(K^\bullet) \longrightarrow H^{i+1}(L^\bullet) \longrightarrow H^{i+1}(M^\bullet) \\ \cdots\cdots\cdots\cdots\cdots\cdots \end{array}$$

is exact.

(ii) For a commutative diagram of exact sequences:

$$\begin{array}{ccccccccc} 0 & \longrightarrow & K^\bullet & \xrightarrow{u} & L^\bullet & \xrightarrow{v} & M^\bullet & \longrightarrow & 0 \\ & & {\scriptstyle f}\downarrow & & {\scriptstyle g}\downarrow & & {\scriptstyle h}\downarrow & & \\ 0 & \longrightarrow & K'^\bullet & \xrightarrow{u'} & L'^\bullet & \xrightarrow{v'} & M'^\bullet & \longrightarrow & 0 \end{array}$$

we obtain the commutative diagram

$$\begin{array}{ccc} H^i(M^\bullet) & \xrightarrow{\partial^i} & H^{i+1}(K^\bullet) \\ {\scriptstyle H^i(h)}\downarrow & & \downarrow{\scriptstyle H^{i+1}(f)} \\ H^i(M'^\bullet) & \xrightarrow[\partial^i]{} & H^{i+1}(K'^\bullet) \end{array}$$

(Sketch of proof) By exactness of the hypothesis, the following diagram is commutative and the horizontal sequences are all exact in $\mathscr{C}$:

$$
\begin{array}{ccccccccc}
0 & \longrightarrow & K^i & \xrightarrow{u^i} & L^i & \xrightarrow{v^i} & M^i & \longrightarrow & 0 \\
 & & \downarrow & & \downarrow & & \downarrow & & \\
0 & \longrightarrow & K^{i+1} & \xrightarrow{u^{i+1}} & L^{i+1} & \xrightarrow{v^{i+1}} & M^{i+1} & \longrightarrow & 0 \\
 & & \downarrow & & \downarrow & & \downarrow & & \\
0 & \longrightarrow & K^{i+2} & \xrightarrow{u^{i+2}} & L^{i+2} & \xrightarrow{v^{i+2}} & M^{i+2} & \longrightarrow & 0
\end{array}
$$

For an element $\overline{x} \in H^i(M^\bullet)$, take a corresponding element $x \in Z^i(M^\bullet) \subset M^i$. Then, by exactness we have $y \in L^i$ such that $v^i(y) = x$. As $v^{i+1} \circ d_L^i(y) = d_M^i \circ v^i(y) = 0$, there exists an element $z \in K^{i+1}$ such that $u^{i+1}(z) = d_L^i(y)$. Here, z is an element in $Z^{i+1}(K^\bullet)$. In fact, we have $u^{i+2} \circ d_K^{i+1}(z) = d_L^{i+1} \circ u^{i+1}(z) = d_L^{i+1} \circ d_L^i(y) = 0$ and the monomorphism u^{i+2} induces $d_K^{i+1}(z) = 0$. Here, we have only to define $\partial^i(\overline{x}) := \overline{z}$. Here we note that $\overline{z}$ denotes the element in $H^{i+1}(K^\bullet)$ corresponding to z.

Definition 3.1.11. For an object $A \in \mathscr{C}$, $0 \longrightarrow A \xrightarrow{\varepsilon} L^\bullet$ is called a *right resolution* if $L^\bullet = (L^i, d^i)$ is a cochain complex such that $0 \longrightarrow A \xrightarrow{\varepsilon} L^0 \xrightarrow{d^0} L^1 \xrightarrow{d^1} \cdots$ is an exact sequence. If, moreover, L^i are all injective objects, we call it an *injective resolution*.

Proposition 3.1.12. *Every object $A \in \mathscr{C}$ has an injective resolution.*

Proof. By Proposition 3.1.4, there is an injective object L^0 such that $A \overset{\varepsilon}{\hookrightarrow} L^0$. Take an injective object L^1 such that $\operatorname{Coker} \varepsilon \hookrightarrow L^1$ and define d^0 to be the composite $L^0 \to \operatorname{Coker} \varepsilon \hookrightarrow L^1$. Next take $\operatorname{Coker} d^0$ as A and define L^2, d^1 in a similar way. Successive procedures give an injective resolution.

When are the cohomologies obtained from two different complexes isomorphic?

Definition 3.1.13. We say that two morphisms $u, v : K^\bullet \to L^\bullet$ in $C_o(\mathscr{C})$ are *homotopic* if there is a collection of morphisms $h = \{h^i\}_{i \geq 0}$, $h^i : K^i \to L^{i-1}$ in $C_o(\mathscr{C})$ satisfying $u^i - v^i = h^{i+1} \circ d_K^i + d_L^{i-1} \circ h^i$ ($\forall i \geq 0$). In this case we write $u \simeq_h v$.

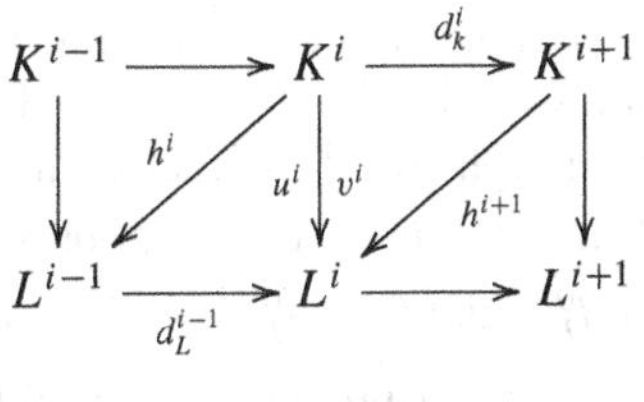

The relation $\simeq_h$ is an equivalence relation. Let $K(\mathscr{C})$ be the category such that $\mathcal{O}b_j(K(\mathscr{C})) = \mathcal{O}b_j(C_o(\mathscr{C}))$ and a morphism is a homotopical equivalence class of morphisms in $C_o(\mathscr{C})$.

Proposition 3.1.14. *If two morphisms $u, v : K^\bullet \to L^\bullet$ in $C_o(\mathscr{C})$ satisfy $u \simeq_h v$, then $H^i(u) = H^i(v) : H^i(K^\bullet) \to H^i(L^\bullet)$ for every $i \geq 0$.*

Proof. As $u^i - v^i = h^{i+1} \circ d_K^i + d_L^{i-1} \circ h^i$, for an element $x \in Z^i(K^\bullet)$ we have $u^i(x) - v^i(x) = d_L^{i-1} \circ h^i(x)$. Therefore, we have $u^i(x) \equiv v^i(x) \pmod{B^i(L^\bullet)}$.

Definition 3.1.15. If for two complexes $K^\bullet, L^\bullet \in C_o(\mathscr{C})$, there exist morphisms $u : K^\bullet \to L^\bullet$ and $v : L^\bullet \to K^\bullet$ such that $u \circ v \simeq_h 1_{L^\bullet}$ and $v \circ u \simeq_{h'} 1_{K^\bullet}$, then we say that $K^\bullet$ and $L^\bullet$ are *homotopically equivalent.*

If $K^\bullet$ and $L^\bullet$ are homotopically equivalent, then $H^i(u) : H^i(K^\bullet) \xrightarrow{\sim} H^i(L^\bullet)$, $H^i(v) : H^i(L^\bullet) \xrightarrow{\sim} H^i(K^\bullet)$ are both isomorphisms in $\mathscr{C}$ and the inverse morphism of each other.

Proposition 3.1.16. *Let $\varphi : A \to B$ be a morphism in $\mathscr{C}$, $0 \longrightarrow A \xrightarrow{\varepsilon_A} K^\bullet$ a right resolution of A and $0 \longrightarrow B \xrightarrow{\varepsilon_B} L^\bullet$ an injective resolution of B. Then, there exists a morphism $u : K^\bullet \to L^\bullet$ in $C_o(\mathscr{C})$ such that the following diagram*

$$\begin{array}{ccccc} 0 & \longrightarrow & A & \xrightarrow{\varepsilon_A} & K^0 \\ & & \Big\downarrow{\scriptstyle\varphi} & \curvearrowright & \Big\downarrow{\scriptstyle u^0} \\ 0 & \longrightarrow & B & \xrightarrow[\varepsilon_B]{} & L_0 \end{array}$$

is commutative. Such u is unique up to isomorphisms.

Proof. (i) First we will define u^0. By Proposition 3.1.3 the morphism $\varepsilon_B \circ \varphi : A \to L^0$ factors through the monomorphism $\varepsilon_A : A \to K^\bullet$ as $\varepsilon_B \circ \varphi = u^0 \circ \varepsilon_A$ because L^0 is an injective object. By this we define $u^0 : K^0 \to L^0$.

(ii) Assume we have defined $u^0, \ldots, u^{i-1}$, then we define u^i as follows: From $u^{i-1} : K^{i-1} \to L^{i-1}$ a morphism $\overline{u^{i-1}} : K^{i-1}/\mathrm{Ker}\, d_K^{i-1} \to L^{i-1}/\mathrm{Ker}\, d_L^{i-1}$ is induced canonically. For the canonical monomorphisms $\overline{d_K^{i-1}} : K^{i-1}/\mathrm{Ker}\, d_K^{i-1} \hookrightarrow K^i$, $\overline{d_L^{i-1}} : L^{i-1}/\mathrm{Ker}\, d_L^{i-1} \hookrightarrow L^i$, as L^i is an injective object, the morphism $\overline{d_L^{i-1}} \circ \overline{u^{i-1}}$ factors through $\overline{d_K^{i-1}}$ as $\overline{d_L^{i-1}} \circ \overline{u^{i-1}} = u^i \circ \overline{d_K^{i-1}}$. By this we define $u^i : K^i \to L^i$.

Next we will prove uniqueness of u. Let $u, v : K^\bullet \to L^\bullet$ morphisms satisfying the condition. We will construct $h = \{h^i\}$ such that $u \simeq_h v$.

(iii) First define $h^0 = 0$. By commutativity of the diagram we have $u^0 \circ \varepsilon_A = \varepsilon_B \circ \varphi = v^0 \circ \varepsilon_A$. Therefore, it follows that $(u^0 - v^0) \circ \varepsilon_A = 0$. By exactness it follows that $u^0 - v^0|_{\mathrm{Ker}\, d_K^0} = 0$, then a morphism $u^0 - v^0$ factors through the natural surjection $p^0 : K^0 \to K^0/\mathrm{Ker}\, d_K^0$ as $u^0 - v^0 = g^1 \circ p^0$. For $g^1 : K^0/\mathrm{Ker}\, d_K^0 \to L^0$ and for the monomorphism $\overline{d_K^0} : K^0/\mathrm{Ker}\, d_K^0 \hookrightarrow K^1$, we have the decomposition $g^1 = h^1 \circ \overline{d_K^0}$, since L^0 is an injective object. By this we define h^1. Then we obtain $u^0 - v^0 = h^1 \circ d_K^0$.

(iv) Assume we have defined $h^0, h^1, \ldots, h^i$. We define h^{i+1} as follows:
As $u^i \circ d_K^{i-1} = d_L^{i-1} \circ u^{i-1}$, $v^i \circ d_K^{i-1} = d_L^{i-1} \circ v^{i-1}$ we have $(u^i - v^i)d_K^{i-1} =$

$d_L^{i-1}(u^{i-1} - v^{i-1})$. By induction hypothesis we have $u^{i-1} - v^{i-1} = h^i \circ d_K^{i-1} + d_L^{i-2} \circ h^{i-1}$, therefore we obtain $(u^i - v^i)d_K^{i-1} = d_L^{i-1} \circ h^i \circ d_K^{i-1}$. Here we define $\alpha^i := u^i - v^i - d_L^{i-1} \circ h^i$, then $\alpha^i \circ d_K^{i-1} = 0$. Hence, α^i factors through the epimorphism $p^i : K^i \to K^i/\mathrm{Ker}\, d_K^i$ as $\alpha^i = \overline{g^{i+1}} \circ p^i$. For this morphism $g^{i+1} : K^i/\mathrm{Ker}\, d_K^i \to L^i$ and monomorphism $\overline{d_K^i} : K^i/\mathrm{Ker}\, d_K^i \hookrightarrow K^{i+1}$, as L^i is an injective object we have the decomposition $g^{i+1} = h^{i+1} \circ \overline{d_K^i}$. By this we define h^{i+1}. Then it follows that $\alpha^i = h^{i+1} \circ d_K^i$, which yields $u^i - v^i = h^{i+1} \circ d_K^i + d_L^{i-1} \circ h^i$.

Corollary 3.1.17. *For every object $A \in \mathscr{C}$, two injective resolutions $K^\bullet$, $L^\bullet$ of A are homotopically equivalent.*

Proof. By Proposition 3.1.16, for the identity map $1_A : A \to A$, there exist $u : K^\bullet \to L^\bullet$ and $v : L^\bullet \to K^\bullet$. On the other hand, for 1_A $v \circ u$ and $1_{K^\bullet} : K^\bullet \to K^\bullet$ are both satisfying the conditions in Proposition 3.1.16, therefore by uniqueness we have $v \circ u \simeq_h 1_{K^\bullet}$. In the same way, we have $u \circ v \simeq 1_{L^\bullet}$.

Proposition 3.1.18. *Let $0 \to A \to B \to C \to 0$ be an exact sequence in a category $\mathscr{C}$ and $K^\bullet$ and $M^\bullet$ injective resolutions of A and C, respectively. Then there exists an injective resolution $L^\bullet$ of B and morphisms u and v in $K(\mathscr{C})$ such that the following diagram is commutative:*

$$\begin{array}{ccccccccc}
 & & 0 & & 0 & & 0 & & \\
 & & \downarrow & & \downarrow & & \downarrow & & \\
0 & \longrightarrow & A & \xrightarrow{f} & B & \xrightarrow{g} & C & \longrightarrow & 0 \\
 & & \downarrow{\scriptstyle \varepsilon_A} & & \downarrow{\scriptstyle \varepsilon_B} & & \downarrow{\scriptstyle \varepsilon_C} & & \\
0 & \longrightarrow & K^\bullet & \xrightarrow[u]{} & L^\bullet & \xrightarrow[v]{} & M^\bullet & \longrightarrow & 0
\end{array}$$

Proof. (i) First we construct L^0 and ε_B. Let $L^0 := K^0 \oplus M^0$. As K^0 is an injective object, for a monomorphism $f : A \to B$ and $\varepsilon_A : A \to K^0$ there exists a morphism $\lambda^0 : B \to K^0$ such that $\varepsilon_A = \lambda^0 \circ f$. Now define $\varepsilon_B := (\lambda^0, \varepsilon_C \circ g) : B \to L^0$. Let $u^0 : K^0 \to L^0$ be the monomorphism to the first factor and $v^0 : L^0 \to M^0$ the projection to the second factor. Then the diagram

$$\begin{array}{ccccccccc}
 & & 0 & & & & 0 & & \\
 & & \downarrow & & & & \downarrow & & \\
0 & \longrightarrow & A & \longrightarrow & B & \longrightarrow & C & \longrightarrow & 0 \\
 & & \downarrow{\scriptstyle \varepsilon_A} & & \downarrow{\scriptstyle \varepsilon_B} & & \downarrow{\scriptstyle \varepsilon_C} & & \\
0 & \longrightarrow & K^0 & \xrightarrow[u_0]{} & L^0 & \xrightarrow[u_1]{} & M^0 & \longrightarrow & 0
\end{array}$$

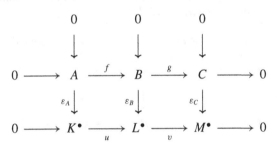

is commutative and exact. By the Five Lemma, a morphism ε_B is a monomorphism. Note that the sequence $0 \to \operatorname{Coker} \varepsilon_A \to \operatorname{Coker} \varepsilon_B \to \operatorname{Coker} \varepsilon_C \to 0$ is exact.

(ii) Assume that $L^0, \ldots, L^{i-1}$ and $d_L^0, \ldots, d_L^{i-2}$ are defined. We will define L^i and d_L^{i-1}. The following is a commutative diagram of exact sequences:

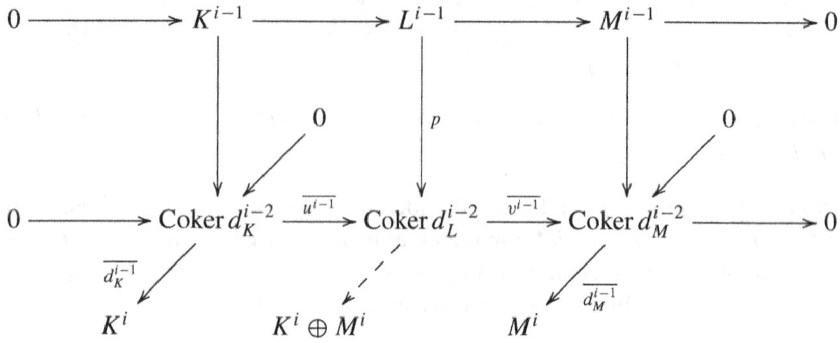

Let $L^i := K^i \oplus M^i$, then, as K^i is an injective object, there exists a morphism $\lambda^i : \operatorname{Coker} d_L^{i-2} \to K^i$ such that $\overline{d_K^{i-1}} = \lambda^i \circ \overline{u^{i-1}}$. Here, let $\overline{d_L^{i-1}} = (\lambda^i, \overline{d_M^i} \circ \overline{v^{i-1}}) : \operatorname{Coker} d_L^{i-2} \to L^i$, and let its composite with the canonical projection $p : L^{i-1} \to \operatorname{Coker} d_L^{i-2}$ be $d_L^{i-1} : L^{i-1} \to L^i$.

Proposition 3.1.19. *For the commutative diagram in a category $\mathscr{C}$:*

$$\begin{array}{ccccccccc} 0 & \longrightarrow & A & \longrightarrow & B & \longrightarrow & C & \longrightarrow & 0 \\ & & \downarrow & & \downarrow & & \downarrow & & \\ 0 & \longrightarrow & A' & \longrightarrow & B' & \longrightarrow & C' & \longrightarrow & 0 \end{array}$$

and the commutative diagrams of injective resolutions:

$$\begin{array}{ccc} A & \longrightarrow & A' \\ \downarrow & & \downarrow \\ K^\bullet & \xrightarrow{\alpha} & K'^\bullet \end{array} \qquad \begin{array}{ccc} C & \longrightarrow & C' \\ \downarrow & & \downarrow \\ M^\bullet & \xrightarrow{\gamma} & M'^\bullet, \end{array}$$

there exist injective resolutions $L^\bullet$ and $L'^\bullet$ of B and B', respectively, such that the diagram

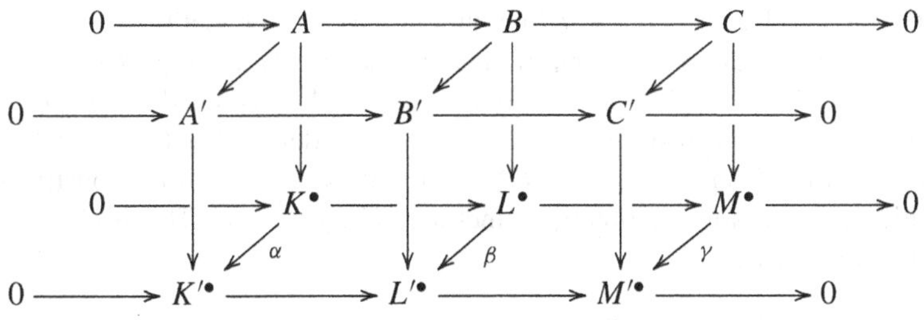

is commutative.

Proof. Define $L^\bullet$ and $L'^\bullet$ as in the proof of Proposition 3.1.18 and define $\beta : L^\bullet \to L'^\bullet$ as $\alpha^i \oplus \gamma^i : K^i \oplus M^i \to K'^i \oplus M'^i$.

3.2 ith Derived Functor

In this section we assume categories $\mathscr{C}$ and $\mathscr{C}'$ to be (Ab), (Mod_R), $(\mathrm{Mod}_{\mathscr{O}_X})$ as in Sect. 3.1.

Definition 3.2.1. A family $T = (T^i)_{i\geq 0}$ of additive functors $T^i : \mathscr{C} \to \mathscr{C}'$ $(i \geq 0)$ is called a *cohomological functor* if the following hold:

(1) For an exact sequence $0 \longrightarrow A \xrightarrow{f} B \xrightarrow{g} C \longrightarrow 0$, there exist morphisms $\partial^i : T^i(C) \to T^{i+1}(A)$ such that the sequence

$$\longrightarrow T^i(A) \xrightarrow{T^i(f)} T^i(B) \xrightarrow{T^i(g)} T^i(C) \xrightarrow{\partial^i} T^{i+1}(A) \xrightarrow{T^{i+1}(f)} \cdots$$

is exact.

(2) For the commutative diagram of exact functors

$$\begin{array}{ccccccccc} 0 & \longrightarrow & A & \longrightarrow & B & \longrightarrow & C & \longrightarrow & 0 \\ & & {\scriptstyle\alpha}\downarrow & & \downarrow{\scriptstyle\beta} & & \downarrow{\scriptstyle\gamma} & & \\ 0 & \longrightarrow & A' & \longrightarrow & B' & \longrightarrow & C' & \longrightarrow & 0 \end{array}$$

the following diagram is commutative:

$$\begin{array}{ccc} T^i(C) & \xrightarrow{\partial^i} & T^{i+1}(A) \\ {\scriptstyle T^i(\gamma)}\downarrow & & \downarrow{\scriptstyle T^{i+1}(\alpha)} \\ T^i(C') & \xrightarrow[\partial'^i]{} & T^{i+1}(A'). \end{array}$$

Definition 3.2.2. Let $F : \mathscr{C} \to \mathscr{C}'$ be an additive functor. Let $0 \longrightarrow A \overset{\varepsilon}{\longrightarrow} L^\bullet$ be an injective resolution of an object $A \in \mathscr{C}$ and $F(L^\bullet)$ a complex such as $0 \to F(L^0) \to F(L^1) \to F(L^2) \to \cdots$. For an object $A \in \mathscr{C}$ we define $R^i F(A) := H^i\big(F(L^\bullet)\big)$. For a morphism $f : A \to B$ in $\mathscr{C}$, we define $R^i F(f) := H^i\big(F(u)\big)$ by using the morphism $u : L^\bullet \to K^\bullet$ of the injective resolutions constructed in Proposition 3.1.16. Now we obtain a functor $R^i F : \mathscr{C} \to \mathscr{C}'$. This is called an ith *derived functor* of F.

The following is an important property of an ith derived functor.

Theorem 3.2.3. *(i) The object $R^i F(A)$ does not depend on a choice of an injective resolution of A.*
(ii) The collection of the functors $RF := (R^i F)_{i \geq 0}$ is a cohomological functor.
(iii) For an injective object M, it follows that $R^i F(M) = 0$ $(i > 0)$.
(iv) A functor F is left exact if and only if $R^0 F \simeq F$.
(v) Any collection $T = (T^i)_{i \geq 0}$ of functors satisfying the conditions (ii), (iii) and $T^0 \cong R^0 F$ satisfies $T^i \cong R^i F$ $(i \geq 0)$.

Proof. (i) Let $K^\bullet$ and $L^\bullet$ be injective resolutions of A. By Corollary 3.1.17 we have morphisms $u : K^\bullet \to L^\bullet$ and $v : L^\bullet \to K^\bullet$ such that $u \circ v \simeq_h 1_{L^\bullet}$, $v \circ u \simeq_{h'} 1_{K^\bullet}$. Since we have $u^i \circ v^i - 1_{L^i} = h^{i+1} \circ d_L^i + d_L^{i-1} \circ h^i$ for each i, by applying the additive functor F we have $F(u^i)F(v^i) - 1_{F(L^i)} = F(h^{i+1})F(d_L^i) + F(d_L^{i-1})F(h^i)$. Therefore we obtain $F(u) \circ F(v) \simeq_{F(h)} 1_{F(L^\bullet)}$. The equality $F(v) \circ F(u) \simeq_{F(h')} 1_{F(K^\bullet)}$ also follows in a similar way. Thus by Proposition 3.1.14 it follows that $H^i\big(F(K^\bullet)\big) \simeq H^i\big(F(L^\bullet)\big)$ for every i.

(ii) By Proposition 3.1.18, for an exact sequence $0 \to A \to B \to C \to 0$ in $\mathscr{C}$, there exist injective resolutions $K^\bullet$, $L^\bullet$, $M^\bullet$ of A, B, C, respectively, with the commutative diagram of exact sequences:

$$
\begin{array}{ccccccccc}
 & & 0 & & 0 & & 0 & & \\
 & & \downarrow & & \downarrow & & \downarrow & & \\
0 & \longrightarrow & A & \longrightarrow & B & \longrightarrow & C & \longrightarrow & 0 \\
 & & \downarrow & & \downarrow & & \downarrow & & \\
0 & \longrightarrow & K^\bullet & \longrightarrow & L^\bullet & \longrightarrow & M^\bullet & \longrightarrow & 0.
\end{array}
$$

As in the proof of Proposition 3.1.18, we may assume that $L^i = K^i \oplus M^i$, the additive functor F satisfies $F(L^i) = F(K^i) \oplus F(M^i)$. Therefore the sequence $0 \to F(K^\bullet) \to F(L^\bullet) \to F(M^\bullet) \to 0$ is exact in $K(\mathscr{C}')$. By Theorem 3.1.10, we conclude that $\{R^i F\}$ is a cohomological functor.

(iii) Let $M \in \mathscr{C}$ be an injective object. Then $0 \longrightarrow M \overset{\varepsilon}{\longrightarrow} M \overset{d^0}{\longrightarrow} 0 \overset{d^1}{\longrightarrow} 0 \to$ is an injective resolution of M. By definition we have $R^i F(M) := H^i\big(0 \longrightarrow F(M) \overset{d^0}{\longrightarrow} F(0) \overset{d^1}{\longrightarrow} F(0) \cdots\big)$. As $F(0) = 0$, we obtain $R^i F(M) = 0$ $(i > 0)$.

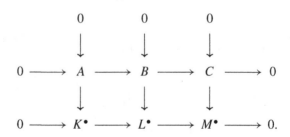

(iv) If F is left exact, for an injective resolution $0 \longrightarrow A \xrightarrow{\varepsilon} L^0 \xrightarrow{d^0} L^1$ of $A \in \mathscr{C}$, the sequence $0 \longrightarrow F(A) \xrightarrow{F(\varepsilon)} F(L^0) \xrightarrow{F(d^0)} F(L^1)$ is exact. By the definition, we have $R^0F(A) = \operatorname{Ker} F(d^0) \cong F(A)$. Conversely, assume $R^0F \simeq F$. For an exact sequence $0 \to A \to B \to C \to 0$ in $\mathscr{C}$, the sequence $0 \to R^0F(A) \to R^0F(B) \to R^0F(C)$ is exact, because RF is a cohomological functor. By the assumption this sequence is $0 \to F(A) \to F(B) \to F(C)$. Therefore F is left exact.

(v) Given $T = (T^i)_{i \geq 0}$, $S = (S^i)_{i \geq 0}$ with properties (ii) and (iii), assume there is a morphism of functors $\varphi : T^0 \to S^0$. It is sufficient to prove that there exists a unique collection $n = (n^i)_{i \geq 0} : T \to S$ of functors satisfying the following conditions:

(a) $n^0 = \varphi$.

(b) For an exact sequence $0 \to A \to B \to C \to 0$ in $\mathscr{C}$, the following is commutative:

$$\begin{array}{ccc} T^i(C) & \xrightarrow{\partial} & T^{i+1}(A) \\ {\scriptstyle n^i(C)}\downarrow & & \downarrow{\scriptstyle n^{i+1}(A)} \\ S^i(C) & \xrightarrow[\partial']{} & S^{i+1}(A). \end{array}$$

First, consider of the case $i = 1$ and construct n^1. (The construction for $i > 1$ is the same.) For an object $A \in \mathscr{C}$, take an injective object M such that $A \hookrightarrow M$ and let $0 \to A \to M \to B \to 0$ be exact. Then we have the following exact sequence:

$$\begin{array}{ccccccc} T^0(M) & \longrightarrow & T^0(B) & \xrightarrow{\partial} & T^1(A) & \longrightarrow & T^1(M) = 0 \\ \downarrow{\scriptstyle n^0(M)} & & \downarrow{\scriptstyle n^0(B)} & & \downarrow & & \\ S^0(M) & \longrightarrow & S^0(B) & \xrightarrow[\partial']{} & S^1(A) & \longrightarrow & . \end{array}$$

As ∂ is an epimorphism, there exists $n^1_M(A) : T^1(A) \to S^1(A)$ such that $n^1_M(A) \circ \partial = \partial' \circ n^0(B)$. If we prove that $n^1_M(A)$ is independent of a choice of M, then the uniqueness of n^1 follows from surjectivity of ∂. Assume for two injective objects M, N there exist monomorphisms $A \hookrightarrow M$, $A \hookrightarrow N$, then the object $M \oplus N$ is also an injective object and the natural monomorphism $A \hookrightarrow M \oplus N$ exists. Therefore if there is a commutative diagram $\begin{array}{ccc} A & \hookrightarrow & M \\ & \searrow & \downarrow \\ & & N \end{array}$, it is sufficient to prove that $n^1_M(A) = n^1_N(A)$.

Let $0 \to A \to M \to B \to 0$, $0 \to A \to N \to C \to 0$ be exact sequences. Then, as T, S are cohomological functors and by the definitions of n^1_M, n^1_N, there is a canonical morphism $f : B \to C$ such that

$$\begin{array}{ccc} T^0(B) & \xrightarrow{n^0(B)} & S^0(B) \\ {\scriptstyle T^0(f)}\downarrow & & \downarrow{\scriptstyle S^0(f)} \\ T^0(C) & \xrightarrow{n^0(C)} & S^0(C) \end{array} \quad \text{is commutative, therefore} \quad \begin{array}{ccc} T^1(A) & \xrightarrow{n^1_M(A)} & S^1(A) \\ {\scriptstyle 1}\downarrow & & \downarrow{\scriptstyle 1} \\ T^1(A) & \xrightarrow[n^1_N(A)]{} & S^1(A) \end{array}$$

is commutative.

3.3 Ext

In this section we study the derived functor of a functor $\mathrm{Hom}_{\mathscr{C}}(A, -)$. Here, we also assume a category $\mathscr{C}$ to be (Ab), (Mod_R), $(\mathrm{Mod}_{\mathscr{O}_X})$ as in Sect. 3.1, and $\mathscr{C}^0$ denotes the dual category of $\mathscr{C}$. First we define a projective object, the dual of an injective object.

Definition 3.3.1. For $X \in \mathscr{C}$, the sequence $0 \to \mathrm{Hom}_{\mathscr{C}}(X, A) \to \mathrm{Hom}_{\mathscr{C}}(X, B) \to \mathrm{Hom}_{\mathscr{C}}(X, C)$ induced from an exact sequence $0 \to A \to B \to C \to 0$ in $\mathscr{C}$ is exact. If the sequence $0 \to \mathrm{Hom}_{\mathscr{C}}(X, A) \to \mathrm{Hom}_{\mathscr{C}}(X, B) \to \mathrm{Hom}_{\mathscr{C}}(X, C) \to 0$ is exact, we call X a *projective object*. As $\mathrm{Hom}_{\mathscr{C}}(A, B) = \mathrm{Hom}_{\mathscr{C}^0}(B, A)$, a projective object in $\mathscr{C}$ is an injective object in $\mathscr{C}^0$.

Proposition 3.3.2. *If $\mathscr{C} =$ (Ab), (Mod_R), then for every object $A \in \mathscr{C}$ there exists a projective object $L \in \mathscr{C}$ and an epimorphism $L \to A$. But for $\mathscr{C} = (\mathrm{Mod}_{\mathscr{O}_X})$ this does not hold.*

Proof. For $\alpha \in A$ let $L_\alpha = \mathbb{Z}$ if $\mathscr{C} =$ (Ab), and let $L_\alpha = R$ if $\mathscr{C} = (\mathrm{Mod}_R)$, then define $L := \bigoplus_{\alpha \in A} L_\alpha$.

We can show an example of the statement on $\mathscr{C} = (\mathrm{Mod}_{\mathscr{O}_X})$. Let $X = \mathbb{A}^1_{\mathbb{C}}$ and $\mathscr{O}_X = \mathbb{Z}_X$ (constant sheaf, i.e., for any open subset $U \subset X$, $\Gamma(U, \mathbb{Z}) = \mathbb{Z}$). Then the statement follows from that a projective $\mathscr{O}_X$-Module is only 0. The proof of this fact is as follows: Let $\mathscr{L}$ be a non-zero projective $\mathscr{O}_X$-Module, then there exists $x \in X$ such that $\mathscr{L}_x \neq 0$. Therefore, there is a non-zero $\mathbb{Z}$-homomorphism $\varphi_x : \mathscr{L}_x \to \mathbb{Z}/(m)$ for some $m \geq 0$. Next define $\mathscr{G}$ as a sheaf with the support only on x with the stalk $\mathbb{Z}/(m)$. Then it is an $\mathscr{O}_X$-Module and φ_x induces a morphism $\varphi : \mathscr{L} \to \mathscr{G}$ of $\mathscr{O}_X$-Modules. By construction, there is an open subset $x \in U \subset X$ such that $\Gamma(U, \mathscr{L}) \xrightarrow{\varphi_U} \Gamma(U, \mathscr{G})$ is not a zero map. Here, fix an open subset $V \subset X$ such that $x \in V$ and $U \not\subset V$ and define a sheaf $\mathbb{Z}_V$ as follows: For an open subset $W \subset X$,

$$\Gamma(W, \mathbb{Z}_V) = \begin{cases} \mathbb{Z}, \ if \ W \subset V \\ 0, \ if \ W \not\subset V. \end{cases}$$

Then $\mathbb{Z}_V$ is an $\mathscr{O}_X$-Module and from the canonical surjective homomorphism $\mathbb{Z} \to \mathbb{Z}/(m)$, we obtain a surjective homomorphism $\psi : \mathbb{Z}_V \to \mathscr{G}$. As $\mathscr{L}$ is projective, the following diagram is commutative:

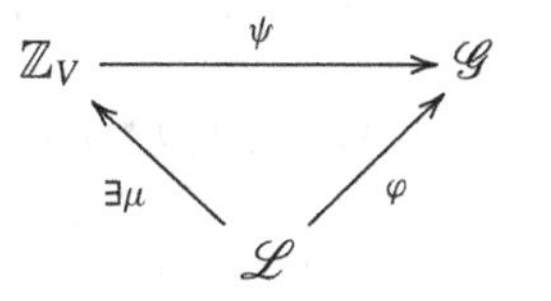

Take the sets of sections on U, then we have

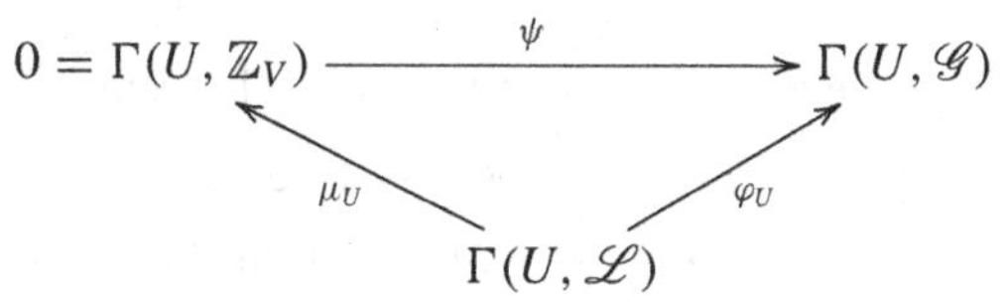

Although φ_U is not a zero map, μ_U is a zero map, which contradicts the commutativity of the diagram.

Definition 3.3.3. Let $A \in \mathscr{C}$. A sequence $L_\bullet \xrightarrow{p} A \longrightarrow 0$ in $\mathscr{C}$ is called a *left resolution* if $L_\bullet = (L_i, d_i)$ is a chain complex in $\mathscr{C}$ (i.e., a sequence $L_i \xrightarrow{d_{i-1}} L_{i-1} \longrightarrow \cdots$ satisfying $d_{i-1} \circ d_i = 0$) such that $\longrightarrow L_2 \xrightarrow{d_1} L_1 \xrightarrow{d_0} L_0 \xrightarrow{p} A \longrightarrow 0$ is exact. In addition, if each L_i is a projective object, then we call the left resolution a *projective resolution*.

Proposition 3.3.4. *Let* $\mathscr{C} =$ (Ab), (Mod$_R$), *then every object* $A \in \mathscr{C}$ *has a projective resolution.*

Proof. By Proposition 3.3.2, we can prove the proposition in the same way as Proposition 3.1.12.

Definition 3.3.5. For $A \in \mathscr{C}$ and $n \in \mathbb{N}$ if there exists an injective resolution $0 \longrightarrow A \xrightarrow{\varepsilon} L^0 \longrightarrow L^1 \cdots \longrightarrow L^n \longrightarrow 0$, we define $\operatorname{inj\,dim} A \leq n$. If there exists a projective resolution $0 \to F_n \to F_{n-1} \to \cdots \to F_0 \to A \to 0$, we define $\operatorname{proj\,dim} A \leq n$.

Definition 3.3.6. Fix an object $A \in \mathscr{C}$. The functor $F : \mathscr{C} \to$ (Ab), $Y \mapsto \operatorname{Hom}_{\mathscr{C}}(A, Y)$ is a left exact covariant functor. The ith derived functor $R^i F(Y)$ is denoted by $\operatorname{Ext}^i(A, Y)$. In particular, if $\mathscr{C} =$ (Mod$_R$), then $\operatorname{Ext}^i(A, Y)$ has an R-module structure. This is denoted by $\operatorname{Ext}^i_R(A, Y)$.

Proposition 3.3.7. *(i) Let $B \in \mathscr{C}$ be an injective object, then we have* $\mathrm{Ext}^i(A, B) = 0$ $(i > 0)$.

(ii) Let $A \in \mathscr{C}$ be a projective object, then we have $\mathrm{Ext}^i(A, B) = 0$ $(i > 0)$.

Proof. (i) is clear from property Theorem 3.2.3 (iii) of the derived functor. (ii) is proved as follows: Let $0 \to B \to L^\bullet$ be an injective resolution of B. Apply $\mathrm{Hom}(A, -)$ on this sequence. Then, as A is projective, the sequence $0 \to \mathrm{Hom}_{\mathscr{C}}(A, B) \to \mathrm{Hom}_{\mathscr{C}}(A, L^0) \to \mathrm{Hom}_{\mathscr{C}}(A, L^1) \to \cdots$ is exact. Therefore

$$\mathrm{Ext}^i(A, B) = H^i\big(\mathrm{Hom}\,(A, L^\bullet)\big) = 0, \quad i > 0.$$

Proposition 3.3.8. *(i) The functor* $\big(\mathrm{Ext}^i(A, -)\big)_{i \geq 0} : \mathscr{C} \to (\mathrm{Ab})$ *is a covariant cohomological functor.*

(i) The functor $\big(\mathrm{Ext}^i(-, B)\big)_{i \geq 0} : \mathscr{C}^0 \to (\mathrm{Ab})$ *is a covariant cohomological functor.*

(Sketch of proof) (i) is clear by the property (ii) in Theorem 3.2.3 of the derived functor. For (ii), let $0 \longrightarrow M \xrightarrow{\psi^0} L \xrightarrow{\varphi^0} K \longrightarrow 0$ be an exact sequence in $\mathscr{C}^0$ (i.e., an exact sequence $0 \longrightarrow K \xrightarrow{\varphi} L \xrightarrow{\psi} M \longrightarrow 0$ in $\mathscr{C}$). Construct $(\partial^i)_{i \geq 0}$ such that $\mathrm{Ext}^i(M, B) \longrightarrow \mathrm{Ext}^i(L, B) \longrightarrow \mathrm{Ext}^i(K, B) \xrightarrow{\partial^i} \mathrm{Ext}^{i+1}(M, B) \longrightarrow \cdots$ is exact and satisfies Definition 3.2.1 (2). For an injective resolution $0 \to B \to Y^\bullet$ of B in $\mathscr{C}$, let $Z^i\big(\mathrm{Hom}_{\mathscr{C}}(*, Y^\bullet)\big)$ and $B^i\big(\mathrm{Hom}_{\mathscr{C}}(*, Y^\bullet)\big)$ be $Z^i(*)$ and $B^i(*)$, respectively. Then for every i the sequences $0 \to Z^i(M) \to Z^i(L) \to Z^i(K)$ and $B^i(L) \to B^i(K) \to 0$ are exact. On the following commutative diagram of exact sequence:

$$\begin{array}{ccccccc}
 & 0 & & 0 & & 0 & \\
 & \downarrow & & \downarrow & & \downarrow & \\
0 \longrightarrow & Z^i(M) & \longrightarrow & Z^i(L) & \longrightarrow & Z^i(K) & \\
 & \downarrow & & \downarrow & & \downarrow & \\
0 \longrightarrow & \mathrm{Hom}_{\mathscr{C}}(M, Y^i) & \longrightarrow & \mathrm{Hom}_{\mathscr{C}}(L, Y^i) & \longrightarrow & \mathrm{Hom}_{\mathscr{C}}(K, Y^i) & \longrightarrow 0 \\
 & \downarrow & & \downarrow & & \downarrow & \\
0 \longrightarrow & Z^{i+1}(M) & \longrightarrow & Z^{i+1}(L) & \longrightarrow & Z^{i+1}(K) & \\
 & \downarrow & & \downarrow & & \downarrow & \\
 & Z^{i+1}(M)/B^{i+1}(M) & \longrightarrow & Z^{i+1}(L)/B^{i+1}(L) & \longrightarrow & Z^{i+1}(K)/B^{i+1}(K) & \\
 & \downarrow & & \downarrow & & \downarrow & \\
 & 0 & & 0 & & 0 &
\end{array}$$

apply the Snake Lemma, and we obtain that there is a morphism $\widetilde{\partial}$ such that

$$\begin{aligned}
0 &\longrightarrow Z^i(M) \longrightarrow Z^i(L) \longrightarrow Z^i(K) \xrightarrow{\widetilde{\partial}} Z^{i+1}(M)/B^{i+1}(M) \\
&\longrightarrow Z^{i+1}(L)/B^{i+1}(L) \longrightarrow Z^{i+1}(K)/B^{i+1}(K)
\end{aligned}$$

is exact.

As $\widetilde{\partial}|_{B^i(K)} = 0$, a morphism $\widetilde{\partial}$ factors through $\partial : Z^i(K)/B^i(K) \to Z^{i+1}(M)/B^{i+1}(M)$ and the sequence

$$\begin{aligned} &Z^i(M)/B^i(M) \longrightarrow Z^i(L)/B^i(L) \longrightarrow Z^i(K)/B^i(K) \\ &\xrightarrow{\partial} Z^{i+1}(M)/B^{i+1}(M) \longrightarrow Z^{i+1}(L)/B^{i+1}(L) \longrightarrow Z^{i+1}(K)/B^{i+1}(K) \end{aligned}$$

is exact.

We can see Ext^i from a different viewpoint.

Theorem 3.3.9. *When* $\mathscr{C} = (\mathrm{Ab}), (\mathrm{Mod}_R)$, *the functor* $\operatorname{Ext}^i(-, B) : \mathscr{C}^0 \to (\mathrm{Ab})$ *is an ith derived functor of* $\operatorname{Hom}_{\mathscr{C}^0}(B, -) : \mathscr{C}^0 \to (\mathrm{Ab})$.

Proof. By Proposition 3.3.7 (ii) for an injective object A in $\mathscr{C}^0$ (i.e., a projective object in $\mathscr{C}$), we have $\operatorname{Ext}^i(A, B) = 0, i > 0$. As $\operatorname{Ext}^0(A, B) = \operatorname{Hom}_{\mathscr{C}^0}(B, A)$, by Proposition 3.3.8 (ii) and Theorem 3.2.3 (v) we obtain $\operatorname{Ext}^i(-, B) \cong R^i\operatorname{Hom}_{\mathscr{C}^0}(B, -)$.

Corollary 3.3.10. *When* $\mathscr{C} = (\mathrm{Ab}), (\mathrm{Mod}_R)$, *let* $X_\bullet \to A \to 0$ *be a projective resolution of an object* $A \in \mathscr{C}$. *Then for every object* $B \in \mathscr{C}$ *it follows that* $\operatorname{Ext}^i(A, B) = H^i(\operatorname{Hom}_{\mathscr{C}}(X_\bullet, B))$.

Proof. A projective resolution of A in $\mathscr{C}$ is an injective resolution in $\mathscr{C}^0$. Therefore by Theorem 3.3.9, we have $\operatorname{Ext}^i(A, B) \cong R^i\operatorname{Hom}_{\mathscr{C}^0}(B, A) = H^i(\operatorname{Hom}_{\mathscr{C}^0}(B, X^\bullet)) = H^i(\operatorname{Hom}_{\mathscr{C}}(X_\bullet, B))$.

Theorem 3.3.11 (p). *Let n be a non-negative integer, then for $A \in \mathscr{C}$ the following are equivalent:*

(i) proj dim $A \leqq n$;
(ii) $\operatorname{Ext}^{n+1}(A, B) = 0$ *for every* $B \in \mathscr{C}$;
(iii) $\operatorname{Ext}^n(A, -)$ *is a right exact functor;*
(iv) *Let* $0 \to X_n \to X_{n-1} \to \cdots \to X_0 \to A \to 0$ *be an exact sequence in* $\mathscr{C}$. *If* $X_0, \ldots, X_{n-1}$ *are projective, then* X_n *is projective.*

Proof. (i) ⇒ (ii) Let $0 \to X_n \to X_{n-1} \to \cdots \to X_0 \to A \to 0$ be a projective resolution. By Corollary 3.3.10 we have $\operatorname{Ext}^{n+1}(A, B) = H^{n+1}(\operatorname{Hom}_{\mathscr{C}}(X_\bullet, B)) = 0$.

(ii) ⇒ (iii) Let $0 \to B' \to B \to B'' \to 0$ be an exact sequence in $\mathscr{C}$, then we have $\operatorname{Ext}^n(A, B') \to \operatorname{Ext}^n(A, B) \to \operatorname{Ext}^n(A, B'') \to \operatorname{Ext}^{n+1}(A, B) = 0$.

(iii) ⇒ (iv) For $n = 0$, condition (iii) implies that $\operatorname{Hom}(A, -)$ is an exact functor, therefore A is a projective object. So the left resolution $0 \to X_0 \to A \to 0$ of length 0 yields that X_0 is projective. Now, let $n > 0$. Let $0 \to B' \to B \to B'' \to 0$ be an exact sequence in $\mathscr{C}$. It is sufficient to prove that $\operatorname{Hom}_{\mathscr{C}}(X_n, B) \to \operatorname{Hom}_{\mathscr{C}}(X_n, B'')$ is surjective. By taking a projective resolution $\to X'_{n+1} \to X'_n \to X_n \to 0$ of X_n, the sequence $\to X'_{n+1} \to X'_n \to X_{n-1} \to X_{n-2} \to \cdots \to A \to 0$ is a projective resolution of A. Here, we have

$$\begin{aligned}\operatorname{Ext}^n(A,B) &= H\big(\operatorname{Hom}_{\mathscr{C}}(X_{n-1},B)\longrightarrow \operatorname{Hom}_{\mathscr{C}}(X'_n,B)\longrightarrow \operatorname{Hom}_{\mathscr{C}}(X'_{n+1},B)\big)\\ &= H\big(\operatorname{Hom}_{\mathscr{C}}(X_{n-1},B)\longrightarrow \operatorname{Hom}_{\mathscr{C}}(X_n,B)\longrightarrow 0\big).\end{aligned}$$

Therefore $\operatorname{Hom}_{\mathscr{C}}(X_{n-1},B)\to\operatorname{Hom}_{\mathscr{C}}(X_n,B)\to\operatorname{Ext}^n(A,B)\to 0$ is an exact sequence. By constructing such a sequence on B'', we obtain a commutative diagram of exact sequences:

$$\begin{array}{ccccccc}\operatorname{Hom}_{\mathscr{C}}(X_{n-1},B) & \longrightarrow & \operatorname{Hom}_{\mathscr{C}}(X_n,B) & \longrightarrow & \operatorname{Ext}^n(A,B) & \longrightarrow & 0\\ \downarrow{\scriptstyle\alpha} & & \downarrow{\scriptstyle\beta} & & \downarrow{\scriptstyle\gamma} & & \\ \operatorname{Hom}_{\mathscr{C}}(X_{n-1},B'') & \longrightarrow & \operatorname{Hom}_{\mathscr{C}}(X_n,B'') & \longrightarrow & \operatorname{Ext}^n(A,B'') & \longrightarrow & 0.\end{array}$$

Here, as X_{n-1} is projective, a morphism α is surjective. By assumption (iii) we have that γ is surjective. Hence β is also surjective and X_n is a projective object.

(iv) $\Rightarrow$ (i) Let $X_{n-1}\xrightarrow{d_{n-2}} X_{n-2}\longrightarrow\cdots\longrightarrow X_0\longrightarrow A\longrightarrow 0$ be a projective resolution of A and $X_n := \operatorname{Ker} d_{n-2}$. Then by (iv), X_n is projective. By this we obtain a projective resolution $0\to X_n\to\cdots\to X_0\to A\to 0$ of length n.

Corollary 3.3.12 (p). *An object A in $\mathscr{C}$ is projective if and only if* $\operatorname{Ext}^1(A,B)=0$ *holds for all $B\in\mathscr{C}$.*

Proof. In the equivalence of (i) and (ii) in Theorem 3.3.11 (p), let $n=0$. Then we have the assertion.

Theorem 3.3.13 (i). *For a non-negative integer n and an object $B\in\mathscr{C}$, the following are equivalent:*

(i) $\operatorname{inj\,dim} B\le n$.
(ii) $\operatorname{Ext}^{n+1}(A,B)=0$ *holds for every $A\in\mathscr{C}$.*
(iii) $\operatorname{Ext}^n(-,B)$ *is a right exact functor.*
(iv) Let $0\to B\to Y^0\to\cdots\to Y^n\to 0$ be an exact sequence in $\mathscr{C}$. If $Y^0,\dots,Y^{n-1}$ are injective objects, then Y^n is also injective.

Corollary 3.3.14 (i). *An object B is injective in $\mathscr{C}$ if and only if* $\operatorname{Ext}^1(A,B)=0$ *holds for all $A\in\mathscr{C}$.*

The proofs of Theorem 3.3.13 (i), Corollary 3.3.14 (i) are similar to the proofs of Theorem 3.3.11 (p), Corollary 3.3.12 (p).

3.4 Sheaf Cohomologies

Let $(X,\mathscr{O}_X)$ be a ringed space. By making use of the previous section, we introduce cohomologies with sheaves on $(X,\mathscr{O}_X)$ as coefficients.

Definition 3.4.1. A set Φ is called the *family of supports* of X if:

(1) Φ is non-empty set consisting of closed subsets of X.
(2) If $F_1, F_2 \in \Phi$, then $F_1 \cup F_2 \in \Phi$.
(3) If $F' \subset F$, $F \in \Phi$ for a closed subset $F' \subset X$, then $F' \in \Phi$.

Example 3.4.2. (i) $\Phi = \{\text{closed subsets of } X\}$.
(ii) $\Phi = \{\text{compact closed subsets of } X\}$.
(iii) Let $F \subset X$ be a closed subset and let $\Phi = \{\text{closed subsets contained in } F\}$.
Then, these are all family of supports of X.

Proposition 3.4.3. *For a family of supports Φ and $\mathscr{F} \in (\mathrm{Mod}_{\mathscr{O}_X})$ define*

$$\Gamma_\Phi(X, \mathscr{F}) := \{s \in \Gamma(X, \mathscr{F}) \mid \mathrm{supp}\,(s) \in \Phi\}$$

and for a morphism $\varphi : \mathscr{F} \to \mathscr{G}$ in $(\mathrm{Mod}_{\mathscr{O}_X})$, define $\Gamma_\Phi(\varphi) : \Gamma_\Phi(X, \mathscr{F}) \to \Gamma_\Phi(X, \mathscr{G})$ as the restriction of $\varphi_X : \Gamma(X, \mathscr{F}) \to \Gamma(X, \mathscr{G})$. Then, $\Gamma_\Phi : (\mathrm{Mod}_{\mathscr{O}_X}) \to$ (Ab) is a left exact functor.

Definition 3.4.4. For a left exact functor $\Gamma_\Phi : (\mathrm{Mod}_{\mathscr{O}_X}) \to (\mathrm{Ab})$, the ith derived functor $R^i\Gamma_\Phi(X, \mathscr{F})$ is called an *ith cohomology* of $\mathscr{F}$ with family of supports Φ and denote it by $H^i_\Phi(X, \mathscr{F})$.

In particular, if Φ is as in Example 3.4.2 (i)–(iii), then $H^i_\Phi(X, \mathscr{F})$ is denoted by $H^i(X, \mathscr{F})$, $H^i_c(X, \mathscr{F})$, $H^i_F(X, \mathscr{F})$, respectively.

By properties in Theorem 3.2.3 of the ith derived functor, we obtain the following:

Theorem 3.4.5. *(i) $\left(H^i_\Phi(X, -)\right)_{i \geq 0}$ is a cohomological functor.*
(ii) For an injective $\mathscr{O}_X$-Module $\mathscr{F}$, we have $H^i_\Phi(X, \mathscr{F}) = 0$ $(i > 0)$.
(iii) $H^0_\Phi(X, \mathscr{F}) = \Gamma_\Phi(X, \mathscr{F})$.

The following characterization of an affine variety in terms of a cohomology is important.

Theorem 3.4.6 (Serre [Se2]). *For an algebraic variety $(X, \mathscr{O}_X)$ the following are equivalent:*

(i) X is an affine variety.
(ii) The vanishing $H^i(X, \mathscr{F}) = 0$ $(i > 0)$ holds for a coherent sheaf $\mathscr{F}$.
(iii) The vanishing $H^1(X, \mathscr{I}) = 0$ holds for a cohenent $\mathscr{O}_X$-ideal sheaf $\mathscr{I}$.

A sheaf cohomology group $H^i_\Phi(X, \mathscr{F})$ is not easy to calculate, but by Theorem 3.4.6, it can be calculated as a Čech cohomology for a coherent sheaf $\mathscr{F}$ on an algebraic variety (for details, see [Se1]).

Definition 3.4.7. For $\mathscr{F} \in (\mathrm{Mod}_{\mathscr{O}_X})$ a functor $F : (\mathrm{Mod}_{\mathscr{O}_X}) \to (\mathrm{Mod}_{\mathscr{O}_X})$, $\mathscr{G} \mapsto \mathscr{H}om_{\mathscr{O}_X}(\mathscr{F}, \mathscr{G})$ is a left exact functor. The ith derived functor $R^iF(\mathscr{G})$ is denoted by $\mathscr{E}xt^i_{\mathscr{O}_X}(\mathscr{F}, \mathscr{G})$.

Similarly as in Corollary 3.3.10, we obtain the following for $\mathscr{C} = (\mathrm{Mod}_{\mathscr{O}_X})$.

Proposition 3.4.8. *For coherent sheaf $\mathscr{F}$ on a quasi-projective variety X, there exists a left resolution $\mathscr{L}_\bullet \to \mathscr{F}$ such that each $\mathscr{L}_i$ is locally free of finite rank (see Sect. 5.1). In addition, for $\mathscr{G} \in (\mathrm{Mod}_{\mathscr{O}_X})$ the following holds:*

$$\mathscr{E}xt^i_{\mathscr{O}_X}(\mathscr{F}, \mathscr{G}) \simeq H^i\big(\mathscr{H}om_{\mathscr{O}_X}(\mathscr{L}_\bullet, \mathscr{G})\big).$$

Proof. The proof is given by using a "very ample invertible sheaf" which is not defined yet in this book. The reader can see it in [Ha2]. For a very ample invertible sheaf $\mathscr{H}$, the sheaf $\mathscr{H} \otimes \mathscr{F}$ is generated by a finite number of global sections, then there exists a surjection $\mathscr{O}_X^{\oplus r} \to \mathscr{H} \otimes \mathscr{F}$. Applying $\otimes \mathscr{H}^{-1}$ on this morphism we obtain a surjection $(\mathscr{H}^{-1})^{\oplus r} \to \mathscr{F}$. Let $\mathscr{L}_0 = (\mathscr{H}^{-1})^{\oplus r}$, and then define $\mathscr{L}_i$ $(i \geq 0)$ successively as in the proof of Proposition 3.1.12.

As for the latter statement, note that by the condition on $\mathscr{L}_i$, a functor $\mathscr{H}om_{\mathscr{O}_X}(\mathscr{L}_i, \ )$ is exact. Therefore if we define

$$T^i = H^i\big(\mathscr{H}om_{\mathscr{O}_X}(\mathscr{L}_\bullet, \ )\big) : (\mathrm{Mod}_{\mathscr{O}_X}) \longrightarrow (\mathrm{Mod}_{\mathscr{O}_X})$$

then $(T^i)_{\geq 0}$ is a cohomological functor. As $\mathscr{H}om_{\mathscr{O}_X}(\ , \mathscr{G})$ is a left exact functor, it follows that $T^0(\mathscr{G}) = \mathscr{H}om_{\mathscr{O}_X}(\mathscr{F}, \mathscr{G})$. If, moreover, $\mathscr{G}$ is injective, we can show that $\mathscr{G}|_U \in (\mathrm{Mod}_{\mathscr{O}_{X|U}})$ is injective for every open subset U. Therefore, $\mathscr{H}om_{\mathscr{O}_X}(\ , \mathscr{G})$ is an exact sequence. For an injective object $\mathscr{G}$, we obtain $T^i(\mathscr{G}) = H^i\big(\mathscr{H}om_{\mathscr{O}_X}(\mathscr{L}_\bullet, \mathscr{G})\big) = 0$. Hence, by Theorem 3.2.3, (v), we have that T^i is isomorphic to the ith derived funcor of $\mathscr{H}om_{\mathscr{O}_X}(\mathscr{F}, \ )$.

Proposition 3.4.9. *If $\mathscr{F}, \mathscr{G}$ are coherent $\mathscr{O}_X$-Modules, then $\mathscr{E}xt^i_{\mathscr{O}_X}(\mathscr{F}, \mathscr{G})$ is also coherent and the following holds for an affine open subset U:*

$$\Gamma\big(U, \mathscr{E}xt^i_{\mathscr{O}_X}(\mathscr{F}, \mathscr{G})\big) = \mathrm{Ext}^i_{\Gamma(U, \mathscr{O}_X)}\big(\Gamma(U, \mathscr{F}), \Gamma(U, \mathscr{G})\big).$$

Proof. As $\mathscr{F}$ is coherent, by Proposition 3.4.8 there exists a resolution $\mathscr{L}_\bullet \to \mathscr{F}$ with each $\mathscr{L}_i$ a locally free sheaf of finite rank such that $\mathscr{E}xt^i_{\mathscr{O}_X}(\mathscr{F}, \mathscr{G}) \simeq H^i\big(\mathscr{H}om_{\mathscr{O}_X}(\mathscr{L}_\bullet, \mathscr{G})\big)$. Here, as $\mathscr{H}om_{\mathscr{O}_X}(\mathscr{L}_i, \mathscr{G})$ is coherent, by Proposition 2.4.5 the sheaf $H^i\big(\mathscr{H}om_{\mathscr{O}_X}(\mathscr{L}_\bullet, \mathscr{G})\big)$ is also coherent. The following discussion needs information which appears later in this book. For an affine open subset U consider the spectral sequence in Example 3.6.10:

$$E_2^{p,q} = H^p\big(U, \mathscr{E}xt^q_{\mathscr{O}_{X|U}}(\mathscr{F}|_U, \mathscr{G}|_U)\big) \Longrightarrow E^{p+q} = \mathrm{Ext}^{p+q}_{\mathscr{O}_{X|U}}(\mathscr{F}|_U, \mathscr{G}|_U).$$

Since U is affine and $\mathscr{E}xt^q_{\mathscr{O}_{X|U}}(\mathscr{F}|_U, \mathscr{G}|_U)$ is coherent, by Theorem 3.4.6 we obtain $E_2^{p,q} = 0$ $(\forall p \neq 0)$. Then, this spectral sequence degenerates at E_2 and $E_2^{0,n} \simeq E^n$. Here, noting that an injective resolution $0 \to \mathscr{G}|_U \to \mathscr{I}^\bullet$ of $\mathscr{G}|_U$ induces the injective resolution $0 \to \Gamma(U, \mathscr{G}|_U) \to \Gamma(U, \mathscr{I}^\bullet)$ of $\Gamma(U, \mathscr{G}|_U)$ we obtain:

$$\begin{aligned} E^n &= \mathrm{Ext}^n_{\mathcal{O}_{X|U}}(\mathcal{F}|_U, \mathcal{G}|_U) \\ &= H^n\big(\mathrm{Hom}_{\mathcal{O}_{X|U}}(\mathcal{F}|_U, \mathcal{I}^\bullet)\big) \\ &= H^n\big(\mathrm{Hom}_{\Gamma(U,\mathcal{O}_X)}(\Gamma(U,\mathcal{F}), \Gamma(U,\mathcal{I}^\bullet))\big) \\ &= \mathrm{Ext}^n_{\Gamma(U,\mathcal{O}_X)}\big(\Gamma(U,\mathcal{F}), \Gamma(U,\mathcal{G})\big). \end{aligned}$$

On the other hand, we have $E_2^{0,n} = \Gamma\big(U, \mathcal{E}xt^n_{\mathcal{O}_{X|U}}(\mathcal{F}|_U, \mathcal{G}|_U)\big)$, which yields the required isomorphism.

3.5 Derived Functors and Duality

As in Sect. 3.1, categories $\mathscr{C}, \mathscr{C}'$ are assumed to be (Ab), (Mod_R), ($\mathrm{Mod}_{\mathcal{O}_X}$). Let $K^+(\mathscr{C}), K^-(\mathscr{C}), K^b(\mathscr{C})$ be subcategories of $K(\mathscr{C})$ consisting of the objects bounded below, bounded above and bounded both below and above, respectively (see Definition 3.1.13). Including $K(\mathscr{C})$ we denote these by $K^*(\mathscr{C})$.

Definition 3.5.1. For objects $K^\bullet, L^\bullet \in K(\mathscr{C})$ a morphism $u : K^\bullet \to L^\bullet$ is called *quasi-isomorphic* if u induces an isomorphism $H^i(K^\bullet) \to H^i(L^\bullet)$ of cohomology for every i. We denote the set of all quasi-isomorphisms of $K(\mathscr{C})$ by Qis.

Proposition 3.5.2. *There exist a unique category $D(\mathscr{C})$ and a unique functor $Q : K(\mathscr{C}) \to D(\mathscr{C})$ satisfying the following:*

(i) For every $f \in$ Qis, $Q(f)$ is an isomorphism in $D(\mathscr{C})$.
(ii) Any functor $F : K(\mathscr{C}) \to D'$ such that $F(f)$ is isomorphic for every $f \in$ Qis factors uniquely through Q.

This category $D(\mathscr{C})$ is called the derived category of $\mathscr{C}$.

Also for categories $K^+(\mathscr{C})$, $K^-(\mathscr{C})$, $K^b(\mathscr{C})$, there exist uniquely $D^+(\mathscr{C})$, $D^-(\mathscr{C})$, $D^b(\mathscr{C})$, respectively [Ha1, I, 3.1]. We denote them by $D^*(\mathscr{C})$.

Definition 3.5.3. A functor $F : K^*(\mathscr{C}) \to K(\mathscr{C}')$ is called a ∂*-functor* if F is commutative with a functor $T : K^*(\mathscr{C}) \to K^*(\mathscr{C})$ (or $K(\mathscr{C}') \to K(\mathscr{C}')$), $T(X^\bullet)^p = X^{p+1}$, $d_{T(X)} = -d_{X^\bullet}$, i.e., $T \circ F = F \circ T$.

Also for a functor $F : D^*(\mathscr{C}) \to D(\mathscr{C}')$, we can define ∂-functor in a similar way.

Definition 3.5.4. For a ∂-functor $F : K^*(\mathscr{C}) \to K(\mathscr{C}')$, a ∂-functor $\mathbb{R}^*F : D^*(\mathscr{C}) \to D(\mathscr{C}')$ is called the *derived functor* of F, if there is a morphism $\xi : Q \circ F \to \mathbb{R}^*F \circ Q$ of functors such that:

For every ∂-functor $G : D^*(\mathscr{C}) \to D(\mathscr{C}')$ and a morphism $\zeta : Q \circ F \to G \circ Q$ of functors, there exists a unique morphism $\eta : \mathbb{R}^*F \to G$ satisfying $\zeta = (\eta \circ Q) \circ \xi$.

We write $\mathbb{R}^{*i} F(X^\bullet) = H^i(\mathbb{R}^* F(X^\bullet))$. For each case of $K^*(\mathscr{C}) = K(\mathscr{C}), K^+(\mathscr{C}), K^-(\mathscr{C}), K^b(\mathscr{C})$, $\mathbb{R}^* F$ denotes $\mathbb{R}F, \mathbb{R}^+ F, \mathbb{R}^- F, \mathbb{R}^b F$. Sometimes we denote it just by $\mathbb{R}F$.

The following is a basic theorem which describes when the derived functor exists.

Theorem 3.5.5 ([Ha1, I.5.1]). *Let $\mathscr{C}$, $\mathscr{C}'$, $K^*(\mathscr{C})$, F be as in Definition 3.5.4.*

Assume there exists a triangulated (for the proof, see [Ha1, I, §1]) subcategory $L \subset K^*(\mathscr{C})$ such that:

(i) For every object $X^\bullet$ in $K^*(\mathscr{C})$, there exist an $I^\bullet \in L$ and a quasi-isomorphism $X^\bullet \to I^\bullet$.
(ii) If an object $I^\bullet \in L$ satisfies $H^i(I^\bullet) = 0$ for every i, then $FI^\bullet$ has the same property.

Then the functor F has the derived functor $\mathbb{R}^* F$ and for every object $I^\bullet \in L$, we have an isomorphism $\xi(I^\bullet) : Q \circ F(I^\bullet) \to \mathbb{R}^* F \circ Q(I^\bullet)$.

The derived functor of this section and the ith derived functor in Sect. 3.2 have the following relation.

Example 3.5.6. Let $F : K^+(\mathscr{C}) \to K(\mathscr{C}')$ be induced from an additive functor $F_0 : \mathscr{C} \to \mathscr{C}'$. Let L be the subcategory consisting of injective objects in $\mathscr{C}$, then L satisfies the conditions of Theorem 3.5.5. Therefore there exists a derived functor $\mathbb{R}^+ F$. For every $K^\bullet \in K(\mathscr{C})$, an object $I^\bullet$ in L that is quasi-isomorphic to $K^\bullet$ is called an *injective resolution of* $K^\bullet$. By Theorem 3.5.5, we have $\mathbb{R}^+ FK^\bullet \simeq FI^\bullet$ in $D(\mathscr{C}')$. In particular we have $\mathbb{R}^{+i} FK^\bullet \simeq H^i(FI^\bullet)$. If, moreover, F_0 is left exact and $K^\bullet = K^0$ (the complex with the degree 0-part being K^0 and the other part being 0), then we have $\mathbb{R}^{+i} FK^\bullet = H^i(FI^\bullet) = R^i F_0 K^0$ (ith derived functor of Sect. 3.2). Here, we note that an injective resolution $I^\bullet$ of $K^\bullet$ is an injective resolution of K^0.

Example 3.5.7 ([Ha1, II §2]). Let $f : X \to Y$ be a morphism of algebraic varieties over k. Let $\mathscr{C} = (\mathrm{Mod}_{\mathscr{O}_X})$ and $\mathscr{C}' = (\mathrm{Mod}_{\mathscr{O}_Y})$. Let $F : K(\mathscr{C}) \to K(\mathscr{C}')$ be a ∂-functor induced from $f_* : \mathscr{C} \to \mathscr{C}'$. Let L be the subcategory consisting of complexes of flabby $\mathscr{O}_X$-Modules, then L satisfies the conditions of Theorem 3.5.5. Therefore the derived functor $\mathbb{R}F_*$ exists. We denote it by $\mathbb{R}f_*$. Here, $\mathscr{I}$ is called *flabby*, if for a pair $V \subset U$ of open subset the restriction map $\rho^U_V : \Gamma(U, \mathscr{I}) \to \Gamma(V, \mathscr{I})$ is surjective. Note that an injective $\mathscr{O}_X$-Module is flabby. An object $I^\bullet$ of L quasi-isomorphic to an object $K^\bullet \subset K(\mathscr{C})$ is called a *flabby resolution* of $K^\bullet$. By Theorem 3.5.5, $\mathbb{R}F_* K^\bullet = f_* I^\bullet$ holds in $D(\mathscr{C}')$, therefore in particular we have $\mathbb{R}^i f_* K^\bullet = H^i(f_* I^\bullet)$. If, moreover, $K^\bullet = K^0$, we obtain $\mathbb{R}^i f_* K^\bullet = R^i f_* K^0$ in a similar way to Example 3.5.6.

Now if we put $Y = \{\text{one point}\}$, then $\mathscr{C}' = (\mathrm{Mod}_k)$ and $f_* = \Gamma$. Then the existence of the derived functor $\mathbb{R}\Gamma$ is proved. For Γ_E (E is a closed subset) also, the derived functor exists, which can be proved similarly.

Example 3.5.8 ([Ha1, I, §6, II, §3]). For $X^\bullet \in K(\mathscr{C})$, $Y^\bullet \in K^+(\mathscr{C})$, we define $\mathrm{Hom}^\bullet(X^\bullet, Y^\bullet) \in K(\mathscr{C})$ as follows:

$$\mathrm{Hom}^n(X^\bullet, Y^\bullet) = \prod_{p\in\mathbb{Z}} \mathrm{Hom}_{\mathscr{C}}(X^p, Y^{p+n})$$

$$d^n = \prod \left(d_X^{p-1} + (-1)^{n+1} d_Y^{p+n}\right).$$

By this we obtain a bi-∂-functor:

$$\mathrm{Hom} : K(\mathscr{C})^0 \times K^+(\mathscr{C}) \longrightarrow K(\mathrm{Ab}).$$

The derived functor of this functor exists and is denoted by $\mathbb{R}\mathrm{Hom}^\bullet : D(\mathscr{C})^0 \times D^+(\mathscr{C}) \to D(\mathrm{Ab})$.

In particular, for $\mathscr{C} = (\mathrm{Mod}_{\mathscr{O}_X})$, define $\mathscr{H}om_{\mathscr{O}_X}(\mathscr{F}^\bullet, \mathscr{G}^\bullet)$ in a similar way. Then, we obtain

$$\mathscr{H}om_{\mathscr{O}_X} : K(\mathscr{C})^0 \times K^+(\mathscr{C}) \longrightarrow K(\mathscr{C})$$

and the derived functor

$$\mathbb{R}\mathscr{H}om : D(\mathscr{C})^0 \times D^+(\mathscr{C}) \longrightarrow D(\mathscr{C})$$

also exists.

Note 3.5.9 (dualizing complex [Ha1, V, §2]). Let X be an algebraic variety, let $\mathscr{C} = (\mathrm{Mod}_{\mathscr{O}_X})$ and let $D_c(\mathscr{C})$ be the subcategory of $D(\mathscr{C})$ consisting of complexes with each cohomology coherent. The subcategory $D_c^+(\mathscr{C})$ of $D^+(\mathscr{C})$ is defined in the same way. An object $R^\bullet \in D_c^+(\mathrm{Mod}_{\mathscr{O}_X})$ is called a *dualizing complex* if it has an injective resolution of finite length and for a functor $D = \mathbb{R}\mathscr{H}om(\ , R^\bullet) : D_c(\mathscr{C})^0 \to D_c(\mathscr{C})$ the canonical morphism $\mathscr{F}^\bullet \to DD\mathscr{F}^\bullet$ is isomorphic. A dualizing complex is not unique [Ha1, V, 3.1], but a normalized dualizing complex $D_X^\bullet$ is locally unique up to isomorphisms in $D_c^+(\mathscr{C})$ [Ha1, V, §6].

Theorem 3.5.10 (Duality theorem for proper morphism [Ha1, VII, 3.4]). *Let $f : X \to Y$ be a proper morphism of algebraic varieties and let $\mathscr{C} = (\mathrm{Mod}_{\mathscr{O}_X})$, $\mathscr{C}' = (\mathrm{Mod}_{\mathscr{O}_Y})$. Then for $\mathbb{R}f_* : D_c(\mathscr{C}) \to D_c(\mathscr{C}')$, $\mathbb{R}\mathscr{H}om_{\mathscr{O}_X}(\ , D_X^\bullet) : D_c(\mathscr{C}) \to D_c(\mathscr{C})$, $\mathbb{R}\mathscr{H}om_{\mathscr{O}_Y}(\ , D_Y^\bullet) : D_c(\mathscr{C}') \to D_c(\mathscr{C}')$, $\forall \mathscr{F}^\bullet \in D_c(\mathscr{C})$, we obtain the following:*

$$\mathbb{R}f_*\mathbb{R}\mathscr{H}om_{\mathscr{O}_X}(\mathscr{F}^\bullet, D_X^\bullet) \simeq \mathbb{R}\mathscr{H}om_{\mathscr{O}_Y}(\mathbb{R}f_*\mathscr{F}^\bullet, D_Y^\bullet).$$

Here, we define a Cohen–Macaulay variety.

Definition 3.5.11. Let R be a Noetherian local ring with the maximal ideal $\mathfrak{m}$. A set of elements $f_1, \dots, f_r$ of $\mathfrak{m}$ is called an *R-regular sequence* if for every i $(1 \le i \le r)$ the R-homomorphism $R/\sum_{j=1}^{i-1} f_j R \to R/\sum_{j=1}^{i-1} f_j R$ multiplying f_i is injective. The maximal length of an R-regular sequence is written as depth R and called the *depth* of R. In general depth $R \leq \dim R$ holds. If in particular the equality holds, the ring R is called a *Cohen–Macaulay ring*.

An algebraic variety or an analytic space X is called *Cohen–Macaulay* if for every $x \in X$, the local ring $\mathscr{O}_{X,x}$ is a Cohen–Macaulay ring.

Proposition 3.5.12. *For an n-dimensional algebraic variety X, the following are equivalent:*

(i) *X is a Cohen–Macaulay variety;*
(ii) $D_X^{\cdot} \cong D_X^{-n}[n]$ *(i.e.,* $D_X^{\cdot}$ *is the complex with* D_X^{-n} *as the −n-part and with 0 as the other parts).*

In this case, $D_X^{-n} \cong \omega_X$ *holds, where* ω_X *is the canonical sheaf that will be defined in Sect. 5.3.*

Proposition 3.5.13 ([Gr, 3.10]). *Let R be a Noetherian local ring and k the residue field. Then the following are equivalent:*

(i) depth $R > n$.
(ii) $\mathrm{Ext}^i_R(k, R) = 0, \forall i \leq n$.
(iii) $H^i_{\mathfrak{m}}(R) = 0, \forall i \leq n$.

By this we obtain:
R is a Cohen–Macaulay ring $\Longleftrightarrow \mathrm{Ext}^i_R(k, R) = 0, i < \dim R \Longleftrightarrow H^i_{\mathfrak{m}}(R) = 0, i < \dim R$.

Theorem 3.5.14 (Local Duality Theorem [Ha1, V, §6]). *Let* $\mathscr{O}_{X,x}$ *be the local ring of an algebraic variety X over k at a point and* $\mathfrak{m}$ *the maximal ideal.*
For an object $M^\bullet \in D_c^+(\mathrm{Mod}_{\mathscr{O}_X})$ *the following holds:*

$$\mathbb{R}\Gamma_{\{x\}}(M^\bullet) \cong \mathrm{Hom}_{\mathscr{O}_X}\big(\mathbb{R}\mathscr{H}om_{\mathscr{O}_X}(M^\bullet, D_X^{\cdot}), I\big),$$

where I is the injective hull of $k = \mathscr{O}_{X,x}/\mathfrak{m}$, *i.e., I is injective and contains k as* $\mathscr{O}_{X,x}$*-submodule and any non-zero* $\mathscr{O}_{X,x}$*-submodule N of I satisfies* $k \cap N \neq 0$.

The following corollary is the goal of this section. This is often used for a resolution of singularities f.

Corollary 3.5.15. *Let X be an n-dimensional Cohen–Macaulay algebraic variety and* $f : X \to Y$ *a proper morphism of algebraic varieties over k. For a point* $y \in Y$ *let* $E := f^{-1}(y)$ *and I the injective hull of* $\mathscr{O}_{Y,y}/\mathfrak{m}_{Y,y}$. *Then for an invertible sheaf (see Definition 5.1.2)* $\mathscr{L}$ *on X, the following holds:*

$$H^i_E(X, \mathscr{L}) \cong \mathrm{Hom}_{\mathscr{O}_Y}\big(R^{n-i} f_*(\omega_X \otimes \mathscr{L}^{-1}), I\big).$$

In particular, if $X - E \xrightarrow{\sim} Y - \{y\}$, *then* $\dim_k H^i_E(X, \mathscr{L}) = \dim_k R^{n-i} f_*(\omega_X \otimes \mathscr{L}^{-1})$.

Proof. For simplicity, denote $\mathrm{Hom}_{\mathscr{O}_{Y,y}}(M^\bullet, I)$ by $(M^\bullet)'$.
As $\Gamma_E = \Gamma_{\{x\}} \circ f_*$, we have $\mathbb{R}\Gamma_E \mathscr{L} \cong \mathbb{R}\Gamma_{\{x\}}\big(\mathbb{R}f_*(\mathscr{L})\big)$. By the local duality theorem: The right-hand side $\cong \big(\mathbb{R}\mathrm{Hom}_{\mathscr{O}_Y}(\mathbb{R}f_*\mathscr{L}, D_Y^{\cdot})\big)'$.

This is isomorphic to $(\mathbb{R}f_*(\mathbb{R}\mathrm{Hom}_{\mathcal{O}_X}(\mathscr{L}, D_X^{\cdot})))'$ by the duality theorem for proper morphism. Here, this is also isomorphic to $(\mathbb{R}f_*(\omega_X \otimes \mathscr{L}^{-1}[n]))'$ by $D_X^{\cdot} = \omega_X[n]$.

If $X \setminus E \simeq Y \setminus \{y\}$, then $\mathscr{M} = R^{n-i}f_*(\omega_X \otimes \mathscr{L}^{-1})$ has the support on $\{y\}$. In general, it is sufficient to prove $\dim \mathscr{M} = \dim \mathscr{M}'$ for $\mathscr{M}$ with the support on $\{y\}$. If $\mathscr{M}$ is a simple $\mathcal{O}_{Y,y}$-module, by $\mathscr{M} \cong \mathcal{O}_{Y,y}/\mathfrak{m}_{Y,y}$, there is an injection $\mathcal{O}_{Y,y}/\mathfrak{m}_{Y,y} \hookrightarrow I$, therefore we have $\mathscr{M}' \neq 0$. Hence, we obtain $1 = \dim_k \mathscr{M} \leq \dim_k \mathscr{M}'$. Let $n \geq 2$ and assume the condition $\dim_k \mathscr{M} \leq \dim_k \mathscr{M}'$ holds up to $\dim \mathscr{M} = n - 1$. If $\dim \mathscr{M} = n$, since $\mathscr{M}$ is not simple, we have an exact sequence of $\mathcal{O}_{Y,y}$-modules:

$$0 \longrightarrow \mathscr{M}_1 \longrightarrow \mathscr{M} \longrightarrow \mathscr{M}_2 \longrightarrow 0,$$

with $\dim_k \mathscr{M}_1$, $\dim_k \mathscr{M}_2 < n$. As $(\ )'$ is an exact functor, we obtain an exact sequence:

$$0 \longrightarrow \mathscr{M}_2' \longrightarrow \mathscr{M}' \longrightarrow \mathscr{M}_1' \longrightarrow 0.$$

Therefore, it follows that

$$\begin{aligned} \dim_k \mathscr{M}' &= \dim_k \mathscr{M}_1' + \dim_k \mathscr{M}_2' \\ &\geq \dim_k \mathscr{M}_1 + \dim_k \mathscr{M}_2 = n. \end{aligned}$$

By this we have $\dim_k \mathscr{M} \leq \dim_k \mathscr{M}'$. On the other hand, from $(\mathscr{M}')' \cong \mathscr{M}$ it follows that $\dim_k \mathscr{M} = \dim_k \mathscr{M}'$.

Theorem 3.5.16 (Serre's criteria [Ma2, 23.8]). *A Noetherian ring R is integrally closed if and only if the following two conditions hold:*

(R_1) *For every prime ideal* $\mathfrak{p} \subset R$ *such that* $\mathrm{ht}\,\mathfrak{p} = 1$ *the local ring* $R_{\mathfrak{p}}$ *is regular.*
(S_2) *For every prime ideal* $\mathfrak{p} \subset R$ *the inequality* $\mathrm{depth}\, R_{\mathfrak{p}} \geq \min\{\mathrm{ht}\,\mathfrak{p}, 2\}$ *holds.*

Corollary 3.5.17. *If a 2-dimensional local domain R is integrally closed, then R is a Cohen–Macaulay ring.*

Proof. Let $\mathfrak{m}$ be the maximal ideal of R, then we have $\mathrm{ht}\,\mathfrak{m} = 2$. Therefore by Theorem 3.5.16 it follows that $\mathrm{depth}\, R = \mathrm{depth}\, R_{\mathfrak{m}} \geq 2 = \dim R$.

3.6 Spectral Sequences

As in Sect. 3.1, categories $\mathscr{C}, \mathscr{C}'$ are assumed to be (Ab), (Mod_R) or $(\mathrm{Mod}_{\mathcal{O}_X})$.

Definition 3.6.1. A family $E = \{E_r^{p,q}, d_r^{p,q}, E^n, F \mid p, q, r, n \in \mathbb{Z},\ r \geq r_0\}$ is called a *spectral sequence* with the value on $\mathscr{C}$ if there exists an integer $r_0 \geq 0$ and if the following are satisfied:

(1) $E_r^{p,q}, E^n \in \mathscr{C}, \forall p, q, n, r \geq r_0$.
(2) $d_r^{p,q} : E_r^{p,q} \to E_r^{p+r,q-r+1}$ is a morphism in $\mathscr{C}$ such that $d_r^{p+r,q-r+1} \circ d_r^{p,q} = 0$ $\forall p, q, r \geq r_0$.
(3) $E_{r+1}^{p,q} \simeq H(E_r^{p-r,q+r-1} \to E_r^{p,q} \to E_r^{p+r,q-r+1}), \forall p, q, r \geq r_0$.
(4) For any integers p, q, there exists k such that for each $r \geq k$ the equality $E_r^{p,q} = E_{r+1}^{p,q}$ holds, which is denoted by $E_\infty^{p,q}$.
(5) F is a descending chain $F : E^n \supset \cdots \supset F^p(E^n) \supset F^{p+1}(E^n) \supset \cdots \supset 0$ of subobjects of E^n such that for some ℓ, m, the equalities $F^\ell(E^n) = E^n$, $F^m(E^n) = 0$ hold and, moreover, for every p, q the equality $E_\infty^{p,q} \cong Gr_F^p E^{p+q} = F^p(E^{p+q})/F^{p+1}(E^{p+q})$ holds.

In this case the spectral sequence E is written as $E_{r_0}^{p,q} \Longrightarrow E^{p+q}$. In particular, if in (4) the integer k can be taken commonly for all p, q, then we say that E is E_k-degenerate.

Proposition 3.6.2. *If a spectral sequence $E_{r_0}^{p,q} \Longrightarrow E^{p+q}$ satisfies $E_2^{p,q} = 0$ for every $p < 0$, $q < 0$, then there exists the following exact sequence:*

$$0 \longrightarrow E_2^{1,0} \longrightarrow E^1 \longrightarrow E_2^{0,1} \longrightarrow E_2^{2,0} \longrightarrow E^2.$$

Proof. By the assumption, if $p < 0$, then we have $E_\infty^{p,n-p} = E_2^{p,n-p} = 0$ and if $q < 0$, then we have $E_\infty^{n-q,q} = E_2^{n-q,q} = 0$ for every n. Therefore it follows that $E^n = F^0(E^n) \supset \cdots \supset F^{n+1}(E^n) = 0$. As for $r \geq 2$ we have $d_r^{1-r,r-1} = 0$, $d_r^{1,0} = 0$, it follows that $E_2^{1,0} = E_\infty^{1,0} \cong Gr_F^1 E^1 = F^1(E^1)$. Since $0 \to F^1(E^1) \to E^1 \to F^0(E^1)/F^1(E^1) \to 0$ is an exact sequence, we obtain an exact sequence $0 \to E_2^{1,0} \to E^1 \to E_\infty^{0,1} \to 0$. On the other hand, by $d_r^{-r,r} = 0$ $(r \geq 2)$, $d_r^{0,1} = 0$ $(r \geq 3)$ we have that $0 \longrightarrow E_\infty^{0,1} = E_3^{0,1} \longrightarrow E_2^{0,1} \xrightarrow{d_2^{0,1}} E_2^{2,0}$ is an exact sequence. By connecting this with the sequence above, we obtain an exact sequence $0 \longrightarrow E_2^{1,0} \longrightarrow E^1 \longrightarrow E_2^{0,1} \xrightarrow{d_2^{0,1}} E_2^{2,0}$. In addition, by $d_r^{2-r,r-1} = 0$ $(r \geq 3)$, $d_r^{2,0} = 0$ $(r \geq 2)$ we have $\operatorname{Coker} d_2^{0,1} = E_3^{2,0} = E_\infty^{2,0} = Gr_F^2 E^2 = F^2(E^2) \hookrightarrow E^2$, which yields the required exact sequence.

Proposition 3.6.3. *For a spectral sequence $E_{r_0}^{p,q} \Longrightarrow E^{p+q}$ the following hold:*

(i) Assume $r \geq 2$, and $E_r^{p,q} = 0$ holds for every $q \neq 0$, then $E_r^{p,0} \cong E^p$ holds for every p.
(ii) Assume $r \geq 2$ and $E_r^{p,q} = 0$ holds for every $p \neq 0$, then $E_r^{0,q} \cong E^q$ holds for every q.

Proof. (i) For $s \geq r$ we have $d_s^{p-s,s-1} = 0$ and $d_s^{p,0} = 0$. Therefore $E_r^{p,0} = E_\infty^{p,0}$ holds. On the other hand, for $t \neq p$ we have $Gr_F^t E^p \cong E_\infty^{t,p-t} = 0$, therefore we also have $F^p(E^p) = E^p$ and $F^{p+1}(E^p) = 0$. Hence, it follows that $E^p = Gr_F^p E^p \cong E_\infty^{p,0} = E_r^{p,0}$. (ii) is proved in the same way.

Let us introduce a typical method to construct a spectral sequence. It is from a filtered complex.

Definition 3.6.4. $(K^\bullet, F)$ is called a *filtered complex* if $K^\bullet \in K(\mathscr{C})$ and a descending chain F:

$$K^\bullet \supset \cdots \supset F^i(K^\bullet) \supset F^{i+1}(K^\bullet) \supset \cdots \supset 0$$

of subcomplexes satisfies $F^M(K^\bullet) = K^\bullet$ and $F^m(K^\bullet) = 0$ for some $M, m \in \mathbb{Z}$. We denote $F^p(K^\bullet)/F^{p+1}(K^\bullet)$ by $Gr_F^p K^\bullet$.

Theorem 3.6.5. *A filtered complex $(K^\bullet, F)$ induces a spectral sequence $E_1^{p,q} = H^{p+q}(Gr_F^p K^\bullet) \Longrightarrow E^{p+q} = H^{p+q}(K^\bullet)$ in a natural way.*

(Sketch of proof) For simplicity, the boundary operator $K^i \to K^{i+1}$ of a complex $K^\bullet$ is denoted by d (without superscripts). We define:

$$Z_r^{p,q} := d^{-1}\big(F^{p+r}(K^{p+q+1})\big) \cap F^p(K^{p+q}),$$
$$B_r^{p,q} := dF^{1+p-r}(K^{p+q-1}) + F^{p+1}(K^{p+q})$$

and define also $E_r^{p,q} := Z_r^{p,q}/Z_r^{p,q} \cap B_r^{p,q}$. From the inclusions $dZ_r^{p,q} \subset Z_r^{p+r,q-r+1}$ and $d(Z_r^{p,q} \cap B_r^{p,q}) \subset Z_r^{p+r,q-r+1} \cap B_r^{p+r,q-r+1}$, the boundary operator d induces a homomorphism $d_r : E_r^{p,q} \to E_r^{p+r,q-r+1}$. Let $E^n := H^n(K^\bullet)$ and let the filtration F on E^n be induced from the filtration F from that of $K^\bullet$. Then these satisfy the conditions of a spectral sequence. Here, we note that $B_1^{p,q} \subset Z_1^{p,q}$ and $E_1^{p,q} = Z_1^{p,q}/B_1^{p,q} = H^{p+q}(Gr_F^p K^\bullet)$.

Definition 3.6.6. Let $(K^\bullet, F)$ be a filtered complex. A filtered complex $(L^\bullet, G)$ is called a *filtered injective resolution* of $(K^\bullet, F)$, if there is a morphism $u : (K^\bullet, F) \to (L^\bullet, G)$ of filtered complexes (i.e., a morphism of complexes satisfying $u(F^pK^i) \subset F^pL^i$) such that the morphism $Gr_F^p K^\bullet \to Gr_G^p L^\bullet$ induced from u is an injective resolution of $Gr_F^p K^\bullet$ for each p.

In this case we note that the morphism $u : K^\bullet \to L^\bullet$ itself also gives an injective resolution of $K^\bullet$. In fact, in general if $0 \to A \to B \to C \to 0$ is an exact sequence in $\mathscr{C}$ and A is injective, then as "B is injective $\Leftrightarrow$ C is injective" we obtain that G^pL^i $(\forall p)$ is injective by the injectivity of $Gr_G^p L^i$ successively.

The fact that the morphism $K^\bullet \to L^\bullet$ is a quasi-isomorphism is proved by using the commutative diagram of exact sequences:

$$\begin{array}{ccccc}
H^{i-1}\big(Gr_F^p(K^\bullet)\big) & \longrightarrow & H^i\big(F^{p+1}(K^\bullet)\big) & \longrightarrow & H^i\big(F^p(K^\bullet)\big) \\
\downarrow & & \downarrow & & \downarrow \\
H^{i-1}\big(Gr_G^p(L^\bullet)\big) & \longrightarrow & H^i\big(G^{p+1}(L^\bullet)\big) & \longrightarrow & H^i\big(G^p(L^\bullet)\big)
\end{array}$$

$$\begin{array}{ccc} \longrightarrow H^i\big(Gr_F^p(K^\bullet)\big) & \longrightarrow & H^{i+1}\big(F^{p+1}(K^\bullet)\big) \\ \downarrow & & \downarrow \\ \longrightarrow H^i\big(Gr_G^p(L^\bullet)\big) & \longrightarrow & H^{i+1}\big(G^{p+1}(L^\bullet)\big). \end{array}$$

Here, as $F^m(K^\bullet) \cong G^m(L^\bullet) \cong 0$ $(\exists m)$, $H^j\big(Gr_F^p(K^\bullet)\big) \simeq H^j\big(Gr_G^p(L^\bullet)\big)$ $(\forall j)$ we obtain $H^i(K^\bullet) \simeq H^i(L^\bullet)$ by induction on p.

Theorem 3.6.7. *For a left exact functor $T : \mathscr{C} \to \mathscr{C}'$ and a filtered complex $(K^\bullet, F)$ on $\mathscr{C}$, there is the following spectral sequence:*

$$E_1^{p,q} = \mathbb{R}^{p+q} T Gr_F^p K^\bullet \Longrightarrow E^{p+q} = \mathbb{R}^{p+q} T K^\bullet.$$

Proof. Let $(L^\bullet, G)$ be a filtered injective resolution of a filtered complex $(K^\bullet, F)$. For a filtered complex $(TL^\bullet, TG)$ on $\mathscr{C}'$ we have a spectral sequence by Theorem 3.6.5:

$$E_1^{p,q} = H^{p+q}(Gr_{TG}^p TL^\bullet) \Longrightarrow E^{p+q} = H^{p+q}(TL^\bullet).$$

As $L^\bullet$ is an injective resolution of $K^\bullet$, we have $H^{p+q}(TL^\bullet) = \mathbb{R}^{p+q} T K^\bullet$. On the other hand, since $Gr_G^p L^\bullet$ is an injective resolution of $Gr_F^p K^\bullet$, it follows that $H^{p+q}(T Gr_G^p L^\bullet) = \mathbb{R}^{p+q} T Gr_F^p K^\bullet$. Here, $0 \to G^{p+1} L^i \to G^p L^i \to Gr_G^p L^i \to 0$ is an exact sequence of injective objects, hence it follows that $R^1 T G^{p+1} L^i = 0$, which yields $T Gr_G^p L^\bullet = Gr_{TG}^p T L^\bullet$.

Example 3.6.8 (Hodge spectral sequence). Let $(X, \mathscr{O}_X)$ be a non-singular analytic space over $\mathbb{C}$ (cf. Definition 4.1.8). Consider a complex of $\mathscr{O}_X$-Modules on X:

$$\Omega^\bullet : 0 \longrightarrow \mathscr{O}_X \longrightarrow \Omega_X^1 \longrightarrow \Omega_X^2 \longrightarrow \cdots.$$

Let F be the stupid filtration of $\Omega^\bullet$, i.e., a filtration F of $K^\bullet$ that is defined as:

$$(F^p K^\bullet)^i = \begin{cases} K^i & i \geq p \\ 0 & i < p. \end{cases}$$

Then $Gr_F^p \Omega_X^\bullet = \Omega_X^p[-p] = \{0 \to \cdots \to 0 \to \overset{p}{\Omega_X} \to 0 \to \cdots\}$; therefore by letting $T = \Gamma$ we obtain $\mathbb{R}^{p+q} \Gamma Gr_F^p \Omega_X^\bullet = \mathbb{R}^{p+q} \Gamma \Omega_X^p[-p] = H^q(X, \Omega_X^p)$.

On the other hand, by the De Rham Theorem, $\Omega_X^\bullet$ is right resolution of the constant sheaf $\mathbb{C}$ on X, which implies that $\mathbb{C}$ and $\Omega_X^\bullet$ are quasi-isomorphic (here, $\mathbb{C}$ is regarded as a complex with $\mathbb{C}$ at degree 0 and 0 at all other degrees). Therefore we have $\mathbb{R}^{p+q} \Gamma \Omega^\bullet = \mathbb{R}^{p+q} \Gamma \mathbb{C} = H^{p+q}(X, \mathbb{C})$. By this we obtain a spectral sequence:

$$E_1^{p,q} = H^q(X, \Omega_X^p) \Longrightarrow E^{p+q} = H^{p+q}(X, \mathbb{C}),$$

which is called the *Hodge spectral sequence*. In particular, if X is a compact Kähler manifold, it degenerates at E_1. In fact by the harmonic integral theory it is known that $H^i(X, \mathbb{C}) \cong \bigoplus_{p+q=i} H^q(X, \Omega_X^p)$ [MK, Theorem 5.4]. Therefore we have $\dim_{\mathbb{C}} E^i = \sum_{p+q=i} \dim_{\mathbb{C}} E_1^{p,q}$.

On the other hand, in general we have

$$(1_{p,q}) \qquad \dim_{\mathbb{C}} E_1^{p,q} \geqq \dim_{\mathbb{C}} E_2^{p,q} \geqq \cdots \geqq \dim_{\mathbb{C}} E_\infty^{p,q} = \dim_{\mathbb{C}} Gr_F^p E^{p+q},$$

which implies that $\dim E^i \leqq \sum_{p+q=i} \dim_{\mathbb{C}} E_1^{p,q}$ and the equality holds if and only if for every (p, q) such that $p+q = i$ all equalities in $(1_{p,q})$ hold. Therefore, in this case the equality in $(1_{p,q})$ holds and the spectral sequence degenerates at E_1.

Example 3.6.9 (*Leray's spectral sequence*). Let $f : X \to Y$, $g : Y \to Z$ be morphisms of ringed spaces. For $\mathscr{F} \in (\mathrm{Mod}_{\mathscr{O}_X})$, take an injective resolution $0 \to \mathscr{F} \to \mathscr{I}^\bullet$ of $\mathscr{F}$. On a complex $f_*\mathscr{I}^\bullet$ we define a filtration F as follows:

$$(F^p f_*\mathscr{I}^\bullet)^i = \begin{cases} 0, & i > -p \\ \operatorname{Ker} d^{-p}, & i = -p \\ f_*\mathscr{I}^i, & i < -p \end{cases}.$$

For a filtered complex $(f_*\mathscr{I}^\bullet, F)$ and a left exact functor $g_* : (\mathrm{Mod}_{\mathscr{O}_Y}) \to (\mathrm{Mod}_{\mathscr{O}_Z})$, we obtain a spectral sequence by Theorem 3.6.7:

$$E_1^{p,q} = \mathbb{R}^{p+q} g_* Gr_F^p f_*\mathscr{I}^\bullet \Longrightarrow E^{p+q} = \mathbb{R}^{p+q} g_*(f_*\mathscr{I}^\bullet).$$

Here, by the definition of the filtration, we have

$$H^i(Gr_F^p f_*\mathscr{I}^\bullet) = \begin{cases} 0, & i \neq -p \\ \operatorname{Ker} d^{-p}/\operatorname{Im} d^{-p-1}, & i = -p \end{cases}.$$

Then, noting that $\operatorname{Ker} d^{-p}/\operatorname{Im} d^{-p-1} = R^{-p} f_*\mathscr{F}$, we obtain $E_1^{p,q} = R^{2p+q} g_* (R^{-p} f_*\mathscr{F})$. On the other hand, $f_*\mathscr{I}^i$ is flabby for each i; by Example 3.5.7 it follows that $E^{p+q} = H^{p+q}\big(g_*(f_*\mathscr{I}^\bullet)\big) = H^{p+q}\big((g \circ f)_*\mathscr{I}^\bullet\big) = R^{p+q}(g \circ f)_*\mathscr{F}$. By this the above spectral sequence becomes

$$E_1^{p,q} = R^{2p+q} g_*(R^{-p} f_*\mathscr{F}) \Longrightarrow E^{p+q} = R^{p+q}(g \circ f)_*\mathscr{F}.$$

Here, by letting $E_r^{p,q}$ be $E_{r+1}^{2p+q,-p}$, the spectral sequence becomes

$$E_2^{p,q} = R^p g_*(R^q f_*\mathscr{F}) \Longrightarrow E^{p+q} = R^{p+q}(g \circ f)_*\mathscr{F}.$$

This is called Leray's spectral sequence.

The above example is a spectral sequence induced from two right exact functors f_*, g_*. In a similar way, by two left exact functors we obtain a spectral sequence. The following is such an example:

Example 3.6.10. Let $(X, \mathcal{O}_X)$ be a ringed space. Consider two left exact functors $\mathcal{H}om_{\mathcal{O}_X}(\mathcal{F}, \ ) : (\mathrm{Mod}_{\mathcal{O}_X}) \to (\mathrm{Mod}_{\mathcal{O}_X})$, $\Gamma(X,) : (\mathrm{Mod}_{\mathcal{O}_X}) \to (\mathrm{Ab})$. For $\mathcal{G} \in (\mathrm{Mod}_{\mathcal{O}_X})$ take an injective resolution $0 \to \mathcal{G} \to \mathcal{I}^\bullet$ of $\mathcal{G}$. Define a filtration F on $\mathcal{H}om_{\mathcal{O}_X}(\mathcal{F}, \mathcal{I}^\bullet)$ as follows:

$$F^p\big(\mathcal{H}om_{\mathcal{O}_X}(\mathcal{F}, \mathcal{I}^\bullet)\big) = \begin{cases} 0, & i > -p \\ \mathrm{Ker}\, d^{-p}, & i = -p \\ \mathcal{H}om_{\mathcal{O}_X}(\mathcal{F}, \mathcal{I}^i), & i < -p \end{cases} .$$

For a filtered complex $\big(\mathcal{H}om_{\mathcal{O}_X}(\mathcal{F}, \mathcal{I}^\bullet), F\big)$ and a left exact functor $\Gamma(X,)$ we obtain a spectral sequence by Theorem 3.6.7:

$$\begin{aligned} E_1^{p,q} &= \mathbb{R}^{p+q}\Gamma\big(X, Gr_F^p \mathcal{H}om_{\mathcal{O}_X}(\mathcal{F}, \mathcal{I}^\bullet)\big) \\ &\Longrightarrow E^{p+q} = \mathbb{R}^{p+q}\Gamma\big(X, \mathcal{H}om_{\mathcal{O}_X}(\mathcal{F}, \mathcal{I}^\bullet)\big). \end{aligned}$$

Here, by the definition of the filtration we have

$$H^i\big(Gr_F^p \mathcal{H}om_{\mathcal{O}_X}(\mathcal{F}, \mathcal{I}^\bullet)\big) = \begin{cases} 0, & i \neq -p \\ \mathrm{Ker}\, d^{-p}/\mathrm{Im}\, d^{-p-1}, & i = -p. \end{cases}$$

Therefore, by noting that $\mathrm{Ker}\, d^{-p}/\mathrm{Im}\, d^{-p-1} = \mathcal{E}xt_{\mathcal{O}_X}^{-p}(\mathcal{F}, \mathcal{G})$, we obtain $E_1^{p,q} = H^{2p+q}\big(X, \mathcal{E}xt_{\mathcal{O}_X}^{-p}(\mathcal{F}, \mathcal{G})\big)$. On the other hand, noting that $\mathcal{H}om_{\mathcal{O}_X}(\mathcal{F}, \mathcal{I}^i)$ is injective for each i, we obtain

$$E^{p+q} = H^{p+q}\big(\mathcal{H}om_{\mathcal{O}_X}(\mathcal{F}, \mathcal{I}^\bullet)\big) = \mathrm{Ext}_{\mathcal{O}_X}^{p+q}(\mathcal{F}, \mathcal{G}).$$

By replacement of suffixes as in Example 3.6.9 we obtain the following spectral sequence:

$$E_2^{p,q} = H^p\big(X, \mathcal{E}xt_{\mathcal{O}_X}^q(\mathcal{F}, \mathcal{G})\big) \Longrightarrow E^{p+q} = \mathrm{Ext}_{\mathcal{O}_X}^{p+q}(\mathcal{F}, \mathcal{G}).$$

Chapter 4
Definition of a Singularity, Resolutions of Singularities

Many things can cause mistakes:
similar symbols, sloppy handwriting,
alcohol last night, teacher's advice...

In this chapter we define singularities on analytic spaces or on algebraic varieties. Then, we introduce the fact that an isolated singularity on an analytic space can be regarded as a singularity on an algebraic variety. We also introduce Hironaka's theorem stating that every algebraic variety over a field of characteristic zero has a resolution of the singularities. In this book our interest is focused on singularities. To this end, we study the resolved space instead of studying the singularity itself, therefore this resolution theorem is essential.

Convention As are defined in Chap. 2, an analytic space and an algebraic variety are certain ringed spaces $(X, \mathscr{O}_X)$. From now on, unless otherwise stated, an analytic space or an algebraic variety is denoted by X instead of $(X, \mathscr{O}_X)$.

The base field k of an algebraic variety is, unless otherwise stated, an arbitrary algebraically closed field.

4.1 Definition of a Singularity

In order to define singularities we need to define the tangent cone and the tangent space.

Definition 4.1.1. Let x be a point of an algebraic variety over k and X an affine neighborhood of x. The neighborhood X is an algebraic set in $\mathbb{A}_k^N$ and by the coordinate transformation of $\mathbb{A}_k^N$, we may assume that x is the origin 0. Let $I \subset k[X_1, \ldots, X_N]$

S. Ishii, *Introduction to Singularities*,
https://doi.org/10.1007/978-4-431-56837-7_4

be the defining ideal of X. Here, ideals I^*, I^1 in $k[X_1, \dots, X_N]$ are defined as follows:

$$I^* := \text{the ideal generated by the leading form of each element of } I,$$
$$I^1 := \text{the ideal generated by the linear form of each element of } I.$$

The closed algebraic set in $\mathbb{A}_k^N$ defined by I^* is called the *tangent cone* of X at x and denoted by C_x. The closed algebraic set in $\mathbb{A}_k^N$ defined by I^1 is called the *tangent space* of X at x and denoted by T_x.

Here, let $f \in k[X_1, \dots, X_N]$ be presented as $f(X_1, \dots, X_N) = \sum_{i=r}^{s} f_i$, ($f_i$ is a homogeneous polynomial of degree i and $f_r \neq 0$). Then the leading form of f is f_r and the linear form of f is f_1. As $0 \in X$, each element of I has the leading form of degree ≥ 1. Therefore, if f has the linear form, then it is the leading form, which yields $I^1 \subset I^*$.

Definition 4.1.1′. Let x be a point in an analytic space X in $\mathbb{C}^N$. We can define the tangent cone C_x and the tangent space T_x of X at x in the same way. Here, we should have Taylor expansion of each element of I. The leading form and the linear form are those of the Taylor series.

By the definition, C_x and T_x are both affine varieties in $\mathbb{A}_k^N$ (or $\mathbb{C}^N$ in the case of an analytic space).

Below, we will see properties of affine varieties C_x and T_x.

Lemma 4.1.2. *Let X be an affine variety or an analytic space in $\mathbb{C}^N$, x a point of X, $\mathscr{O}_X$ the structure sheaf and $\mathfrak{m} \subset \mathscr{O}_{X,x}$ the maximal ideal. Then, about the tangent cone and the tangent space, the following hold:*

(i) $T_x \cong \mathrm{Hom}_k(\mathfrak{m}/\mathfrak{m}^2, k)$ as vector spaces;
(ii) Let $A(C_x)$ be the affine coordinate ring of C_x. Then, $A(C_x) \cong \bigoplus_{n\geq 0} \mathfrak{m}^n/\mathfrak{m}^{n+1}$ and the dimension at the origin of C_x is the dimension of X at x.

By this lemma, T_x and C_x are independent of the choice of embeddings $x = 0 \in X \subset \mathbb{A}_k^N$ up to isomorphism.

Proof. (i) Let M be the defining ideal $(X_1, \dots, X_N)$ of the origin of $\mathbb{A}_k^N$, then we have $\mathrm{Hom}_k(k^N, k) \cong M/M^2$. This isomorphism sends the kernel of the dual $\mathrm{Hom}_k(k^N, k) \xrightarrow{\alpha} \mathrm{Hom}_k(T_x, k)$ of the injection $T_x \hookrightarrow k^N$ to the vector subspace $I^1/I^1 \cap M^2$. Noting that $I^1 + M^2 = I + M^2$, we obtain $\mathrm{Hom}_k(T_x, k) \cong (M/M^2)/(I^1/I^1 \cap M^2) \cong M/(I^1 + M^2) \cong M/(I + M^2) \cong \frac{M/I}{(M/I)^2} \cong \mathfrak{m}/\mathfrak{m}^2$.

(ii) As I^* is a homogeneous ideal, it follows that $A(C_x) = k[X_1, \dots, X_N]/I^* \cong (\bigoplus_{n\geq 0} M^n/M^{n+1}) / \bigoplus_{n\geq 0} (I^* \cap M^n / I^* \cap M^{n+1}) \cong \bigoplus_{n\geq 0} M^n/(I^* \cap M^n + M^{n+1})$. By the definition of I^*, we have $I^* \cap M^n + M^{n+1} = I \cap M^n + M^{n+1}$, therefore $A(C_x) \cong \bigoplus_{n\geq 0} M^n/(I \cap M^n + M^{n+1}) = \bigoplus_{n\geq 0} \mathfrak{m}^n/\mathfrak{m}^{n+1}$. For the statement on the dimension, let A be the localization of the ring $\bigoplus_{k\geq 0} \mathfrak{m}^k/\mathfrak{m}^{k+1}$ by the maximal ideal $\bigoplus_{k\geq 1} \mathfrak{m}^k/\mathfrak{m}^{k+1}$ and let the maximal ideal be $\mathscr{M}$. Then

$\mathscr{M}^k/\mathscr{M}^{k+1} \cong \mathfrak{m}^k/\mathfrak{m}^{k+1}$. Hence, we have $\dim_k \mathscr{O}_{X,x}/\mathfrak{m}^k = \sum_{l=0}^{k-1} \dim \mathfrak{m}^l/\mathfrak{m}^{l+1} = \dim_k A/\mathscr{M}^k$. By using the dimension theorem [AM, Th. 11.14], we obtain $\dim \mathscr{O}_{X,x} = \big(\text{degree of the polynomial } \dim \mathscr{O}_{X,x}/\mathfrak{m}^k\ (k \gg 0) \text{ in } k\big) = \dim A$.

Let us see examples of C_x and T_x:

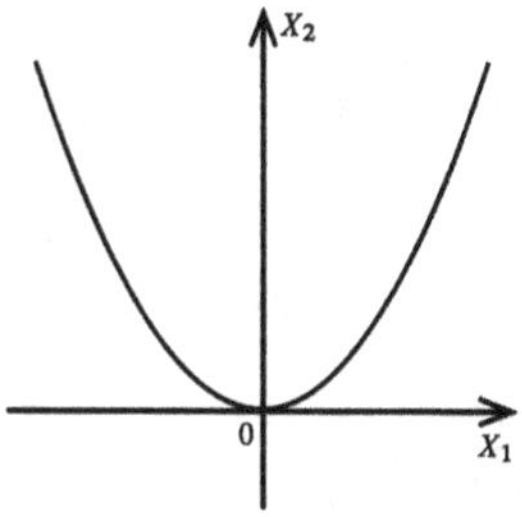

Example 4.1.3

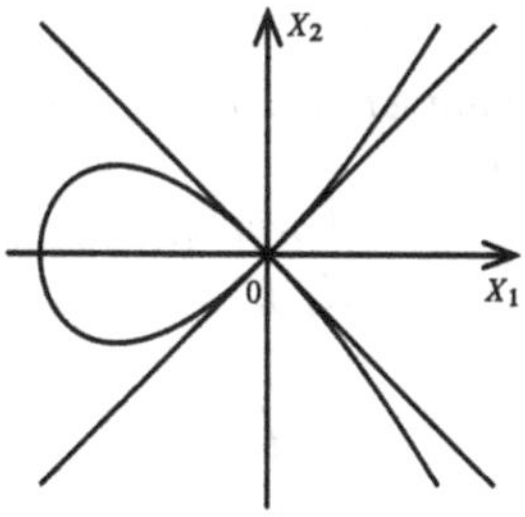

Example 4.1.4

Example 4.1.3. Let the defining ideal I of $X \subset \mathbb{A}^2_{\mathbb{C}}$ be $(X_2 - X_1^2)$. Then $I^* = (X_2)$ and $I^1 = (X_2)$, which yield $C_0 = T_0 =$(X_1-axis).

Example 4.1.4. Let the defining ideal of $X \subset \mathbb{A}^2_{\mathbb{C}}$ be $I = ((X_1 + X_2)(X_1 - X_2) + X_1^3)$, then $I^* = ((X_1 + X_2)(X_1 - X_2))$ and $I^1 = (0)$. Therefore, the tangent cone C_0 is the union of lines $X_1 + X_2 = 0$ and $X_1 - X_2 = 0$. On the other hand, the tangent space T_0 is the whole space $\mathbb{A}^2_{\mathbb{C}}$.

Example 4.1.5. Let the defining ideal of $X \subset \mathbb{A}^2_{\mathbb{C}}$ be $I = ({X_2}^2 - {X_1}^3)$, then $I^* = ({X_2}^2)$ and $I^1 = (0)$. Therefore the tangent cone C_0 is a double X_1-axis and the tangent space T_0 is the whole space $\mathbb{A}^2_{\mathbb{C}}$.

Remark 4.1.6. For a hypersurface, as we have seen in the examples, I^* is generated by the leading form of generators of I. But it does not hold in general.

On the other hand, I^1 is generated by the linear form of generators of I.

As $\mathbb{A}^2_{\mathbb{C}}$ is of dimension 2 over the complex number field, it is of real dimension 4. So, strictly speaking, the pictures above are not correct. The reader can think these pictures as the sections of the varieties by the real plane.

On defining singularities, the following theorem is important.

Theorem 4.1.7. *Let X be an algebraic variety over k or an analytic space, x a point of X, n the dimension of X at x and $\mathfrak{m} \subset \mathscr{O}_{X,x}$ the maximal ideal. Then the following hold, where if X is an analytic space, we regard k as $\mathbb{C}$:*

(i) $C_x \subseteq T_x$.
(ii) $\dim T_x \geqq n$.
(iii) *The number of elements of minimal system of generators of $\mathfrak{m}$ over $\mathscr{O}_{X,x}$ is $\geqq n$.*
(iv) $\dim_k \mathfrak{m}/\mathfrak{m}^2 \geq n$.

(v) *Let* $m := \dim_k \mathfrak{m}/\mathfrak{m}^2$, *then there is a surjective homomorphism of graded rings:*

$$\varphi : k[X_1, \ldots, X_m] \longrightarrow \bigoplus_{k \geq 0} \frac{\mathfrak{m}^k}{\mathfrak{m}^{k+1}}.$$

(vi) *The number of elements of the minimal system of generators of* $\Omega_{X,x}$ *over* $\mathcal{O}_{X,x}$ *is* $\geqq n$.

(vii) *For a neighborhood* U *of* x, *assume* $U \subset \mathbb{A}_k^N$ *and let* U *be defined by* $f_1(X_1, \ldots, X_N), \ldots, f_d(X_1, \ldots, X_N)$. *Then, the following holds:*

$$\operatorname{rank} \begin{pmatrix} \frac{\partial f_1}{\partial X_1}(x), & \frac{\partial f_2}{\partial X_1}(x) & \cdots\cdots & \frac{\partial f_d}{\partial X_1}(x) \\ \vdots & & & \vdots \\ \frac{\partial f_1}{\partial X_N}(x) & & \cdots\cdots & \frac{\partial f_d}{\partial X_N}(x) \end{pmatrix} \leqq N - n.$$

The matrix of left-hand side is called the Jacobian matrix of $U \subset \mathbb{A}_k^N$ *at* x *and denoted by* $J(x)$.

(viii) *In a statement in (i)–(vii), the equality holds if and only if the equality holds in another statement. Here, for (v) "the equality holds" means that* φ *is an isomorphism.*

Proof. (i) As $I^1 \subset I^*$, it is clear that $C_x \subseteq T_x$.

(ii) By (i), we have $\dim_x C_x \leq \dim_x T_x$; on the other hand, by Lemma 4.1.2 (ii) the left-hand side $= n$.

(iii) As $\dim_x X = n$, the maximal ideal $\mathfrak{m}$ has height n, therefore the number of generators is greater than or equal to n [AM, Cor. 11.16].

(iv) By Nakayama's lemma, $\dim_k \mathfrak{m}/\mathfrak{m}^2$ is equal to the number of systems of generators of $\mathfrak{m}$, hence the statement follows from (iii).

(v) The graded ring $\bigoplus_{k \geq 0} \mathfrak{m}^k/\mathfrak{m}^{k+1}$ is generated by the degree 1 part $\mathfrak{m}/\mathfrak{m}^2$. Then we obtain the required surjective homomorphism of graded rings by associating X_i of $k[X_1, \ldots, X_m]$ to a generator of $\mathfrak{m}/\mathfrak{m}^2$.

(vi) By Nakayama's lemma, it is sufficient to prove that $\dim_k \Omega_{X,x}/\mathfrak{m}\Omega_{X,x} \geqq n$. The map

$$\Omega_{X,x}/\mathfrak{m}\Omega_{X,x} \longrightarrow \mathfrak{m}/\mathfrak{m}^2; \ \overline{df} \longmapsto \overline{f - f(x)}$$

is an isomorphism of vector spaces, therefore the required statement follows from (iv).

(vii) Let $x = (x_1, \ldots, x_N) \in U \subset \mathbb{A}_k^N$. By a coordinate transformation $x \mapsto 0$, the defining ideal of U is represented as:

$$I = (f_1(X_1 + x_1, \ldots X_N + x_N), \ldots, f_d(X_1 + x_1, \ldots X_N + x_N)),$$

where, noting that I^1 is generated by the linear terms of generators, and the linear term of $f_i(X+x)$ is $\sum_{j=1}^{N} \frac{\partial f_i}{\partial X_j}(x)X_j$. Hence we have

$$I^1 = \left(\sum_{j=1}^{N} \frac{\partial f_1}{\partial X_j}(x)X_j, \ldots, \sum_{j=1}^{N} \frac{\partial f_d}{\partial X_j}(x)X_j \right).$$

Therefore the linear subspace T_x in $\mathbb{A}_k^N$ generated by I^1 is of dimension m by Lemma 4.1.2 (i) and it coincides with

$$N - \operatorname{rank} \begin{pmatrix} \frac{\partial f_1}{\partial X_1}(x) & \cdots & \frac{\partial f_d}{\partial X_1}(x) \\ \vdots & & \vdots \\ \frac{\partial f_1}{\partial X_N}(x) & \cdots & \frac{\partial f_d}{\partial X_N}(x) \end{pmatrix}.$$

By (ii), this is greater than or equal to n.

(vii) Next we check the equivalence of the equalities in each statement. As we see, the inequalities in (ii), (iii), (iv), (vi), (vii) are all equivalent to $m \geqq n$; the equalities are equivalent. In (i), by the inclusion relation and by Lemma 4.1.2, it follows that $m \geq n$. If the equality holds in (i), then obviously we have $m = n$. Conversely, if $m = n$, then as T_x is irreducible, the equality in (i) holds. Regarding (v), the subjectivity in (v) and Lemma 4.1.2 (ii) implies $m \geq n$. If the homomorphism in (v) is isomorphism, then $m = n$. Conversely, assume $m = n$. If $\operatorname{Ker} \varphi \neq \{0\}$, then by Krull's principal ideal theorem [AM, Cor. 11.17], we have the dimension of $\bigoplus_{k \geq 0} \mathfrak{m}^k/\mathfrak{m}^{k+1}$ is $< n$, which is a contradiction.

Definition 4.1.8. If for a point x of an algebraic variety or an analytic space X the equalities in Theorem 4.1.7 hold, we say that x is a *smooth point* or non-singular point of X. A point which is not smooth is called a *singular point*. Let $X_{\text{sing}} = \{x \in X \mid x \text{ is a singular point of } X\}$ and call this set the *singular locus* of X. On the other hand, the set of non-singular points is denoted by X_{reg}. In particular, for a point $x \in X$, if x has an open neighborhood U such that $U_{\text{sing}} = \{x\}$, then x is called an *isolated singularity* of X. If all points of X are non-singular, then X is called *non-singular*.

By Lemma 4.1.2, the notion of singular or non-singular is independent of the choice of embeddings $X \subset \mathbb{A}_k^N$.

By the definition, Examples 4.1.4 and 4.1.5 are both varieties with the singular point at the origin. On the other hand, $\mathbb{A}_k^n$ is a non-singular variety.

Proposition 4.1.9. *If an algebraic variety (or analytic space) X is irreducible and reduced, the singular locus X_{sing} is an algebraic closed set (or analytic closed set) of X and the subset X_{reg} is a dense subset of X.*

Proof. For the first statement, it is sufficient to prove that for each point $x \in X$ there is an open neighborhood U such that the singular locus U_{sing} is an algebraic closed set (or analytic closed set). By Theorem 4.1.7 (vii), there is an open neighborhood U for each point x such that $U_{\text{sing}} = \{y \in U \mid \text{all } (N-n)\text{-minors of } \left(\frac{\partial f_j}{\partial X_i}(y)\right) \text{ are } 0\}$, which shows the required statement. For the second assertion, see for example [Ha, I. Th. 5.3].

Definition 4.1.10. If an n-dimensional local ring $\mathcal{O}$ and its maximal ideal satisfy the equalities in (iii)–(v) in Theorem 4.1.7, $\mathcal{O}$ is called *regular*. If $\mathcal{O}$ is an integral domain and integrally closed in the quotient field, then $\mathcal{O}$ is called *normal*. A point x of an algebraic variety or an analytic space X is called a *normal point* if the local ring $\mathcal{O}_{X,x}$ is normal. If each point of X is normal, then X is called *normal*.

Regarding normality, the following is well known and often used.

Theorem 4.1.11. *(i) A regular local ring is normal. Therefore, a non-singular variety is normal. The converse does not hold in general, but for the 1-dimensional case the converse also holds.*

(ii) The set of non-normal points of an algebraic variety or an analytic space X is an algebraic closed set or an analytic closed set, respectively.

(iii) If X is a normal algebraic variety or a normal analytic space, then the singular locus has a codimension greater than or equal to 2. In particular, in a non-singular variety Y, a closed subvariety X is defined by one equation; the converse holds, i.e., if $\operatorname{codim}_X X_{\text{sing}} \geqq 2$ *holds, then X is normal.*

(iv) An irreducible reduced algebraic variety or analytic space X has a morphism called the normalization $\nu : \overline{X} \to X$ as follows:

(a) $\overline{X}$ is normal.

(b) If $\mu : Y \to X$ is a surjective morphism from a normal algebraic variety or analytic space Y, there exists a unique $g : Y \to \overline{X}$ such that $\mu = \nu \circ g$.

(v) Let X be an algebraic variety or analytic space, and $U \subset X$ an open subset such that $\operatorname{codim}_X(X \setminus U) \geqq 2$. *If U is normal, then the following are equivalent:*

(a) X is normal.

(b) $\mathcal{O}_X = j_*\mathcal{O}_X|_U$, where $j : U \hookrightarrow X$ is the inclusion map.

(Sketch of proof) (i) follows from [Ma1, Th. 36, p. 121]. (iii) follows from the fact that for a normal ring R, the localization R_P by any prime ideal P is again normal and in particular if P is a prime ideal of height 1, R_P is regular by the last statement of (i) (this argument needs a scheme theoretic discussion). (iv) is proved by using the fact that the integral closure of finitely generated k-algebra is also finitely generated k-algebra [ZS, Vol. 1, Ch V, Th. 9, p. 267]. Here, we note that $\nu : \overline{X} \to X$ is a finite morphism. For (ii), we should note that $x \in X$ is normal if and only if $\mathcal{O}_{X,x} \xrightarrow{\nu^*} (\nu_*\mathcal{O}_{\overline{X}})_x$ is surjective. As ν is a finite morphism, the sheaf $\nu_*\mathcal{O}_{\overline{X}}$ is a coherent $\mathcal{O}_X$-Module, therefore Coker φ is also coherent. Then the support is algebraic closed or analytic closed.

(v-1) ⇒ (v-2): For an affine open subset $V \subset X$, the ring $R = \Gamma(V, \mathcal{O}_X)$ is normal. Then, $R = \bigcap_{\substack{\mathfrak{p}:\textit{prime ideal}\\ \mathrm{ht}\,\mathfrak{p}=1}} R_{\mathfrak{p}}$. On the other hand, by the assumption on U, $\Gamma(U \cap V, \mathcal{O}_X) \subset R_{\mathfrak{p}}$ holds for evey prime ideal $\mathfrak{p}$ such that $\mathrm{ht}\,\mathfrak{p} = 1$. Therefore, we have $\Gamma(U \cap V, \mathcal{O}_X) \subset \Gamma(V, \mathcal{O}_X)$. The converse inclusion is obvious. (v-2) ⇒ (v-1): Let $\nu : \overline{X} \to X$ be the normalization, then $\nu_* \mathcal{O}_{\overline{X}}|_U = \mathcal{O}_X|_U$. Therefore $\nu_* \mathcal{O}_{\overline{X}} \subset j_* \mathcal{O}_X|_U = \mathcal{O}_X$, which yields $\nu_* \mathcal{O}_{\overline{X}} = \mathcal{O}_X$ and ν is isomorphic.

4.2 Algebraization Theorem

In this section we assume the base field is $\mathbb{C}$. In the previous section, we introduced singularities. Now we look at singular points, for example

$$y^2 - x^2 - x^3 = 0 \tag{4.1}$$

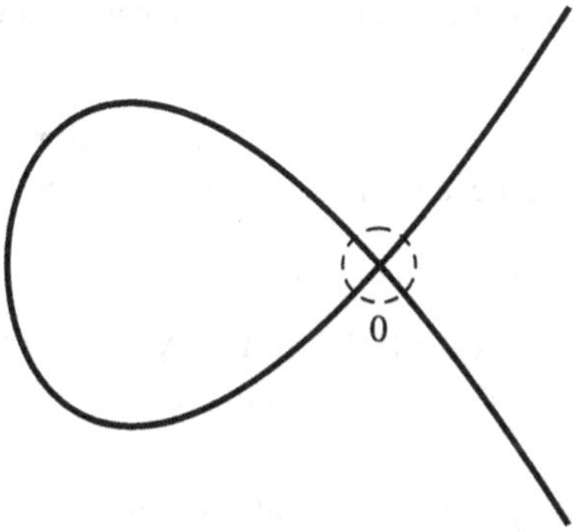

$$y^2 - x^2 = 0 \tag{4.2}$$

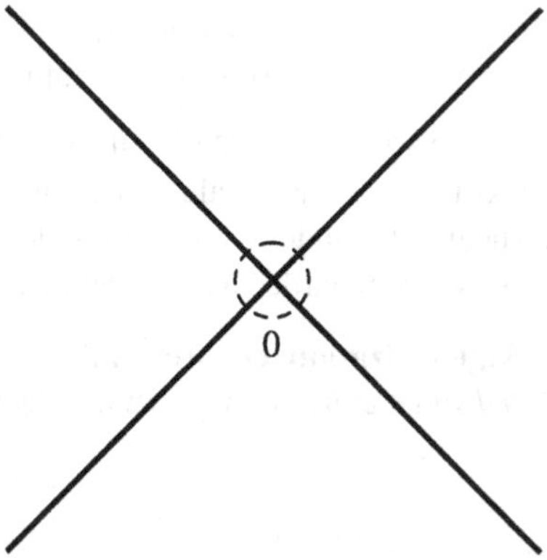

These varieties over $\mathbb{C}$ are of course different varieties. But at very small neighborhoods of the origin, both look like two lines $y = \pm x$ intersecting at the origin. In such a case we want to regard the singularities as the same. In this section we formulate it exactly. On the other hand, an analytic space defined by

$$\sin(y - x)\sin(y + x) = 0 \tag{4.3}$$

is not an algebraic variety, but if one looks at a very small neighborhood of the origin, then we can regard it as the same as (4.1) and (4.2).

Definition 4.2.1. On the set $\mathscr{A}$ of pairs (X, x) consisting of an analytic space X and its point x, we define a relation $\sim$ as follows:

$$(X, x) \sim (Y, y) \iff \text{there exist a neighborhood } U \subset X \text{ of } x, \text{ a neighborhood } V \subset Y \text{ of } y \text{ and an isomorphism } f : U \simeq V \text{ such that } f(x) = y.$$

The relation becomes an equivalence relation; let the quotient set $\mathscr{G} := \mathscr{A}/\sim$. An element of $\mathscr{G}$ is called a *germ of the analytic space* and we denote it by a representative (X, x).

Definition 4.2.2. Let R be a Noetherian local ring and $\mathfrak{m}$ the maximal ideal. Let $\widehat{R} := \varprojlim_{n\to\infty} R/\mathfrak{m}^n$ and call it the *completion* of R by $\mathfrak{m}$. Actually it is the completion of R by $\mathfrak{m}$-adic topology, but we do not step into topology here.

Theorem 4.2.3 (Hironaka, Rossi [HR], Artin [A4]). *Let (X, x), (Y, y) be germs of analytic spaces, then the following are equivalent:*

(i) $(X, x) = (Y, y)$.
(ii) *There exists an isomorphism of $\mathbb{C}$-algebras* $\widehat{\mathscr{O}}_{X,x} \simeq \widehat{\mathscr{O}}_{Y,y}$.
(iii) *There exists an isomorphism of $\mathbb{C}$-algebras* $\mathscr{O}_{X,x} \simeq \mathscr{O}_{Y,y}$.

Proof. The equivalence of (i) and (ii) is proved in [HR] for the isolated singularities' case and [A4, Cor. 1.6] for the general case. (i) $\Rightarrow$ (iii), (iii) $\Rightarrow$ (ii) are trivial.

Examples (4.1)–(4.3) have the same completion $\mathbb{C}[[x, y]]/(x + y)(x - y)$ at the origin. Therefore, by Theorem 4.2.3 the germs at the origin coincide.

Now consider an algebraic variety X over $\mathbb{C}$. The variety X has the structure sheaf $\mathscr{O}_X^{\mathrm{alg}}$ under the Zariski topology as an algebraic variety. On the other hand, there is also the structure sheaf $\mathscr{O}_X^{\mathrm{hol}}$ under the usual topology induced from $\mathbb{C}^N$ as an analytic space. By this X is also regarded as an analytic space.

Theorem 4.2.4 (Artin's Algebraization Theorem). *For a germ (X, x) of an analytic space, if x is an isolated singularity, there exists an algebraic variety A over $\mathbb{C}$ and a point of A such that*

$$(X, x) = (A, P).$$

Proof. By [A5, Theorem 3.8] there exist an algebraic variety over $\mathbb{C}$ and its point P such that $\widehat{\mathscr{O}}^{\text{hol}}_{X,x} \cong \widehat{\mathscr{O}}^{\text{alg}}_{A,P}$. As $\widehat{\mathscr{O}}^{\text{alg}}_{A,P} = \widehat{\mathscr{O}}^{\text{hol}}_{A,P}$, by Theorem 4.2.3 we obtain $(X, x) = (A, P)$.

By this, when we think of an isolated singularity on an analytic space, we may assume that it is an isolated singularity on an algebraic variety.

Henceforth, unless otherwise stated, an isolated singularity (X, x) means that X is an algebraic variety and $\mathscr{O}_X$ is the structure sheaf of the algebraic variety.

4.3 Resolutions of Singularities and Blow-Up

In this section we consider an algebraic variety over an algebraically closed field k. In the first part we introduce Hironaka's theorem which shows the existence of resolutions of singularities of an algebraic variety when the characteristic of k is zero. We then show how to construct resolutions of singularities for special singularities.

For the definition of resolutions we prepare some notions on morphisms.

Definition 4.3.1. (1) A morphism $f : Y \to X$ of algebraic varieties is called *projective* if there exists a closed immersion $i : Y \hookrightarrow \mathbb{P}^N \times X$ such that f factors as $f = p_2 \circ i$, where $p_2 : \mathbb{P}^N \times X \to X$ is the projection onto the second factor.

(2) A morphism $f : Y \to X$ of algebraic varieties is called *locally projective* if there exists an open covering $\{U_i\}$ of X such that $f|_{f^{-1}(U_i)}$ is projective for every i.

(3) A morphism $f : Y \to X$ of algebraic varieties is called a *proper morphism* if for every morphism $g : Z \to X$, the projection $p_2 : Y \times_X Z \to Z$ is a closed map. In particular, if X is a point, and the trivial map $f : Y \to X$ is proper, then we call Y *complete*.

By the definition, a proper map is itself a closed map. On the other hand, a locally projective morphism is proper (cf. [Ha2, II,4]).

Definition 4.3.2. Let X be an algebraic variety (then $X \setminus X_{\text{sing}}$ is an open dense subset of X). A morphism $f : Y \to X$ (or Y) is called a *resolution of the singularities* of X, if the following hold:

(1) The morphism f is proper.
(2) The restriction $f|_{Y \setminus f^{-1}(X_{\text{sing}})} : Y \setminus f^{-1}(X_{\text{sing}}) \to X \setminus X_{\text{sing}}$ of f is isomorphic.
(3) The variety Y is non-singular.

A morphism $f : Y \to X$ is called a *weak resolution* if it satisfies (1), (3) and the following:

(2)′ The morphism f is isomorphic on open subsets of Y and X.

The morphism $f : Y \to X$ (or Y) is called a *partial resolution* of the singularities of X, if it satisfies (1), (2)′ and the following:

(4) The algebraic variety Y is normal.

Definition 4.3.3. Let $f : Y \to X$ be a partial resolution and let an open subset $X_0 \subset X$ be such that the restriction $f|_{f^{-1}(X_0)}$ is isomorphic. For an irreducible closed subset Z of X such that $Z \cap X_0 \neq \emptyset$, the closure $Z' := \overline{f^{-1}(Z \cap X_0)}$ is called the *strict transform* of Z.

Theorem 4.3.4 (Hironaka's Resolution Theorem [H] [AHV]). *An algebraic variety X over an algebraically closed field k of characteristic zero has a resolution of the singularities $f : Y \to X$ satisfying the following:*

(i) The inverse image $f^{-1}(X_{\text{sing}})$ is a closed subset of codimension one and with simple normal crossings.
(ii) For every automorphism $\varphi : X \xrightarrow{\sim} X$, there exists an isomorphism $\widetilde{\varphi} : Y \xrightarrow{\sim} Y$ such that $f \circ \widetilde{\varphi} = \varphi \circ f$.

Here, a closed subset $E \subset Y$ of codimension one has simple normal crossings, if every irreducible component of E is non-singular and is defined by the product $f_{i_1} \cdots f_{i_k} = 0$, where $f_1, \ldots, f_n$ are generators of the maximal ideal $\mathfrak{m} \subset \mathcal{O}_{Y,x}$, ($n = \dim X = \dim Y$).

For the positive characteristic case, people expect the same statement as in Theorem 4.3.4, but this has not yet been proved.

By this theorem, we obtain a resolution of the singularities of an algebraic variety over $\mathbb{C}$. We will describe the properties of singularities of an algebraic variety in terms of resolutions of the singularities. We will discuss it in Chap. 5 and after. In this section we think how to construct a resolution concretely. First we introduce a blow-up by an ideal sheaf on an algebraic variety. Hironaka's resolution is obtained by these blow-ups. It is not easy to get a resolution of a given singularity, but for special cases such as Examples 4.3.9–4.3.14 we obtain resolutions by ad-hoc techniques.

Definition 4.3.5 (Blow-up). Let X be an affine integral variety, I an ideal of the affine coordinate ring $A(X)$ of X and $g_0, \ldots, g_s$ generators of I. Define $U_i (i = 0, \ldots, s)$ as the affine variety with the affine coordinate ring $A(X)\left[\frac{g_0}{g_i}, \ldots \frac{g_s}{g_i}\right]$ which is a subring of the quotient field of $A(X)$. Let $U_{ij} = \{x \in U_i \mid g_j/g_i(x) \neq 0\}$, then we have $U_{ij} \simeq U_{ji}$ ($\forall i \neq j$). By these isomorphisms we can patch $\{U_i\}$ together and obtain an algebraic variety $B_I(X)$. For each i, a morphism $b_i : U_i \to X$ is induced from the natural inclusion of affine coordinate rings and it is compatible with the patching of $B_I(X)$. Then, we obtain a canonical morphism $b : B_I(X) \to X$ and it is called the *blow-up* of X by an ideal I. This is independent of the choice of generators $g_0, \ldots, g_s$ of I up to isomorphisms over X.

In particular, if I is reduced, we call $V(I)$ the *center of the blow-up*. Therefore the blow-up at the center $x \in X$ is the blow-up by the maximal ideal defined by $M_x \subset A(X)$. We denote it by $B_x(X)$.

The following are basic properties of a blow-up.

Proposition 4.3.6. *Let X be an affine integral variety over k, $b : B_I(X) \to X$ the blow-up by an ideal I. Then the following hold:*

(i) *The blown up variety $B_I(X)$ is also an integral variety.*
(ii) *The morphism b is projective.*
(iii) *The restriction $b|_{B_I(X)\setminus b^{-1}(V)} : B_I(X) \setminus b^{-1}(V) \xrightarrow{\sim} X \setminus V$ of the morphism b is isomorphic, where V is the closed subvariety defined by the ideal I.*
(iv) $\operatorname{codim}_{B_I(X)} b^{-1}(V) = 1$.
(v) *Take an affine integral variety Z such that it contains X as a closed subvarity. Let $A(Z) \xrightarrow{\varphi} A(X)$ be the surjective homomorphism corresponding to the inclusion morphism $X \subset Z$ and take an ideal J in $A(Z)$ such that $\varphi(J) = I$. Then, the restriction of the blow-up $\overline{b} : B_J(Z) \to Z$ onto the strict transform of X in $B_J(Z)$ is isomorphic to $b : B_I(X) \to X$ over X.*

(Sketch of proof)

(i) It is clear from the definition of a blow-up.
(ii) Let $g_0 \cdots g_s$ be generators of I. By the surjective homomorphism:

$$A(X)\left[\frac{g_0}{g_i}, \dots, \frac{g_s}{g_i}\right] \twoheadleftarrow A(X)\left[\frac{X_0}{X_i}, \dots, \frac{X_s}{X_i}\right] : \frac{g_j}{g_i} \leftarrow\!\shortmid \frac{X_j}{X_i}$$

we have a closed immersion $U_i \hookrightarrow X \times W_i$ ($W_i \cong \mathbb{A}_k^s$). The patching of $\{U_i\}$ in $B_I(X)$ and that of $\{X \times W_i\}$ in $X \times \mathbb{P}_k^s$ are compatible, therefore we obtain a closed immersion $B_I(X) \hookrightarrow X \times \mathbb{P}^s$. The morphism b is the restriction of the projection $p_1 : X \times \mathbb{P}_k^s \to X$ to the first factor, which shows that b is a projective morphism.
(iii) The open subset $X \setminus V(I)$ is covered by $X \setminus V(g_i)$ ($i = 0, \dots, s$) whose inverse image is isomorphic to $U_i - V(g_i)$.
(iv) On $A(X)\left[\frac{g_0}{g_i}, \dots, \frac{g_s}{g_i}\right]$, we have $(g_0, \dots, g_s) = (g_i)$. The statement about the codimension follows from Krull's principal ideal theorem [AM, Cor. 11.17].
(v) Let $\overline{g}_0, \dots, \overline{g}_s$ be generators of J and $g_0, \dots, g_s$ their images by φ, then the canonical surjections $A(X)\left[\frac{g_0}{g_i}, \dots \frac{g_s}{g_i}\right] \twoheadleftarrow A(Z)\left[\frac{\overline{g}_0}{\overline{g}_i}, \dots, \frac{\overline{g}_s}{\overline{g}_i}\right]$ have compatibility with the patching of $B_I(X)$ and $B_J(Z)$, and we obtain a closed immersion $B_I(X) \hookrightarrow B_J(Z)$. This immersion has compatibility with the blow-ups b, $\overline{b}$. Here $B_I(X)$ is integral, therefore it is the closure of $\overline{b}^{-1}(X \setminus V)$.

The following proposition is a kind of inverse of Proposition 4.3.6.

Proposition 4.3.7 ([Ha2, II, 7.17]). *Let Y and X be integral algebraic varieties and assume X to be affine. Let $\varphi : Y \to X$ be a projective morphism isomorphic on open subsets. Then, there is an ideal $I \subset A(X)$ such that φ is isomorphic to $b : B_I(X) \to X$.*

Next, blow-ups of a general algebraic variety are defined by blow-ups of affine varieties.

Definition 4.3.8. Let X be an arbitrary algebraic variety over k and $\mathscr{I} \subset \mathscr{O}_X$ an $\mathscr{O}_X$-ideal sheaf. Then the *blow-up* $b : B_{\mathscr{I}}(X) \to X$ of X by $\mathscr{I}$ is defined as follows: For an affine covering $\{U_i\}$ of X, let $I_i := \Gamma(U_i, \mathscr{I})$, then I_i is an ideal of the affine ring of U_i. For each i, construct the blow-ups $b_i : B_{I_i}(U_i) \to U_i$, then there are isomorphisms $b_i^{-1}(U_i \cap U_j) \overset{\varphi_{ij}}{\underset{\sim}{\longrightarrow}} b_j^{-1}(U_i \cap U_j)$ satisfying $b_j \circ \varphi_{ij} = b_i$ $(\forall i \neq j)$. By these isomorphisms φ_{ij} we can patch $\{B_{I_i}(U_i)\}_i$ together and obtain an algebraic variety $B_{\mathscr{I}}(X)$. By the compatibility of $\{b_i\}$ and the patching we obtain $b : B_{\mathscr{I}}(X) \to X$. We call it the *blow-up of X by* $\mathscr{I}$.

Here, we note that $B_{\mathscr{I}}(X) \xrightarrow{b} X$ is independent of choices of an affine covering $\{U_i\}$. On the other hand, by the definition and Proposition 4.3.6 (ii), the morphism b is locally projective.

Example 4.3.9 (blow-up with the center at the origin of $\mathbb{A}_k^n$*).* Let us consider the blow-up $b : B_0(\mathbb{A}_k^n) \to \mathbb{A}_k^n$ with the center at the origin of $\mathbb{A}_k^n$. As the defining ideal of 0 is $M = (X_1, \dots, X_n) \subset k[X_1, \dots, X_n]$ the affine coordinate ring of U_i $(i = 1, \dots, n)$ is $k\left[X_i, \frac{X_1}{X_i}, \dots, \frac{X_n}{X_i}\right]$ and it is isomorphic to the polynomial ring of n variables, therefore $U_i \cong \mathbb{A}_k^n$ $(i = 1, \dots, n)$. Of course $B_0(\mathbb{A}_k^n)$ is non-singular. As is seen in the proof of Proposition 4.3.6 (iv), the inverse image $b^{-1}(0)$ is defined by $X_i = 0$ in U_i, and the affine ring of $U_i \cap b^{-1}(0)$ is $k\left[\frac{X_1}{X_i}, \dots, \frac{X_n}{X_i}\right]$. Looking at the patching of them, we can see that it is the same as the patching of $\mathbb{P}^{n-1}$. Consequently we have $b^{-1}(0) \cong \mathbb{P}^{n-1}$.

Example 4.3.10 (blow-up of a hypersurface of $\mathbb{A}_k^n$*).* Let $X \subset \mathbb{A}_k^n$ be defined by an irreducible polynomial $f \in k[X_1, \dots, X_n]$ (an affine subvariety defined by one equation is called a *hypersurface*). Assume $0 \in X$ and we take the blow-up $B_0(X)$ of X with the center at 0. First consider the blow-up $\overline{b} : B_0(\mathbb{A}_k^n) \to \mathbb{A}_k^n$ of $\mathbb{A}_k^n$ and then study the inverse image $\overline{b}^{-1}(X)$ of X. As X is defined by f, by the definition of the inverse image, $\overline{b}^{-1}(X)$ is defined by f in $B_0(\mathbb{A}_k^n)$. The polynomial f is irreducible in $k[X_1, \dots, X_n]$, but in the affine coordinate ring $k\left[X_i, \frac{X_1}{X_i}, \dots, \frac{X_n}{X_i}\right]$ of U_i it is factored as follows:

$$f(X_1, \dots, X_n) = f\left(X_i \frac{X_1}{X_i}, \dots, X_i, \dots, X_i \frac{X_n}{X_i}\right) = X_i^m g_i\left(\frac{X_1}{X_i}, \dots, X_i, \dots, \frac{X_n}{X_i}\right)$$

$(m > 0$, g_i is a polynomial in n variables). On U_i, the inverse image $\overline{b}^{-1}(X)$ is the union of $V(X_i^m)$ and $V(g_i)$. As $V(X_i^m) \subset \overline{b}^{-1}(0)$, $V(g_i) \not\subset \overline{b}^{-1}(0)$, by Proposition 4.3.6 (v), it turns out that $V(g_i)$ gives $B_0(X)$ on U_i. Next, look at the inverse image $b^{-1}(0)$ of 0 by the morphism $b : B_0(X) \to X$. It is a subset of $\overline{b}^{-1}(0) \cong \mathbb{P}_k^{n-1}$. In the open subset $\overline{b}^{-1}(0) \cap U_i \cong \mathbb{A}_k^{n-1}$, $b^{-1}(0)$ is defined by $g_i\left(\frac{X_1}{X_i}, \dots, 0, \dots, \frac{X_n}{X_i}\right)$. Here let the leading term of f be g, then for every i, we have $g\left(\frac{X_1}{X_i}, \dots, 1, \dots, \frac{X_n}{X_i}\right) = g_i\left(\frac{X_1}{X_i}, \dots, 0, \dots, \frac{X_n}{X_i}\right)$. Therefore, $b^{-1}(0)$ is defined in $\overline{b}^{-1}(0) \cong \mathbb{P}^{n-1}$ by a homogeneous polynomial g.

Example 4.3.11 (*A_1-singularity*). Let $X \subset \mathbb{A}_k^3$ be a hypersurface defined by a polynomial ${X_1}^2 + {X_2}^2 + {X_3}^2$. By Theorem 4.1.7 (vii) we can see that X has an isolated singularity at the origin 0. We study the blow-up at the origin $b : B_0(X) \to X$. On the affine cover $U_i (i = 1, 2, 3)$ of $B_0(\mathbb{A}_k^3)$, f is factored as

$$X_1^2\left(1 + \left(\frac{X_2}{X_1}\right)^2 + \left(\frac{X_3}{X_1}\right)^2\right),$$

$$X_2^2\left(\left(\frac{X_1}{X_2}\right)^2 + 1 + \left(\frac{X_3}{X_2}\right)^2\right),$$

$$X_3^2\left(\left(\frac{X_1}{X_3}\right)^2 + \left(\frac{X_2}{X_3}\right)^2 + 1\right).$$

Therefore $B_0(X)$ is defined in $U_i \cong \mathbb{A}_k^3$ by $1 + Y_1^2 + Y_2^2$ (where Y_i is a coordinate in $\mathbb{A}_k^3$). Hence in particular it is non-singular. The inverse image $b^{-1}(0)$ is in $\overline{b}^{-1}(0) \cong \mathbb{P}_k^2$ defined by $X_1^2 + X_2^2 + X_3^2$ and it is isomorphic to $\mathbb{P}_k^1$. By this we obtain a resolution of A_1-singularity. For other Du Val singularities, we obtain resolutions by successive blow-ups with the centers at points (see Sect. 7.5). A singularity which can be resolved by finite blow-ups by points is called an *absolute isolated singularity*.

Example 4.3.12 (*Simple elliptic singularity*). Take the blow-up of the $\widetilde{E}_6$-singularity $x^3 + y^3 + z^3 + \lambda xyz = 0$ at the origin of $\mathbb{A}_{\mathbb{C}}^3$. It is also a resolution of the singularity. How about the $\widetilde{E}_7$-singularity? Let $X \subset \mathbb{A}_{\mathbb{C}}^3$ be defined by $x^4 + y^4 + z^2 + \lambda xyz = 0$. For simplicity assume $\lambda = 0$. First, consider the blow-up $b : B_0(\mathbb{A}_{\mathbb{C}}^3) \to \mathbb{A}_{\mathbb{C}}^3$. On each affine ring, the defining polynomial is as follows:

$$\mathbb{C}\left[x, \frac{y}{x}, \frac{z}{x}\right] : x^2 + x^2\left(\frac{y}{x}\right)^4 + \left(\frac{z}{x}\right)^2, \tag{4.4}$$

$$\mathbb{C}\left[y, \frac{x}{y}, \frac{z}{y}\right] : y^2\left(\frac{x}{y}\right)^4 + y^2 + \left(\frac{z}{y}\right)^2, \tag{4.5}$$

$$\mathbb{C}\left[z, \frac{x}{z}, \frac{y}{z}\right] : z^2\left(\frac{x}{z}\right)^2 + z^2\left(\frac{y}{z}\right)^2 + 1. \tag{4.6}$$

The variety defined by (4.6) is non-singular. The variety defined by (4.4) has singularities on $x = \frac{z}{x} = 0$ and the variety defined by (4.5) has singularities on $y = \frac{z}{y} = 0$. In both cases the singular locus is $\mathbb{A}_{\mathbb{C}}^1$ and by the patching, the singular locus of $B_0(X)$ is $\mathbb{P}_{\mathbb{C}}^1$. Hence, we will not have a resolution of the singularities by successive blow-ups of finite points. Also for the $\widetilde{E}_8$-singularity, we have a similar phenomenon. But if we use the normalization, we can construct a resolution.

Theorem 4.3.13 (**Zariski [Z]**). *Let k be an algebraically closed field of characteristic 0. Let X be a reduced 2-dimensional variety over k. Then there is a sequence of morphisms $X_r \xrightarrow{\varphi_r} X_{r-1} \longrightarrow \cdots \longrightarrow X_1 \xrightarrow{\varphi_1} X$ such that:*

(i) Each morphism φ_i is either the normalization or a blow-up at an isolated singular point of X_{i-1}.
(ii) X_r is non-singular.

Example 4.3.14. The singularity $\widetilde{E}_7$ is not resolved by the blow-up at the origin as was seen in Example 4.3.12. This singularity is resolved by the blow-up $b : B_I(\mathbb{A}^3_{\mathbb{C}}) \to \mathbb{A}^3_{\mathbb{C}}$ of the ideal $I = (x^2, xy, y^2, z)$. In fact, on each affine coordinate ring of $B_I(\mathbb{A}^3_{\mathbb{C}})$ the defining polynomial of $B_I(X)$ is as follows:

$$\mathbb{C}\left[x, \frac{y}{x}, \frac{z}{x^2}\right]; \quad 1 + \left(\frac{y}{x}\right)^4 + \left(\frac{z}{x^2}\right)^2, \tag{4.7}$$

$$\mathbb{C}\left[x, \frac{y}{x}, \frac{z}{xy}\right]_{\frac{y}{x}}; \quad \left(\frac{x}{y}\right)^2 + \left(\frac{y}{x}\right)^2 + \left(\frac{z}{xy}\right)^2, \tag{4.8}$$

$$\mathbb{C}\left[y, \frac{x}{y}, \frac{z}{y^2}\right]; \quad \left(\frac{x}{y}\right)^4 + 1 + \left(\frac{z}{y^2}\right)^2, \tag{4.9}$$

$$\mathbb{C}\left[x, y, z, \frac{x^2}{z}, \frac{xy}{z}, \frac{y^2}{z}\right] \cong \frac{\mathbb{C}[x, y, z, u, v, w]}{(x^2 - zu,\ xy - zv,\ y^2 - zw,\ v^2 - uw,\ vy - wx,\ uy - vx)} \; ; u^2 + w^2 + 1. \tag{4.10}$$

The affine coordinate rings of the affine open subsets of (4.7) and (4.9) are isomorphic to $\mathbb{A}^3_{\mathbb{C}}$. In case (4.8), it is isomorphic to $\mathbb{A}^3_{\mathbb{C}} \setminus$ (one coordinate plane). On each chart looking at the defining polynomial of $B_I(X)$, we can see that $B_I(X)$ is non-singular. Regarding (4.10), we need to calculate the Jacobian matrix J of $B_I(X)$ in $\mathbb{A}^6_{\mathbb{C}}$.

The points where the rank of the following matrix

$$J = \begin{pmatrix} 2x & y & 0 & 0 & -w & -v & 0 \\ 0 & x & 2y & 0 & v & u & 0 \\ -u & -v & -w & 0 & 0 & 0 & 0 \\ -z & 0 & 0 & -w & 0 & y & 2u \\ 0 & -z & 0 & 2v & y & -x & 0 \\ 0 & 0 & -z & -u & -x & 0 & 2w \end{pmatrix}$$

is less than 4 are singular points of $B_I(X)$. The singularities are on $z = 0$ (therefore they are also on $x = y = 0$); we have only to find the points where the following matrix J_0 has rank < 4:

$$J_0 = \begin{pmatrix} 0 & 0 & 0 & 0 & -w & -v & 0 \\ 0 & 0 & 0 & 0 & v & u & 0 \\ -u & -v & -w & 0 & 0 & 0 & 0 \\ 0 & 0 & 0 & -w & 0 & 0 & 2u \\ 0 & 0 & 0 & 2v & 0 & 0 & 0 \\ 0 & 0 & 0 & -u & 0 & 0 & 2w \end{pmatrix}$$

Here, if we assume $w \neq 0$, there exists the following submatrix

$$\begin{pmatrix} 0 & 0 & -w & 0 \\ -w & 0 & 0 & 0 \\ 0 & -w & 0 & 2u \\ 0 & -u & 0 & 2w \end{pmatrix}$$

in J_0.

Hence if $u \neq w$, rank $J_0 = 4$. If $u = w$, then by $v^2 = uw$ it follows that $v \neq 0$ and it yields rank $J_0 = 4$. For the case $u \neq 0$, the discussion is similar. Therefore the points that satisfy rank $J_0 < 4$ are only on the subset $u = v = w = 0$. However, such points do not satisfy the equation $u^2 + w^2 + 1 = 0$; those points do not exist on $B_I(X)$.

4.4 Toric Resolutions of Singularities

A toric variety is an almost homogeneous space (cf. Definition 4.4.4) of the algebraic torus $(k^*)^n$. The structure of a toric variety is completely determined by the fan in a real Euclidean space. It has an orbit of $(k^*)^n$ as an open dense subset, therefore it is a rational variety (i.e., an integral algebraic variety whose function field is rational). So the category of toric varieties is very restricted; however, it has very good properties. One good property is, for example, that there exist resolutions of singularities no matter what is the characteristic of the base field. This good property is inherited by a hypersurface (Theorem 4.4.23).

In this section the base field k is an algebraically closed field of arbitrary characteristic.

First we prepare an algebraic group and its action.

Definition 4.4.1. An algebraic variety G over k is called an *algebraic group* if the following three conditions hold:

(1) There exists a morphism $\mu : G \times G \to G$ of algebraic varieties over k satisfying $\mu(\mu(g_1, g_2), g_3) = \mu(g_1, \mu(g_2, g_3))$.
(2) There exists an element $e \in G$ such that for an arbitrary $g \in G$ the equality $\mu(g, e) = \mu(e, g) = g$ holds.
(3) There exists an automorphism $\beta : G \xrightarrow{\sim} G$ over k such that the equality $\mu(\beta(g), g) = \mu(g, \beta(g)) = e$ holds for every $g \in G$.

Example 4.4.2. (i) The affine line $\mathbb{A}_k^1$ is a commutative algebraic group by the operation $\mu : \mathbb{A}_k^1 \times \mathbb{A}_k^1 \to \mathbb{A}_k^1 : (a, b) \mapsto a + b$.

(ii) Let $T_k^1 = \mathbb{A}_k^1 \setminus \{0\}$, then by the operation $\mu : T_k^1 \times T_k^1 \to T_k^1 : (a, b) \mapsto a \cdot b$, T_k^1 is also a commutative algebraic group.
If we put $T_k^n := \underbrace{T_k^1 \times T_k^1 \times \cdots \times T_k^1}_{n \text{ times}}$ and define the operation by $\mu : T_k^n \times T_k^n \to T_k^n : ((a_i)_i, (b_i)_i) \mapsto (a_i b_i)_i$, then T_k^n is also a commutative algebraic group. This is called the *n-dimensional torus*.

(iii) Let R be the polynomial ring over k with the variables $x_{ij} (i, j = 1, \ldots, n)$. Let $\Delta(x_{ij}) := \sum_{\sigma \in S_n} \operatorname{sign} \sigma x_{1\sigma(1)} \cdots x_{n\sigma(n)}$ and let the affine variety which has R_Δ as the affine coordinate ring be $GL(n, k)$. Obviously the subset $GL(n, k) \subset \mathbb{A}_k^{n^2}$ is an open dense subset. We write a point of $GL(n, k)$ by $(a_{ij}) = \begin{pmatrix} a_{11} & \cdots & a_{1n} \\ \vdots & & \vdots \\ a_{n1} & \cdots & a_{nn} \end{pmatrix}$. For two points $(a_{ij}), (b_{ij}) \in GL(n, k)$, the operation μ is defined by $\mu((a_{ij}), (b_{ij})) = \left((\sum_{k=1}^n a_{ik} b_{kj})_{ij}\right)$, then $GL_k(n)$ is an algebraic group over k.

Definition 4.4.3. Let G be an algebraic group over k and X an algebraic variety over k. We say that G *acts on* X if the following two conditions hold:

(1) There exists a morphism $\sigma : G \times X \to X$ of algebraic varieties over k such that the equality $\sigma(\mu(g, g'), x) = \sigma(g, \sigma(g', x))$ holds for every $g, g' \in G, x \in X$.
(2) The equality $\sigma(e, x) = x$ holds for every $x \in X$.

In this case $\sigma(g, x)$ is represented simply as gx.

Definition 4.4.4. Assume that an algebraic group G acts on an algebraic variety X over k. For a point $x \in X$, we define $O_G(x) = \{gx \mid g \in G\} \subset X$ and call it the *orbit* of x.

(1) If there is a point $x \in X$ such that $X = O_G(x)$, then we call X a *homogeneous space*.
(2) If there is a point $x \in X$ such that $X = \overline{O_G(x)}$, then we call X an *almost homogeneous space*.

Every two points of homogeneous space correspond to each other by an automorphism. Therefore, by Proposition 4.1.9, every point corresponds to a non-singular point by an automorphism, which yields that a homogeneous space is non-singular. On the other hand, an almost homogeneous space has a dense orbit $O_G(x)$ which is non-singular, but it is possible that there are singularities. But any points of one orbit correspond by automorphisms, so we can see that the same singularities are ranging along the orbit.

Example 4.4.5. (i) Let an action of $GL(n+1,k)$ on $\mathbb{P}_k^n$ be defined by

$$\sigma((a_{ij}),(x_0:\cdots:x_n)) = {}^t\!\left((a_{ij})\begin{pmatrix}x_0\\ \vdots \\ x_n\end{pmatrix}\right).$$

Then every two points of $\mathbb{P}_k^n$ correspond each other by this action, therefore $\mathbb{P}_k^n$ is a homogeneous space.

(ii) Let an action of T_k^n on $\mathbb{P}_k^n$ be defined by $\sigma((a_1,\dots,a_n),(x_0:\cdots:x_n)) := (x_0 : a_1x_1 : \cdots : a_nx_n)$. Then, $O_{T_k^n}((1:1:\cdots:1)) \subset \mathbb{P}_k^n$ is an open dense subset, which implies that $\mathbb{P}_k^n$ is an almost homogeneous space.

Now we are on the stage of defining a toric variety.

Definition 4.4.6. An algebraic variety X is called a *toric variety* if the following two conditions hold:

(1) X is a normal n-dimensional variety over k.
(2) X is an almost homogeneous space by the action of T_k^n.

Definition 4.4.7. For two toric varieties X, Y, a morphism $f : X \to Y$ of algebraic varieties over k is called a *toric morphism*, if there is a homomorphism $\varphi : T^m \to T^n$ such that $f|_{T^m} = \varphi : T^m \to T^n$ and the following holds:

$$\begin{array}{ccc} T^m \times X & \xrightarrow{\ \sigma\ } & X \\ {\scriptstyle \varphi\times f}\downarrow & & \downarrow{\scriptstyle f} \\ T^n \times Y & \xrightarrow{\ \sigma'\ } & Y \end{array} \quad \text{is commutative,}$$

where σ, σ' are morphisms giving the actions.

By this we obtain the category consisting of toric varieties as objects and toric morphisms as morphisms.

Next we introduce the category of finite fans. This category will be proved to be equivalent to the category of toric varieties.

Let M be the free $\mathbb{Z}$-module of rank n and let N be the dual $\mathbb{Z}$-module of M, i.e., $N = \mathrm{Hom}_{\mathbb{Z}}(M,\mathbb{Z})$. By this we have a canonical $\mathbb{Z}$-bilinear map $(\ ,\ ) : N \times M \to \mathbb{Z}$ and it is canonically extended to a $\mathbb{R}$-bilinear map $N_{\mathbb{R}} \times M_{\mathbb{R}} \to \mathbb{R}$ (where $M_{\mathbb{R}} = M \otimes_{\mathbb{Z}} \mathbb{R}$, $N_{\mathbb{R}} = N \otimes_{\mathbb{Z}} \mathbb{R}$).

Definition 4.4.8. A subset $\sigma \subset N_{\mathbb{R}}$ is called a *strongly convex rational polyhedral cone* if there exist a finite number of elements $n_1,\dots,n_s$ of N such that $\sigma = \mathbb{R}_{\geq 0}n_1 + \cdots + \mathbb{R}_{\geq 0}n_s := \{a_1n_1 + \cdots + a_sn_s : a_i \in \mathbb{R},\ a_i \geq 0\ \forall i\}$ and, moreover, the condition $\sigma \cap (-\sigma) = \{0\}$ is satisfied. A strongly convex rational polyhedral cone is sometimes called just a "cone in N".

The *dimension of a cone* σ in N is the dimension of the $\mathbb{R}$-vector subspace of $N_{\mathbb{R}}$ spanned by σ.

A cone σ in N is called *simplicial* if it is written as $\sigma = \mathbb{R}_{\geq 0}n_1 + \cdots + \mathbb{R}_{\geq 0}n_s$ for $s = \dim \sigma$.

Let σ be a cone in N. If there exists $m \in M$ such that $(m, x) \geqq 0, \forall x \in \sigma$ and $\tau = \{y \in \sigma \mid (m, y) = 0\}$, we call τ a *face* of σ and denote by $\tau < \sigma$. A face τ is clearly generated by a subset of generators $n_1, \ldots, n_s$ of σ over $\mathbb{R}_{\geq 0}$.

Example 4.4.9. Let $n = 2$ and define σ as follows, then it is a cone over N:

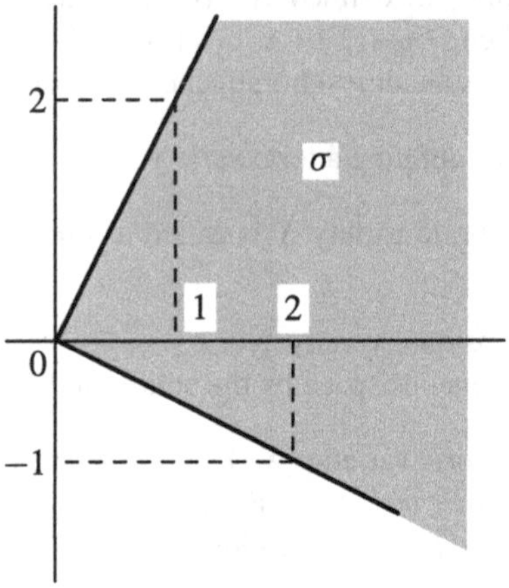

The following two are not cones in N:

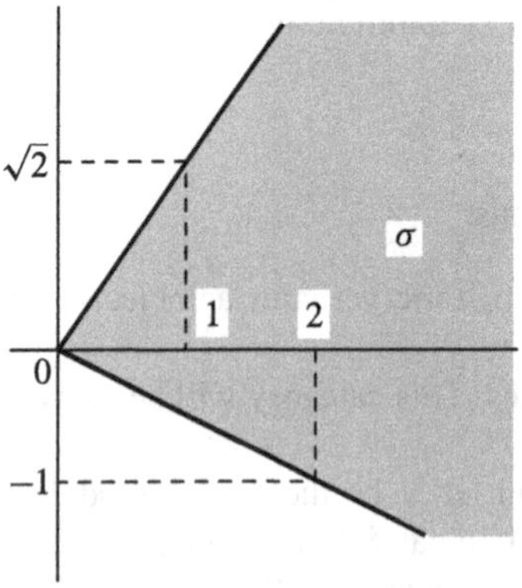

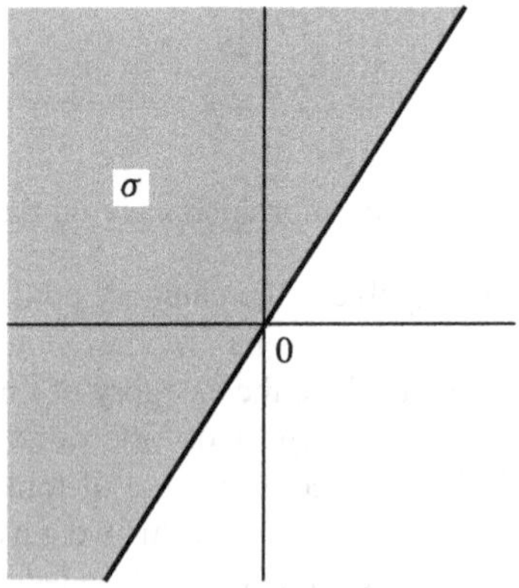

Definition 4.4.10. A pair (N, Δ) is called a *finite fan* if Δ is a finite set of cones in N such that:

(a) If $\sigma \in \Delta$, then every face $\tau < \sigma$ satisfies $\tau \in \Delta$.
(b) If $\sigma, \tau \in \Delta$, then the intersection satisfies $\sigma \cap \tau < \sigma, \sigma \cap \tau < \tau$.

Definition 4.4.11. Let (N, Δ) and (N', Δ') be finite fans. We call $\varphi : (N, \Delta) \to (N'\Delta')$ a *morphism of finite fans* if $\varphi : N \to N'$ is a homomorphism of $\mathbb{Z}$-modules and its canonical extension $\varphi : N_{\mathbb{R}} \to N'_{\mathbb{R}}$ satisfies the following property:
For every $\sigma \in \Delta$, there is $\sigma' \in \Delta'$ such that $\varphi(\sigma) \subset \sigma'$.
By this we obtain the category of finite fans.

The following is an important theorem to connect toric varieties and finite fans:

Theorem 4.4.12. *(i) Let N, M be as above. There is a bijective map:*

$$\eta : (\textit{set of cones in } N) \xrightarrow{\ \sim\ } \begin{pmatrix} \textit{set of subsemigroups } S \\ \textit{of } M \textit{ satisfying (a) (b) (c)} \end{pmatrix}.$$

(a) S is a finitely generated subsemigroup of M containing 0.
(b) S generates the group M.
(c) S is saturated, i.e., for every $m \in M$ such that $am \in S$ for a positive integer a, it follows that $m \in S$.

(ii) There exists an equivalence of categories:

$$\begin{array}{ccc} \Phi : (\textit{category of finite fans}) & \xrightarrow{\ \sim\ } & (\textit{category of toric varieties}) \\ \Psi & & \Psi \\ (N, \Delta) & \longmapsto & T_N(\Delta). \end{array}$$

(iii) For the correspondence $(N, \Delta) \longmapsto T_N(\Delta)$ we have the following isomorphism:

$$\begin{array}{ccc} (\textit{set of faces of } \Delta) & \longrightarrow & (\textit{set of orbits of } T_N(\Delta)) \\ \Psi & & \Psi \\ \sigma & \longmapsto & \mathrm{orb}\,(\sigma), \end{array}$$

where $\sigma < \tau \iff \overline{\mathrm{orb}\,(\sigma)} \supset \mathrm{orb}\,(\tau)$. *We also have* $\dim \mathrm{orb}\,(\sigma) = \mathrm{rank}\, N - \dim \sigma$.

(Sketch of proof) The maps η, Φ are defined as follows: First for $\sigma \in \Delta$, define $\sigma^\vee := \{m \in M_{\mathbb{R}} \mid (m, x) \geqq 0, \forall x \in \sigma\}$ and $\eta(\sigma) := \sigma^\vee \cap M$. Let U_σ be the affine variety with the affine coordinate ring $k[\sigma^\vee \cap M]$ generated by the subsemigroup $\sigma^\vee \cap M$.

Here, $k[\sigma^\vee \cap M] := k[X_1^{m_1} X_2^{m_2} \cdots X_n^{m_n}]_{(m_1 \cdots m_n) \in \sigma^\vee \cap M}$, where $X_1, \ldots, X_n$ are variables. For a face $\tau < \sigma$, the corresponding affine variety U_τ is canonically regarded as an open subset of U_σ. Now, for cones $\sigma, \sigma' \in \Delta$ we patch U_σ and $U_{\sigma'}$ by identifying the open subsets $U_{\sigma\cap\sigma'}$ of both affine varieties. We patch $\{U_\sigma\}$ for all the cones $\sigma \in \Delta$ in this way and construct a variety $T_N(\Delta)$. For $\sigma \in \Delta$ define $P(\sigma)$ as the ideal in $k[\sigma^\vee \cap M]$ generated by $(\sigma^\vee \cap M) \setminus (\sigma^\perp \cap M)$. Then we define orb (σ) the closed subvariety of U_σ defined by the ideal $P(\sigma)$. For a precise discussion, see [Od1, Theorem 4.1].

Example 4.4.13. 1. Let $N = \mathbb{Z}^n$ and let Δ be the set of the positive cone $\sigma = \sum_{i=1}^n \mathbb{R}_{\geq 0} e_i$ and all faces of σ. Then, we have $T_N(\Delta) = \mathbb{A}_k^n$. The open dense orbit of T_k^n is $(\mathbb{A}_k^1 \setminus \{0\})^n$. If τ is a face of σ of dimension r, then $\overline{\text{orb}\,(\tau)}$ is the intersection of r coordinate planes of $\mathbb{A}^n$. In particular, for $n = 2$, the fan Δ and the corresponding orbits are as follows:

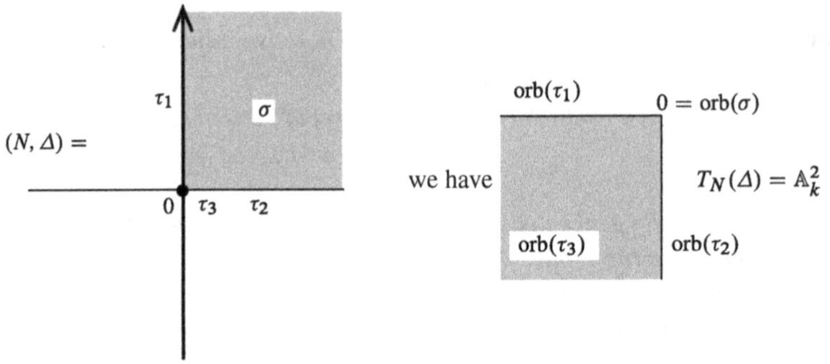

2. For the following fan:

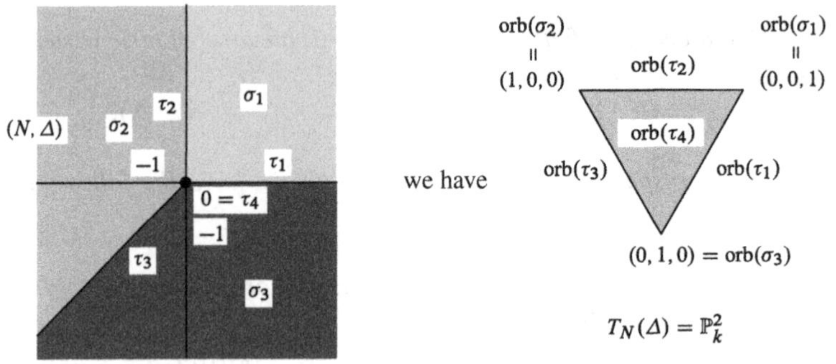

Example 4.4.14. Assume that two finite fans (N, Δ) and (N', Δ') satisfy the following:

(1) $N \subset N'$ is a subgroup of finite index.
(2) In $N_\mathbb{R} = N'_\mathbb{R}$, the equality $\Delta = \Delta'$ holds.

In this case, a canonical morphism $\varphi : (N, \Delta) \to (N', \Delta')$ of finite fans is induced and by Theorem 4.4.12 (ii), we obtain a morphism $T(\varphi) : T_N(\Delta) \to T_{N'}(\Delta')$ of toric varieties. This morphism is the quotient map by the finite group N'/N. In particular, if $T_N(\Delta)$ is non-singular, the singular points of $T_N(\Delta')$ are quotient singularities (cf. 7.4).

We introduce a morphism which is used often.

Example 4.4.15. Assume two finite fans (N, Δ) (N', Δ') satisfy the following:

(i) $N = N'$.
(ii) Δ' is a subdivision of Δ, i.e., every cone $\sigma \in \Delta$ can be described as $\sigma = \bigcup_{\substack{\sigma' \in \Delta' \\ \sigma' \subset \sigma}} \sigma'$.

In this case we have a morphism $\psi : (N', \Delta') \to (N, \Delta)$ of finite fans and also a morphism $T(\psi) : T_{N'}(\Delta') \to T_N(\Delta)$ of toric varieties. This is isomorphic on the open subsets $T_{N'} \simeq T_N$. We will see that the morphism is also proper. In particular, if $T_N(\Delta')$ is non-singular, then the morphism $T(\psi)$ is a weak resolution of the singularities of $T_N(\Delta)$.

As is seen in Theorem 4.4.12, a toric variety is completely determined by a finite fan. The properties of a toric variety are interpreted as those of a finite fan, which is very good since properties of finite fans are combinatorial and sometimes those are more intuition friendly. Here, we will see the non-singularity property.

Theorem 4.4.16. *A toric variety $T_N(\Delta)$ is non-singular if and only if the corresponding finite fan (N, Δ) satisfies the following:*

For every cone $\sigma \in \Delta$, write it by generators as $\sigma = \mathbb{R}_{\geq 0}\mathbf{e}_1 + \cdots + \mathbb{R}_{\geq 0}\mathbf{e}_r$. The generators $\mathbf{e}_1, \ldots, \mathbf{e}_r$ can be extended to a basis of N (such Δ is called a unimodular fan).

Proof. First, assume that Δ is unimodular. For every $\sigma \in \Delta$, if we write it as $\sigma = \sum_{i=1}^r \mathbb{R}_{\geq 0}\boldsymbol{e}_i$, then there is a basis $\{\boldsymbol{e}_1, \ldots, \boldsymbol{e}_r, \boldsymbol{e}_{r+1}, \ldots, \boldsymbol{e}_n\}$ of N. Let $\{\boldsymbol{e}_1^*, \ldots, \boldsymbol{e}_n^*\}$ be its dual basis for M. Then, as $\sigma^\vee \cap M = \sum_{i=1}^r \mathbb{Z}_{\geq 0}\boldsymbol{e}_i^* + \sum_{j=r+1}^n \mathbb{Z}\boldsymbol{e}_j^*$, the affine coordinate ring of U_σ is isomorphic to $k[X_1, \ldots, X_r, X_{r+1}^{\pm 1}, \ldots, X_n^{\pm 1}]$. Hence, we have an isomorphism $U_\sigma \cong \mathbb{A}_k^r \times T_k^{n-r}$, which implies that U_σ is non-singular.

Conversely, assume that U_σ is non-singular for every $\sigma \in \Delta$. Let N_σ be the $\mathbb{Z}$-submodule of N generated by $N \cap \sigma$, let r be the rank of N_σ and let the finite fan consisting of all faces σ be $\langle \sigma \rangle$. Then we have $U_\sigma \cong T_{N_\sigma}(\langle \sigma \rangle) \times T_k^{n-r}$. Therefore $T_{N_\sigma}(\langle \sigma \rangle)$ is non-singular and $\dim \sigma = \operatorname{rank} N_\sigma$. By this, it is sufficient to prove that σ is unimodular for the case $\dim \sigma = \operatorname{rank} N = n$. In this case, as $\sigma^\perp \cap M = 0$, $\mathfrak{m} := P(\sigma)$ is a maximal ideal of $R := k[\sigma^\vee \cap M]$, where $P(\sigma)$ is as in the proof of Theorem 4.4.12(i). The point x defined by this ideal is of course a non-singular point of U_σ, therefore the local ring $R_\mathfrak{m}$ at x has the maximal ideal $\mathfrak{m}R_\mathfrak{m}$ generated by n elements. Represent an element of R as $\sum_{\boldsymbol{m} \in \sigma^\vee \cap M} a_m X^m$ $(a_{\boldsymbol{m}} \in k,\ \boldsymbol{m} = (m_1, \ldots, m_n)\ X^m := X_1^{m_1} \cdots X_n^{m_n})$ by using variables $X_1, \ldots, X_n$. Then the vector space $\mathfrak{m}/\mathfrak{m}^2 = \mathfrak{m}R_\mathfrak{m}/\mathfrak{m}^2R_\mathfrak{m}$ over k is generated by $X^{\boldsymbol{m}_1}, X^{\boldsymbol{m}_2}, \ldots, X^{\boldsymbol{m}_n}$ $(\boldsymbol{m}_i \in \sigma^\vee \cap M)$. This implies that $\boldsymbol{m}_1, \ldots, \boldsymbol{m}_n$ are the all of the elements $\boldsymbol{m}$ in $\sigma^\vee \cap M$ such that it is not represented as $\boldsymbol{m} = \boldsymbol{m}' + \boldsymbol{m}''(\boldsymbol{m}', \boldsymbol{m}'' \neq 0, \boldsymbol{m}', \boldsymbol{m}'' \in \sigma^\vee \cap M)$. Hence, a semigroup $\sigma^\vee \cap M$ is generated by $\boldsymbol{m}_1, \ldots, \boldsymbol{m}_n$. Therefore $M = \sigma^\vee \cap M + (-\sigma^\vee) \cap M$ is generated by those elements as a group and $\sigma^\vee$ is unimodular, which implies that the dual σ is unimodular.

Example 4.4.17. Let $N = \mathbb{Z}^3$, $\sigma = \mathbb{R}_{\geq 0}(1, 2, 1) + \mathbb{R}_{\geq 0}(2, 1, 1) + \mathbb{R}_{\geq 0}(-1, 5, 1)$ and let Δ be the fan that consists of all faces of σ. Then, the affine toric variety $T_N(\Delta)$ is non-singular.

As we saw in Example 4.4.15, for a finite fan (N, Δ), if we construct a unimodular subdivision Δ', then we obtain a weak resolution of $T_N(\Delta)$. We have a stronger statement as follows:

Theorem 4.4.18. *Let k be an algebraically closed field of arbitrary characteristic. For every finite fan (N, Δ), there exists a unimodular subdivision Δ' of Δ, and the corresponding morphism $\varphi : T_N(\Delta') \to T_N(\Delta)$ is a resolution of the singularities.*

Proof. (Procedure 1) For a cone to be unimodular, every face should be at least simplicial. By the following procedure, we can make a subdivision Σ of Δ such that all cones of Σ are simplicial. We make a subdivision so that the smallest-dimensional non-simplicial cones belonging to Δ become simplicial. Let $\sigma \in \Delta$ be a smallest-dimensional cone which is not simplicial. Since if $\dim \sigma = 1, 2$, clearly σ is simplicial, therefore we can assume $\dim \sigma \geq 3$ and the cones of dimension less than $\dim \sigma$ are all simplicial. Let λ be a 1-dimensional face of σ. Subdivide every cone σ' in Δ containing λ by replacing by small cones $\lambda + \tau$, where τ is a proper face of σ'. Now we obtain a new fan Δ_1. Then, σ is subdivided into simplicial cones in Δ_1.

Next we continue the same procedure and finally obtain a fan Σ whose cones are all simplicial.

(Procedure 2) The fan Σ obtained above will be further subdivided as follows: For a cone $\sigma = \sum_{i=1}^{r} \mathbb{R}_{\geq 0} e_i \in \Sigma$ define $N_\sigma = N \cap \left(\sum_{i=1}^{r} \mathbb{R} e_i\right)$ and $\operatorname{mult}(\sigma) := \left|N_\sigma / \sum_{i=1}^{r} \mathbb{Z} e_i\right|$. Then, it is clear that the condition $\operatorname{mult}(\sigma) = 1$ is equivalent to the fact that σ is unimodular. On the other hand, $\operatorname{mult}(\sigma)$ is the number of the integral points in the paralleltope $P_\sigma = \left\{\sum_{i=1}^{r} \alpha_i e_i \mid 0 \leqq \alpha_i < 1\right\}$. If σ is unimodular, then leave it. If σ is not unimodular, then there exists a point $\boldsymbol{x} \in N$ such that $\boldsymbol{x} \neq 0$ and $\boldsymbol{x} \in P_\sigma$. Now define a subdivision of σ, by small cones $\sigma_i = \mathbb{R}_{\geq 0} e_1 + \cdots + \mathbb{R}_{\geq 0} e_{i-1} + \mathbb{R}_{\geq 0} \boldsymbol{x} + \mathbb{R}_{\geq 0} e_{i+1} + \cdots + \mathbb{R}_{\geq 0} e_r$ $(i = 1, \ldots r)$. Then clearly σ_i is simplicial and $\operatorname{mult}(\sigma)$ is equal to the volume of P_σ. We obtain $\operatorname{mult}(\sigma_i) < \operatorname{mult}(\sigma)$. By doing this procedure on each non-unimodular cone, then eventually every cone σ has $\operatorname{mult}(\sigma) = 1$.

By the above procedure, we obtain a unimodular subdivision Δ' of the fan Δ. The corresponding morphism $\varphi : T_N(\Delta') \to T_N(\Delta) = X$ is isomorphic on X_{reg}. In fact for a cone $\sigma \in \Delta$, if every point of $\operatorname{orb}(\sigma)$ is a non-singular point of X, then U_σ is a non-singular affine variety. Then, σ is simplicial, therefore it is left unchanged under Procedure 1, i.e., $\sigma \in \Sigma$ and the affine coordinate ring of U_σ in $T_N(\Sigma)$ is the same $k[\sigma^\vee \cap M]$ as before. Since U_σ is non-singular, N_σ coincides with the $\mathbb{Z}$-submodule generated by σ (see proof of Theorem 4.4.16). Then an integer point of P_σ is only 0. Therefore U_σ does not change under the second procedure.

Example 4.4.19. Let $N = \mathbb{Z}^2$, $\sigma = \mathbb{R}_{\geq 0}(2, 1) + \mathbb{R}_{\geq 0}(1, 3)$ and let Δ be the set of σ and all faces of σ. Then $T_N(\Delta)$ has a singular point. Subdivide Δ as $\sigma_1 = \mathbb{R}_{\geq 0}(2, 1) + \mathbb{R}_{\geq 0}(1, 1)$, $\sigma_2 = \mathbb{R}_{>0}(1, 1) + \mathbb{R}_{\geq 0}(1, 2)$, $\sigma_3 = \mathbb{R}_{\geq 0}(1, 2) + \mathbb{R}_{\geq 0}(1, 3)$. Let Δ' be the finite fan consisting of all faces of these cones, then $T_N(\Delta') \to T_N(\Delta)$ is a resolution of the singularity.

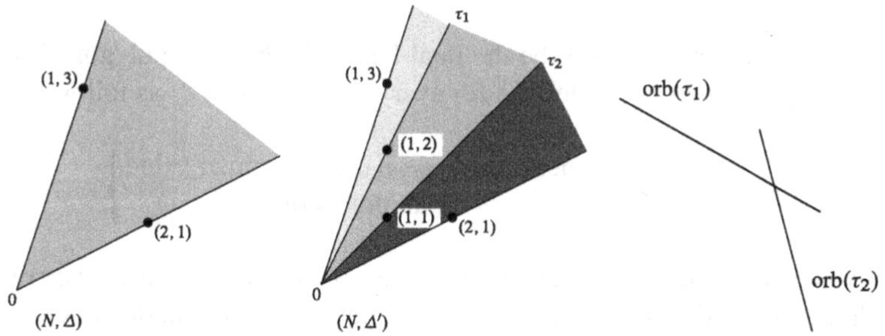

By this any 2-dimensional toric variety has a resolution whose fiber of the singular point (the closed orbit 0) is the union of irreducible curves.

Next we introduce a weighted blow-up which is also a toric morphism.

Example 4.4.20. Let the fan (N, Δ) be as in Example 4.4.13 which gives the toric structure $T_N(\Delta)$ on $\mathbb{A}^n_k$. By using $\boldsymbol{w} = (w_1, w_2, \ldots, w_n) \in \sigma \cap N$ we construct a fan $\Delta(\boldsymbol{w})$ consisting of all faces of $\sigma_i = \mathbb{R}_{\geq 0}\boldsymbol{e}_1 + \cdots + \mathbb{R}_{\geq 0}\boldsymbol{e}_{i-1} + \mathbb{R}_{\geq 0}\boldsymbol{w} + \mathbb{R}_{\geq 0}\boldsymbol{e}_{i+1} + \cdots + \mathbb{R}_{\geq 0}\boldsymbol{e}_n$ $(i = 1, \ldots, n)$. Then, $\Delta(\boldsymbol{w})$ is a subdivision of Δ. Such a subdivision is called a *star-shaped subdivision* by w. The corresponding morphism $\psi : T_N(\Delta(\boldsymbol{w})) \to T_N(\Delta)$ is called the *weighted blow-up* with weight $\boldsymbol{w}$. Let $\boldsymbol{f}_i (i = 1, \ldots, n)$ be the dual basis of $\boldsymbol{e}_i (i = 1, \ldots, n)$ and $a_{ij} = g.c.d.(w_i, w_j)$. Then, the affine coordinate ring of U_{σ_i} is $k[\sigma_i^\vee \cap M]$, where $\sigma_i^\vee = \mathbb{R}_{\geq 0}\boldsymbol{f}_i + \sum_{j \neq i} \mathbb{R}_{\geq 0}(-\frac{w_j}{a_{ij}}\boldsymbol{f}_i + \frac{w_i}{a_{ij}}\boldsymbol{f}_j)$.

The singularities on a toric variety are restricted, but considering a hypersurface on a toric variety, we can treat singularities of more variety. We are going to consider a hypersurface on the affine space $\mathbb{A}_k^{n+1}$.

Definition 4.4.21. For a polynomial $f \in k[X_0, \ldots, X_n]$ take $M = \mathbb{Z}^{n+1}$ and $M_\mathbb{R} = M \otimes_\mathbb{Z} \mathbb{R}$. We define the *Newton polygon* $\Gamma_+(f)$ of f in $M_\mathbb{R}$. For $\boldsymbol{m} = (m_0, \ldots, m_n) \in M$ define $X^m = X_0^{m_0} \cdots X_n^{m_n}$. By using this we represent f as $f = \sum_{m \in M} a_m X^{\boldsymbol{m}}$, $a_m \in k$. Let $\operatorname{supp}(f) := \{\boldsymbol{m} \in M \mid a_{\boldsymbol{m}} \neq 0\}$ and define

$$\Gamma_+(f) := \text{convex hull of}\left(\bigcup_{\boldsymbol{m} \in \operatorname{supp}(f)} (\boldsymbol{m} + \mathbb{R}^{n+1}_{\geq 0})\right).$$

Definition 4.4.22. Under the notation above, f is called *non-degenerate* if for every compact face γ of $\Gamma_+(f)$ the equations $\frac{\partial f_\gamma}{\partial X_i} = 0$ $(i = 0, \ldots, n)$ do not have common zeros on $(\mathbb{A}^1_k \setminus \{0\})^{n+1}$, where $f_\gamma := \sum_{\boldsymbol{m} \in \gamma} a_{\boldsymbol{m}} X^{\boldsymbol{m}}$.

Theorem 4.4.23. *Let X be a hypersurface in $\mathbb{A}_k^{n+1}$ defined by a non-degenerate polynomial $f \in k[X_0, \dots, X_n]$, which has an isolated singularity at the origin. We regard the affine space $\mathbb{A}_k^{n+1}$ as the toric variety $T_N(\Delta)$, where $N := \mathrm{Hom}_{\mathbb{Z}}(M, \mathbb{Z})$ and Δ is as in Example 4.4.13. Then, there is a subdivision Δ' of Δ such that the restriction $\psi|_{X'} : X' \to X$ of the toric morphism $\psi : T_N(\Delta') \to T_N(\Delta)$ on the strict transform X' of X is a weak resolution of the singularity $(X, 0)$.*

Proof. First construct Δ'. Let Σ be the dual fan of the Newton polygon $\Gamma_+(f)$. It means $\Sigma = \{\gamma^*\}$, where γ^* is defined for every face γ of $\Gamma_+(f)$ as follows:

$$\gamma^* := \left\{ \boldsymbol{n} \in |\Delta| \;\middle|\; \begin{array}{c} \text{function } \boldsymbol{n}|_{\Gamma_+(f)} \text{ has the minimal value} \\ \text{for all points of } \gamma \end{array} \right\},$$

where $|\Delta|$ means the union of all cones in Δ. Then γ^* are all strongly convex rational polyhedral cones and Σ is a subdivision of Δ. Let Δ' be a unimodular subdivision of Σ. We will show that this Δ' is a required fan.

For an arbitrary $\tau \in \Delta'$ define a subset τ^* in $M_{\mathbb{R}}$ as follows:

$$\tau^* := \left\{ \boldsymbol{m} \in \Gamma_+(f) \;\middle|\; \begin{array}{l} \text{for any } \boldsymbol{n} \in \tau \\ (\boldsymbol{n}, \boldsymbol{m}) = \min\limits_{m' \in \Gamma_+(f)} (\boldsymbol{n}, \boldsymbol{m}') \end{array} \right\}.$$

Then τ^* is a face of $\Gamma_+(f)$. Here, we note that for $\tau \in \Delta'$ the face $\tau^* \subset \Gamma_+(f)$ is compact if and only if $\operatorname{orb} \tau \subset \psi^{-1}(0)$. Here, $\psi : T_N(\Delta') \to T_N(\Delta)$ is a morphism induced from the subdivision Δ'. As $\psi^{-1}(0)$ is covered by an $\operatorname{orb} \tau$ as above, it is sufficient to prove the following lemma for the proof of the fact that X' is non-singular along $\psi^{-1}(0)$, which completes the theorem.

Lemma 4.4.24. *For $\tau \in \Delta'$, if τ^* is compact, then X' is non-singular along* $\operatorname{orb} \tau$ *and intersects* $\operatorname{orb} \tau$ *transversally.*

Proof. By the assumption, the equations $\frac{\partial f_{\tau^*}}{\partial X_i} = 0$ $(i = 0, \dots, n)$ have no common zeros on $\left(\mathbb{A}_k^1 \setminus \{0\}\right)^{n+1}$. Then the equations $X_i \frac{\partial f_{\tau^*}}{\partial X_i} = 0$ $(i = 0, \dots, n)$ have no common zeros on $\left(\mathbb{A}_k^1 \setminus \{0\}\right)^{n+1}$. For simplicity, we prove the lemma for 1-dimensional τ; for the general case the proof is similar. Let $\tau = \mathbb{R}_{\geq 0}\boldsymbol{p}$ $(\boldsymbol{p} \in N)$. Take $\sigma \in \Delta'$ such that $\tau < \sigma$, $\dim \sigma = n + 1$ and let the generators of σ be $\boldsymbol{p} = (a_0^0, a_1^0, \dots, a_n^0)$, $(a_0^1 a_1^1, \dots, a_n^1), \dots, (a_0^n, \dots, a_n^n) \in N$. As $U_\sigma \cong \mathbb{A}_k^{n+1}$, we can take a coordinate system $Y_0, \dots, Y_n$ of U_σ corresponding to the generators. Then, we have a coordinate transformation $X_i = Y_0^{a_i^0} Y_1^{a_i^1} \cdots Y_n^{a_i^n}$ $(i = 0, \dots, n)$. On the other hand, $\overline{\operatorname{orb} \tau}$ is defined by the equation $Y_0 = 0$ on U_σ. Now, looking at the polynomial $f(X)$ on U_σ,

$$f(X) = Y_0^m (g^0(Y_1, \dots, Y_n) + Y_0 g^1(Y_0, \dots, Y_n)). \tag{4.11}$$

Let $g := g^0 + Y_0 g^1$, then at the neighborhood U_τ of $\operatorname{orb} \tau$ the strict transform X' of X is defined by $g = 0$ and we obtain that $f_{\tau^*} = Y_0^m g^0(Y_1, \dots, Y_n)$.

To show that X' is non-singular along $\operatorname{orb}\tau$, we assume the contrary, i.e., let $P = (0, y_1, y_2, \dots, y_n), y_i \neq 0\ (i = 1, 2, \dots, n)$ be a point of $\operatorname{orb}\tau$ and be a singular point of X'. Then for $i \neq 0$ we have

$$\frac{\partial g}{\partial Y_i}(0, y_1, \dots, y_n) = \frac{\partial g^0}{\partial Y_i}(y_1, \dots, y_n) = 0. \tag{4.12}$$

We have also

$$g(0, y_1, \dots, y_n) = g^0(y_1, \dots, y_n) = 0. \tag{4.13}$$

Now for $i \neq 0$ the value of $Y_i \frac{\partial f_{\tau^*}}{\partial Y_i}(1, y_1, \dots, y_n)$ is $y_i \frac{\partial g^0}{\partial Y_i}(y_1, \dots, y_n)$ which is 0 by (4.12). On the other hand, this is rewritten as

$$Y_i \sum_{j=0}^{n} \frac{\partial X_j}{\partial Y_i} \frac{\partial f_{\tau^*}}{\partial X_j}(1, y_1, \dots, y_n) = \sum_{j=0}^{n} a_j^i X_j \frac{\partial f_{\tau^*}}{\partial X_j}(1, y_1, \dots, y_n) = 0.$$

For $i = 0$, the value of $Y_0 \frac{\partial f_{\tau^*}}{\partial Y_0}(1, y_1, \dots, y_n)$ is 0 by (4.13) and this is rewritten as $\sum_{j=0}^{n} a_j^0 X_j \frac{\partial f_{\tau^*}}{\partial X_j}(1, y_1, \dots, y_n) = 0$.

Now we rewrite the point $(1, y_1, \dots, y_n)$ under the X-coordinates (it means that we consider the coordinates of the image of the point by ψ), then by the coordinate transformation, it is a point in $(\mathbb{A}_k^1 \setminus \{0\})^{n+1}$. The fact that the matrix (a_{ij}) is unimodular yields that the point is the common zero of the equations $X_j \frac{\partial f_{\tau^*}}{\partial X_j} = 0$ $(j = 0, \dots, n)$ on $(\mathbb{A}_k^1 \setminus \{0\})^{n+1}$, which is a contradiction.

In order to show that X' intersects $\operatorname{orb}\tau$ transversally, it is sufficient to prove that the hypersurface defined by $g|_{\operatorname{orb}\tau} = 0$ is non-singular. Here, $g|_{\operatorname{orb}\tau} = g^0(Y_1, \dots, Y_n)$ and if it gives a singular point at $(y_1, \dots, y_n) \in \operatorname{orb}\tau\ (y_i \neq 0)$, then conditions (4.12) and (4.13) hold, which yield a contradiction by the same argument as above.

Corollary 4.4.25. *Assume that the closed subvariety in $\mathbb{A}_k^{n+1}$ defined by a non-degenerate polynomial f has an isolated singularity at the origin. If all 1-dimensional cones in $\Delta' \setminus \Delta$ are in the interior of $\mathbb{R}_{\geq 0}^{n+1}$, then $X' \to X$ is a resolution of the singularity.*

Proof. The set of the points where the morphism $\psi : T_N(\Delta') \to T_N(\Delta)$ is not isomorphic is $\bigcup_{\tau \in \Delta' \setminus \Delta} \operatorname{orb}\tau$. As a 1-dimensional cone $\tau \in \Delta' \setminus \Delta$ is in the interior of $\mathbb{R}_{\geq 0}^{n+1}$, the equality $\psi(\operatorname{orb}\tau) = \{0\}$ holds. On the other hand, for every cone $\sigma \in \Delta' \setminus \Delta$, there is a 1-dimensional cone $\tau < \sigma$ such that $\tau \in \Delta' \setminus \Delta$, therefore $\psi(\operatorname{orb}\sigma) \subset \psi(\overline{\operatorname{orb}\tau}) = \{0\}$ by Theorem 4.4.12 (iii). By this we conclude that ψ is isomorphic outside the origin of $T_N(\Delta) = \mathbb{A}_k^{n+1}$.

Example 4.4.26. Let X be a hypersurface of $\mathbb{A}_k^3$ defined by $f = X_0 X_1 X_2 + X_0^p + X_1^q + X_2^r \left(\frac{1}{p} + \frac{1}{q} + \frac{1}{r} < 1\right)$. Then X has an isolated singularity at the origin. This singularity is called a ***Tpqr*-*singularity***. In this case Σ is as follows. It is a 3-dimensional fan, while the page of this book is 2-dimensional. So we look at the picture of a section cut out of the fan by a sphere centered at the origin. Therefore, a vertex in the picture shows a 1-dimensional cone and the coordinate corresponding to the vertex shows the primitive vector on the 1-dimensional cone.

Σ;

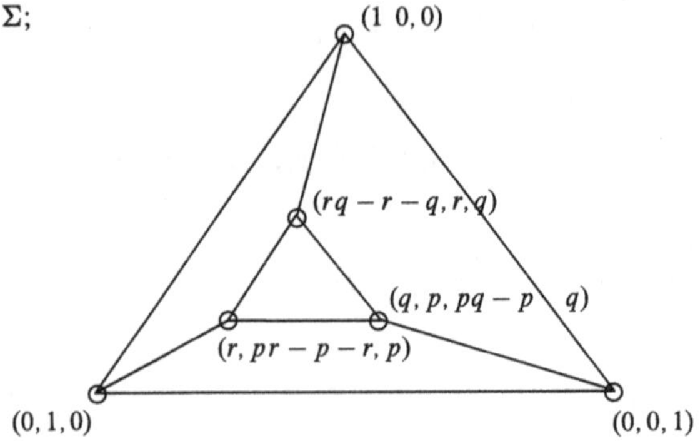

Σ;

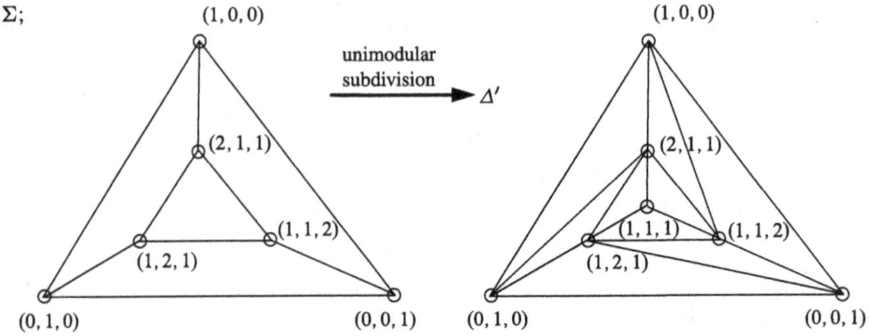

For example, if $(p, q, r) = (4, 4, 4)$, we have the picture of Σ as follows:
By this we obtain a resolution of the T_{444}-singularity, but as is seen easily, unimodular subdivisions are not unique. On the other hand, the T_{444}-singularity can be resolved by the blow-up at the origin, which is much easier than above (see Example 4.3.10). Our toric resolution by the Newton polygon states also the transversality of the strict transform of X and orbits, therefore it needs more procedures than just making a resolution of the singularity of X. But we can definitely get a resolution by this method. This method is very useful to calculate examples.

Chapter 5
Divisors and Sheaves on a Variety

A gorgeous proof can be replaced by a simple proof.

In this chapter we introduce divisors, divisorial sheaves and an equivalence relation of divisors. We also introduce canonical divisors and the canonical sheaf. Later on, we will compare the canonical sheaf of the neighborhood of a singular point and the canonical sheaf of the resolution variety, and by this we will measure the complexity of the singularity.

In this chapter a variety is always an integral algebraic variety over an algebraically closed field k.

5.1 Locally Free Sheaves, Invertible Sheaves, Divisorial Sheaves

Definition 5.1.1. Let $\mathscr{F}$ be a coherent $\mathscr{O}_X$-Module. An element $a \in \Gamma(U, \mathscr{F})$ is called a *torsion element* of $\mathscr{F}$, if there exist an open subset $V \subset U$ and non-zero element $g \in \Gamma(V, \mathscr{O}_X)$ such that $ga = 0$ on V. Here for every open subset $U \subset X$ we define:

$$\Gamma(U, Tor_{\mathscr{O}_X}(\mathscr{F})) := \{a \in \Gamma(U, \mathscr{F}) \mid a \text{ is a torsion element of } \mathscr{F}\}.$$

Then $Tor_{\mathscr{O}_X}(\mathscr{F})$ is a $\mathscr{O}_X$-submodule of $\mathscr{F}$. This is called the *torsion part* of $\mathscr{F}$. A coherent $\mathscr{O}_X$-Module $\mathscr{F}$ satisfies $Tor_{\mathscr{O}_X}(\mathscr{F}) = 0$, then $\mathscr{F}$ is called a *torsion-free* $\mathscr{O}_X$-Module. Clearly the quotient module sheaf $\mathscr{F}/Tor_{\mathscr{O}_X}(\mathscr{F})$ is a torsion free $\mathscr{O}_X$-Module.

S. Ishii, *Introduction to Singularities*,
https://doi.org/10.1007/978-4-431-56837-7_5

Definition 5.1.2. A coherent $\mathcal{O}_X$-Module $\mathcal{F}$ is called *locally free* if there is an open covering $X = \bigcup U_i$ of X, such that $\mathcal{F}|_{U_i} \cong \mathcal{O}_X^{\oplus r_i}|_{U_i}$ holds for every U_i. Here, as X is irreducible, $U_i \cap U_j \neq \emptyset$ holds for every pair U_i, U_j. Hence the value r_i must be common, and this value is called the *rank* of $\mathcal{F}$. In particular, a locally free sheaf of rank one is called an *invertible sheaf.*

By the following proposition we can see the property that one stalk of a coherent sheaf is locally free is extended to a neighborhood.

Proposition 5.1.3. *For a coherent $\mathcal{O}_X$-Module $\mathcal{F}$ the following hold:*

(i) Assume that $\mathcal{F}_x$ is a free $\mathcal{O}_{X,x}$-module for $x \in X$, i.e., there is an isomorphism $\mathcal{F}_x \cong \mathcal{O}_{X,x}^{\oplus r}$. Then there exists an open neighborhood U of x such that there is an isomorphism $\mathcal{F}|_U \cong \mathcal{O}_X^{\oplus r}|_U$.
(ii) If $\mathcal{F}_x$ is a free $\mathcal{O}_{X,x}$-module for every point $x \in X$, then $\mathcal{F}$ is locally free.

Proof. For the proof of (i), we may assume that X is affine. Let the affine ring be A and let the defining ideal of x be M. Let $F := \Gamma(X, \mathcal{F})$, then $\mathcal{F}_x = F \otimes_A A_M$. Take free generators $\alpha_i = \frac{f_i}{g_i}$ $(i = 1, \cdots, r,\ f_i \in F, g_i \in A - M)$ of $\mathcal{F}_x$ over $\mathcal{O}_{X,x}$. Let $g := \prod_i g_i \in A - M$ and replace X by $X_g = \{y \in X \mid g(y) \neq 0\}$, then we may assume that $\alpha_i \in F$. Define a homomorphism $\varphi : \mathcal{O}_X^{\oplus r} \to \mathcal{F}$ by $(h_1, \cdots, h_r) \mapsto \sum h_i\alpha_i$. As $\mathcal{O}_X^{\oplus r}$ and $\mathcal{F}$ are coherent, the kernel Ker φ and the cokernel Coker φ are also coherent. Therefore the supports Supp (Ker φ), Supp (Coker φ) are both closed subsets of X not containing x. Hence, on an appropriate open neighborhood of x, the homomorphism φ is an isomorphism. The statement (ii) follows from (i).

Proposition 5.1.4. *For every coherent $\mathcal{O}_X$-Module $\mathcal{F}$ there is a non-empty open subset U of X such that $\mathcal{F}|_U$ is locally free.*

Proof. For a point $x \in X$ we define $d(x) := \dim_k \mathcal{F}_x \otimes \boldsymbol{k}(x)$, $\boldsymbol{k}(x) = \mathcal{O}_{X,x}/\mathfrak{m}_{X,x}$. Then the subset $U := \{x \in X \mid d(x) \text{ is minimal}\}$ becomes an open subset. Indeed, by Nakayama's lemma, for every point $x \in U$ the stalk $\mathcal{F}_x$ is generated by $d = d(x)$ elements $a_1, \cdots, a_d$. Take an affine open neighborhood U_x such that $a_i \in \Gamma(U_x, \mathcal{F})$ and define a homomorphism $\eta : \mathcal{O}_X^{\oplus d}|_{U_x} \to \mathcal{F}|_{U_x}, (h_1, \cdots, h_d) \mapsto \sum h_i a_i$. Then we have Coker $\eta \otimes \boldsymbol{k}(x) = 0$. If we replace U_x by an appropriately smaller one, we have Coker $\eta = 0$ which implies the η is surjective i.e., $\mathcal{F}|_{U_x}$ is generated by $\{a_i\}$. Therefore we have $U_x \subset U$, which yields that U is an open subset. Next, to show that $\{a_i\}$ are free bases over $\mathcal{O}_X|_{U_x}$, assume a linear relation $\sum g_i a_i = 0, g_i \in \Gamma(U_x, \mathcal{O}_X)$. For every $y \in U_x$, this must be the trivial relation modulo $\mathfrak{m}_{X,y}$; then $g_i \in \mathfrak{m}_{X,y}$ $(\forall y \in U_x)$. Now, the function g_i has a value 0 on every point of an affine integral variety U_x. By the Hilbert Nullstellensatz we obtain $g_i = 0$.

Definition 5.1.5. If a coherent $\mathcal{O}_X$-Module $\mathcal{F}$ is locally free of rank r on an open subset U, then we call r the *rank* of $\mathcal{F}$.

In particular, for a 1-dimensional non-singular variety, the following hold:

Proposition 5.1.6. *(i) Let R be a 1-dimensional normal local ring. If F is a finitely generated R-module, then the following are equivalent:*

$$F \text{ is a free } R\text{-module} \Longleftrightarrow F \text{ is a torsion free } R\text{-module}.$$

(ii) Let X be a 1-dimensional non-singular variety over k and $\mathscr{F}$ a coherent $\mathscr{O}_X$-Module on X. Then the following are equivalent:

$$\mathscr{F} \text{ is locally free} \Longleftrightarrow \mathscr{F} \text{ is torsion free}.$$

Proof. (i) The implication $\Rightarrow$ is trivial. For the implication $\Leftarrow$ we should note that the torsion freeness implies that for $0 \neq f \in R$ the homomorphism $F \xrightarrow{\times f} F$ is surjective. Then, depth $F \geq 1$. On the other hand, as 1-dimensional normal is equivalent to 1-dimensional regular we have projdim $F < \infty$ and from the Auslander–Buchsbaum formula [Ma2, Th19.1] it follows that projdim $F +$ depth $F = \dim R = 1$. Therefore projdim $F = 0$ holds, i.e., F is projective. A projective module over a local ring is a free module. Indeed, if F is projective and $\dim F \otimes R/\mathfrak{m} = d$, then for the canonical homomorphism $\varphi : R^d \longrightarrow\!\!\!\!\rightarrow F$ and $1_F : F \to F$ there exists $\psi : F \to R^d$ such that $1_F = \varphi \circ \psi$. Then, there exists T such that $R^d \cong F \oplus T$. Since $R^d \otimes R/\mathfrak{m} \simeq F \otimes R/\mathfrak{m}$, we have $T \otimes R/\mathfrak{m} = 0$ which yields $T = 0$ by Nakayama's lemma.

(ii) follows from (i), Proposition 5.1.3 (ii) and $(Tor_{\mathscr{O}_X}(\mathscr{F}))_x = Tor_{\mathscr{O}_{X,x}}(\mathscr{F}_x)$.

For the case $\dim X \geq 2$, a torsion free $\mathscr{O}_X$-Module is not necessarily locally free, but we have the following:

Proposition 5.1.7. *Let X be a normal integral variety over k. For a torsion free coherent $\mathscr{O}_X$-Module $\mathscr{F}$ there exists an open subset $X_0 \subset X$ such that:*

(i) $\operatorname{codim}_X(X \setminus X_0) \geq 2$.
(ii) $\mathscr{F}|_{X_0}$ *is locally free.*

Proof. Since the problem is local, we may assume that X is an affine variety. It is sufficient to prove that for any irreducible closed subset Z of codimension 1, there exists an open subset W such that $Z \cap W \neq \emptyset$ and $\mathscr{F}|_W$ is locally free. Let A be the affine coordinate ring of X and $\mathfrak{p}$ the defining ideal of Z, then $A_\mathfrak{p}$ is a normal 1-dimensional local ring, since $\mathfrak{p}$ has height 1 and A is a normal ring. As $F := \Gamma(X, \mathscr{F})$ is a torsion free A-module, $F \otimes_A A_\mathfrak{p}$ is also a torsion free $A_\mathfrak{p}$-module. Therefore by Proposition 5.1.6, it is a free $A_\mathfrak{p}$-module. Let $\alpha_i = \frac{f_i}{h_i}$ ($i = 1, \cdots, r$, $f_i \in F$, $h_i \in A \setminus \mathfrak{p}$) be generators of $F \otimes_A A_\mathfrak{p}$ and let $h = \prod_{i=1}^r h_i$, then, by $h \in A \setminus \mathfrak{p}$ we have $Z \cap X_h \neq \emptyset$. Then by replacing X by X_h, we may assume $\alpha_i \in F$. Here, let $\mathscr{F}'$ be the $\mathscr{O}_X$-submodule of $\mathscr{F}$ generated by $\alpha_1, \cdots, \alpha_r$. Then $A \hookrightarrow A_\mathfrak{p}$ is injective and $\{\alpha_i\}$ are linearly independent over $A_\mathfrak{p}$ in $F \otimes_A A_\mathfrak{p}$, therefore $\{\alpha_i\}$ are free generators of $F' = \Gamma(X, \mathscr{F}')$ over A. Therefore, $\mathscr{F}'$ is a free $\mathscr{O}_X$-Module. We will prove that $\mathscr{F}' = \mathscr{F}$ on an open subset. Let $\mathscr{L} := \mathscr{F}/\mathscr{F}'$, $L := \Gamma(X, \mathscr{L})$ and let $l_1, \cdots, l_s$ be generators of L over A. By the definition $L \otimes_A A_\mathfrak{p} = 0$, there exists $g_j \in A \setminus \mathfrak{p}$ such that $g_j l_j = 0$ $(j = 1, \cdots, s)$. Let $g := \prod_{j=1}^{s} g_j$, then by $g \in A \setminus \mathfrak{p}$ we have $Z \cap X_g \neq \emptyset$ and, moreover, $l_j = 0$ $(j = 1, \cdots, s)$ in $\Gamma(X_g, \mathscr{L}) = L \otimes_A A_g$. Hence it follows that $\mathscr{L}|_{X_g} = 0$ and $\mathscr{F} = \mathscr{F}'$ on X_g.

We will introduce a reflexive sheaf, for which we prepare some notions.

Definition 5.1.8. For a coherent $\mathscr{O}_X$-Module $\mathscr{F}$, we define $\mathscr{F}^* := \mathscr{H}om_{\mathscr{O}_X}(\mathscr{F}, \mathscr{O}_X)$ and call it the *dual* of $\mathscr{F}$.

Remark 5.1.9. For any $\mathscr{F}$, the dual $\mathscr{F}^*$ becomes a torsion free $\mathscr{O}_X$-Module. A canonical homomorphism $\varphi : \mathscr{F} \to \mathscr{F}^{**}$ is defined as follows:

We associate $a \in \mathscr{F}$ to $\varphi(a) : \mathscr{F}^* \to \mathscr{O}_X$ $(f \mapsto f(a))$. Here if $\mathscr{F}$ is locally free, then φ is isomorphic.

Lemma 5.1.10. *Let $\mathscr{F}$ be a coherent $\mathscr{O}_X$-Module such that $\mathscr{F}|_{X_0}$ is locally free on an open subset $X_0 \subset X$ such that* $\operatorname{codim}_X(X \setminus X_0) \geq 2$. *Then, the morphism $\varphi : \mathscr{F} \to \mathscr{F}^{**}$ in Remark 5.1.9 induces an isomorphism $j_*(\mathscr{F}|_{X_0}) \xrightarrow{\sim} \mathscr{F}^{**}$, where $j : X_0 \hookrightarrow X$ is the inclusion map.*

Proof. For every open subset $U \subset X$, we define $U_0 = U \cap X_0$. For every element $a \in \Gamma(U, j_*(\mathscr{F}|_{X_0})) = \Gamma(U_0, \mathscr{F})$ a map $\widetilde{\varphi}_a : \Gamma(U, \mathscr{F}^*) \to \Gamma(U, \mathscr{O}_X)$, $f \mapsto f(a)$ is well-defined. In fact, as $a \in \Gamma(U_0, \mathscr{F})$, $f : \mathscr{F}|_U \to \mathscr{O}_X|_U$, we obtain $f(a) \in \Gamma(U_0, \mathscr{O}_X)$. Here, since $\operatorname{codim}_X(X \setminus X_0) \geq 2$ and X is normal, it follows that $\Gamma(U_0, \mathscr{O}_X) = \Gamma(U, \mathscr{O}_X)$. By this, from φ a morphism $j_*(\mathscr{F}|_{X_0}) \xrightarrow{\widetilde{\varphi}} \mathscr{F}^{**}$ is induced. We have a commutative diagram on U with $\widetilde{\varphi}$:

$$\begin{array}{ccc} \Gamma(U, j_*(\mathscr{F}|_{X_0})) & \xrightarrow{\widetilde{\varphi}_U} & \Gamma(U, \mathscr{F}^{**}) \\ \| & & \downarrow \rho \\ \Gamma(U_0, \mathscr{F}) & \xrightarrow{\varphi_{U_0}} & \Gamma(U_0, \mathscr{F}^{**}). \end{array}$$

Here, by $\mathscr{F}|_{X_0} \xrightarrow{\sim} \mathscr{F}^{**}|_{X_0}$, the morphism φ_{U_0} is an isomorphism. As $\mathscr{F}^{**}$ is torsion free, the morphism ρ is injective, and therefore $\widetilde{\varphi}_U$ is isomorphic.

Theorem 5.1.11. *Let $\mathscr{L}$ be a torsion free coherent $\mathscr{O}_X$-Module on a normal variety X, then the following are equivalent:*

(i) $\mathscr{L} \xrightarrow{\sim} \mathscr{L}^{**}$.
(ii) If a torsion free $\mathscr{O}_X$-Module $\mathscr{M}$ satisfies $\mathscr{L} \subset \mathscr{M}$ and $\operatorname{codim}_X(\operatorname{Supp} \mathscr{M}/\mathscr{L}) \geq 2$, *then $\mathscr{M} = \mathscr{L}$.*
(iii) Let $X_0 \subset X$ be an open subset such that $\operatorname{codim}_X(X \setminus X_0) \geq 2$ *and assume that $\mathscr{L}|_{X_0}$ is locally free. Then the equality $j_*(\mathscr{L}|_{X_0}) = \mathscr{L}$ holds, where $j : X_0 \hookrightarrow X$ is the inclusion map.*

Proof. (i) $\Rightarrow$ (ii). By Proposition 5.1.7 and the assumption in (ii), there is an open subset $U \subset X$ such that $\operatorname{codim}_X(X \setminus U) \geq 2$ and $\mathscr{L}|_U = \mathscr{M}|_U$ is locally free. For the inclusion map $j : U \hookrightarrow X$, we have $j_*(\mathscr{L}|_U) = j_*(\mathscr{M}|_U)$. Hence by Lemma 5.1.10 and the assumption in (i), the inclusion relations $\mathscr{L} \subset \mathscr{M} \subset j_*(\mathscr{M}|_U) = j_*(\mathscr{L}|_U)$ are all identities.
(ii) $\Rightarrow$ (iii). It is sufficient to take $\mathscr{M} = j_*(\mathscr{L}|_{X_0})$.
(iii) $\Rightarrow$ (i) follows from Lemma 5.1.10.

Definition 5.1.12. A torsion free coherent $\mathscr{O}_X$-Module $\mathscr{L}$ on a normal variety X satisfying the conditions (one of) (i), (ii), (iii) in Theorem 5.1.11 is called a *reflexive $\mathscr{O}_X$-Module*. If, moreover, it is of rank one, then it is called a *divisorial $\mathscr{O}_X$-Module* (or *divisorial sheaf*).

5.2 Divisors

Definition 5.2.1. Let K be a field. A map $\nu : K^* = K \setminus \{0\} \to \mathbb{Z}$ is called a *discrete valuation* if the following are satisfied:

(1) $\nu(xy) = \nu(x) + \nu(y)$;
(2) $\nu(x + y) \geq \min(\nu(x), \nu(y))$.

Here, a subset $\mathscr{O} := \{x \in K^* \mid \nu(x) \geq 0\} \cup \{0\}$ becomes a subring of K and it is called the *discrete valuation ring* for ν. It is easy to see that $\mathscr{O}$ is a local ring and the subset $\mathfrak{m} := \{x \in K^* \mid \nu(x) > 0\} \cup \{0\}$ becomes the maximal ideal.

Example 5.2.2. Let R be a 1-dimensional regular local ring, $\mathfrak{m}$ the maximal ideal and $Q(R)$ the quotient field of R. Let x be a generator of $\mathfrak{m}$. Define a map $\nu : Q(R) \setminus \{0\} \to \mathbb{Z}$ as follows:

$$\nu(a) = \begin{cases} r, & a \in R, \quad a = u \cdot x^r \quad (u \text{ is a unit}) \\ \nu(\alpha) - \nu(\beta), & a = \alpha/\beta \quad \alpha, \beta \in R. \end{cases}$$

Then ν is a discrete valuation and R is a discrete valuation ring for ν.

In particular, let X be a normal affine variety, A the affine coordinate ring, $Z \subset X$ an irreducible closed subset of codimension 1 and let $\mathfrak{p} = I(Z)$. Then, the localization $A_{\mathfrak{p}}$ is a 1-dimensional normal local ring, therefore by Theorem 4.1.11 it is regular. The discrete valuation defined as above is denoted by ν_Z and called the *discrete valuation associated to Z*. More generally, let X be a normal variety and $Z \subset X$ an irreducible closed subset of codimension 1. Take an open subset $U \subset X$ such that $U \cap Z \neq \emptyset$. Define a discrete valuation ν_Z associated to Z as the discrete valuation in U associated to $Z \cap U$. We note that this is independent of the choice of affine open subset U.

As X is integral, for every point $x \in X$ the quotient field $Q(\mathscr{O}_{X,x})$ of the local ring $\mathscr{O}_{X,x}$ is common and coincides with the quotient field of affine coordinate ring of any affine open subset of X. This field is denoted by $k(X)$ and called the *rational function field* of X.

Define the sheaf $\mathscr{F}$ by $\Gamma(U, \mathscr{F}) = k(X)$ for every open subset $U \subset X$. This sheaf is denoted also by $k(X)$. Clearly $\mathscr{O}_X$ is a subsheaf of $k(X)$.

A discrete valuation ring has the following characterization.

Proposition 5.2.3. *Let R be a 1-dimensional Noetherian local ring. Then, the following are equivalent:*

(i) R is regular.
(ii) R is a discrete valuation ring.

For the proof see [AM, Proposition 9.2], .

Definition 5.2.4. A 1-codimensional irreducible closed subset Γ on a normal variety X is called a *prime divisor* on X. A formal sum $\sum n_\Gamma \Gamma$ ($n_\Gamma \in \mathbb{Z}$, $n_\Gamma = 0$ except for finitely many Γ) is called a *divisor* (or *Weil divisor*) on X. We call the set of all divisors on X the *divisor group* of X and denote it by $\mathrm{Div}\,(X)$. Actually, it has a free group structure by the addition $\sum n_\Gamma \Gamma + \sum m_\Gamma \Gamma = \sum (n_\Gamma + m_\Gamma)\Gamma$. For a rational function $f \in k(X)$ on X define $(f) = \sum \nu_\Gamma(f)\Gamma$, then by the following lemma the equation $\nu_\Gamma(f) = 0$ holds except for finitely many Γ. Therefore we have $(f) \in \mathrm{Div}\,(X)$. Such a divisor (f) is called a *principal divisor* on X.

If a divisor $D = \sum n_\Gamma \Gamma$ satisfies $n_\Gamma \geq 0$ $(\forall \Gamma)$, then we call D an *effective divisor* and denote it by $D \geq 0$. In particular, if there is a prime divisor Γ such that $n_\Gamma > 0$, then we write $D > 0$. If $D_1 - D_2 \geq 0$ (resp. > 0), we write $D_1 \geq D_2$ (resp. $D_1 > D_2$).

Lemma 5.2.5. *Let X be a normal variety. Then for every $f \in k(X)^*$ there are only a finite number of prime divisors Γ of X such that $\nu_\Gamma(f) \neq 0$.*

Proof. It is sufficient to prove that for a discrete valuation ν the number of prime divisors Γ such that $\nu_\Gamma(f) > 0$ is finite, because if $\nu(f) < 0$ then $\nu(f^{-1}) > 0$. Assume $\nu_\Gamma(f) > 0$, then there exists an affine open subset U of X such that $f \in A_\mathfrak{p}$; here A is the affine coordinate ring of U and $\mathfrak{p} = I(U \cap \Gamma) \subset A$. Then, by replacing U by a smaller one, we may assume that $f \in A$. There are only a finite number of prime divisors in the closed subset $X \setminus U$. On the other hand, let Γ' be a prime divisor on U such that $\nu_{\Gamma'}(f) > 0$, then $f \in \mathfrak{p}' = I(U \cap \Gamma')$. As A is a Noetherian ring, there are at most a finite number of prime ideals of height one with f.

Definition 5.2.6. Let X be a normal variety. For $D = \sum n_\Gamma \Gamma \in \mathrm{Div}\,(X)$ we define a sheaf $\mathscr{O}_X(D)$ as follows:

For an open subset $U \subset X$,

$$\Gamma(U, \mathscr{O}_X(D)) := \left\{ f \in k(X)^* \;\middle|\; \begin{array}{c} \nu_\Gamma(f) \geq -n_\Gamma \\ \Gamma : \text{prime divisor of U} \end{array} \right\} \cup \{0\}.$$

Then this is a $\mathscr{O}_X$-submodule of the constant sheaf $k(X)$. Therefore it is obviously torsion free.

A divisor D and the sheaf $\mathcal{O}_X(D)$ have the following relation:

Theorem 5.2.7. *On a normal variety X the map Φ that associates $D \in \mathrm{Div}(X)$ to $\mathcal{O}_X(D)$ gives the following bijection:*

$$\Phi : \mathrm{Div}(X) \xrightarrow{\sim} \left\{ \begin{matrix} \textit{divisorial } \mathcal{O}_X - \textit{submodule} \\ \mathcal{L} \subset k(X) \end{matrix} \right\}.$$

Proof. First note that $\mathcal{O}_X(D) \cong \mathcal{O}_X$ outside D, then we can see that the rank of $\mathcal{O}_X(D)$ is 1. Take an open subset $X_0 \subset X$ such that $\mathrm{codim}_X(X \setminus X_0) \geq 2$ and $\mathcal{O}_X(D)|_{X_0}$ is invertible. For the inclusion map $j : X_0 \hookrightarrow X$ we will prove that $j_*(\mathcal{O}_X(D)|_{X_0}) = \mathcal{O}_X(D)$, which shows that $\mathcal{O}_X(D)$ is divisorial by Theorem 5.1.11 (iii). For every open subset $V \subset X$, we have $\mathrm{codim}_V(V \setminus X_0) \geq 2$, therefore the set of prime divisors of V and the set of prime divisors of $V \cap X_0$ correspond bijectively. Hence, by the definition of $\mathcal{O}_X(D)$, we obtain

$$\Gamma(V, \mathcal{O}_X(D)) = \Gamma(V \cap X_0, \mathcal{O}_X(D)) = \Gamma(V, j_*(\mathcal{O}_X(D)|_{X_0}))$$

as required.

Conversely, for every divisorial $\mathcal{O}_X$-submodule $\mathcal{L} \subset k(X)$, By Proposition 5.1.7, there is an open subset X_0 such that $\mathrm{codim}(X \setminus X_0) \geq 2$ and $\mathcal{L}|_{X_0}$ is invertible. Then there is an open covering $\{U_i\}_{i=1}^s$ of X_0 such that $\varphi_i : \mathcal{O}_X|_{U_i} \xrightarrow{\sim} \mathcal{L}|_{U_i}$. Here, let $f_i = \varphi_i(1)$ and for a prime divisor Γ take an open subset U_i such that $U_i \cap \Gamma \neq \emptyset$ and then define $n_\Gamma := \nu_\Gamma(f_i)$.

On the intersection $U_i \cap U_j \cap \Gamma \neq \emptyset$ we have $f_j = uf_i$ (u is a unit of $\Gamma(U_i \cap U_j, \mathcal{O})$), therefore the definition of n_Γ does not depend on choice of U_i. By Lemma 5.2.5, it follows that $D = \sum n_\Gamma \Gamma \in \mathrm{Div}(X)$. This correspondence $\mathcal{L} \mapsto \sum n_\Gamma \Gamma$ gives the inverse of Φ.

Proposition 5.2.8. *On a normal variety X, the following is the relation between the sum of divisors and divisorial $\mathcal{O}_X$-Modules:*

$$\mathcal{O}_X(D_1 + D_2) = \left(\mathcal{O}_X(D_1) \otimes \mathcal{O}_X(D_2)\right)^{**}.$$

In particular, for $\mathcal{L} = \mathcal{O}_X(D)$, $m \in \mathbb{N}$, we denote $\mathcal{O}_X(mD)$ by $\mathcal{L}^{[m]}$. Obviously we have $\mathcal{L}^{[m]} = (\mathcal{L}^{\otimes m})^{**} = j_*(\mathcal{L}|_{X_{\mathrm{reg}}}^{\otimes m})$, where $j : X_{\mathrm{reg}} \hookrightarrow X$ is the inclusion map.

Now we define divisors with a good property.

Definition 5.2.9. A divisor D such that the corresponding sheaf $\mathcal{O}_X(D)$ is invertible is called a *Cartier divisor*. The set of Cartier divisors is denoted by $\mathrm{CDiv}(X)$.

Definition 5.2.10. For $D, D' \in \mathrm{Div}(X)$, if there exists $f \in k(X)$ such that $D' = D + (f)$, then we write $D \sim D'$ and say that D and D' are *linearly equivalent*. Actually, $\sim$ is an equivalence relation in $\mathrm{Div}(X)$. Define $\mathcal{Cl}(X) := \mathrm{Div}(X)/\sim$, then $\mathcal{Cl}(X)$ has the structure of abelian group induced from $\mathrm{Div}(X)$.

In particular, if X is a complete non-singular curve, a divisor $D \in \mathrm{Div}(X)$ is represented as $D = \sum_{P \in X} n_P P$, where the degree of D is defined by $\sum_P n_P$ and denoted by $\deg D$. Then, as $\deg(f) = 0$ for a rational function $f \in k(X)$, for two divisors such that $D \sim D'$ we have $\deg D = \deg D'$. Therefore, if we define $\deg \mathscr{O}(D) := \deg D$, the following theorem gives the equality $\deg \mathscr{L} = \deg \mathscr{L}'$ for invertible sheaves $\mathscr{L}$ and $\mathscr{L}'$ such that $\mathscr{L} \cong \mathscr{L}'$.

Theorem 5.2.11. *For a normal variety X the correspondence $\Phi : D \mapsto \mathscr{O}_X(D)$ in Theorem 5.2.7 gives a bijection:*

$$\widetilde{\Phi} : \mathscr{C}\ell(X) \xrightarrow{\sim} \{\text{divisorial } \mathscr{O}_X\text{-Modules}\} \big/ \simeq .$$

Here $\simeq$ is the equivalence relation of isomorphisms of $\mathscr{O}_X$-Modules.

Proof. If $D' = D + (f)$, then by $\mathscr{O}_X(D') = f^{-1} \cdot \mathscr{O}_X(D)$ we have $\mathscr{O}_X(D') \cong \mathscr{O}_X(D)$. Conversely, for two divisorial $\mathscr{O}_X$-submodules $\mathscr{L}_1, \mathscr{L}_2 \subset k(X)$ if $\mathscr{L}_1 \cong \mathscr{L}_2$, then there exists $f \in k(X)$ such that $\mathscr{L}_1 = f \cdot \mathscr{L}_2$. Therefore if we put $\mathscr{L}_i = \mathscr{O}_X(D_i)$ $i = 1, 2$, then we have $D_1 \sim D_2$. Now it is sufficient to show that every divisorial $\mathscr{O}_X$-Module $\mathscr{L}$ is isomorphic to a subsheaf of the constant sheaf $k(X)$. Let X_0 be the open subset in Proposition 5.1.7. The canonical map

$$\mathscr{L}|_{X_0} \longrightarrow \mathscr{L} \otimes_{\mathscr{O}_X} k(X)|_{X_0}, \quad f \longmapsto f \otimes 1$$

is injective. Now compose it with the following isomorphism:

$$\eta : \mathscr{L} \otimes_{\mathscr{O}_X} k(X)|_{X_0} \xrightarrow{\sim} k(X)|_{X_0}.$$

We have $\mathscr{L}|_{X_0} \hookrightarrow k(X)|_{X_0}$, therefore we also obtain

$$\mathscr{L} = j_*(\mathscr{L}|_{X_0}) \hookrightarrow k(X) = j_*(k(X)|_{X_0}),$$

where $j : X_0 \hookrightarrow X$ is the inclusion map and the isomorphism η is obtained as follows:

Take an open subset $U \subset X_0$ such that $\mathscr{L}|_U \simeq \mathscr{O}_X|_U$, then we have $\mathscr{L} \otimes_{\mathscr{O}_X} k(X)|_U \simeq k(X)|_U$. As X is irreducible, if a sheaf is locally a constant sheaf, then it is constant whole on X_0, which gives an isomorphism η.

Next we introduce divisors with coefficients in rational numbers.

Definition 5.2.12. An element of the $\mathbb{Q}$-linear space $\mathrm{Div}(X) \otimes_{\mathbb{Z}} \mathbb{Q}$ is called a $\mathbb{Q}$-*divisor*, and an element of $C\mathrm{Div}(X) \otimes_{\mathbb{Z}} \mathbb{Q}$ is called a $\mathbb{Q}$-*Cartier divisor*.

We consider $C\mathrm{Div}(X) \otimes \mathbb{Q}$ and $\mathrm{Div}(X)$ as a linear subspace and subgroup of $\mathrm{Div}(X) \otimes_{\mathbb{Z}} \mathbb{Q}$, respectively. We have the inclusion $C\mathrm{Div}(X) \otimes \mathbb{Q} \cap \mathrm{Div}(X) \supset C\mathrm{Div}(X)$, but in general the inclusion is not an identity. Let us look at an example.

Example 5.2.13. Let X be a hypersurface in $\mathbb{A}_k^3$ defined by $xy + z^2 = 0$. As the singular point of X is only the origin, X is normal (Theorem 4.1.11 (iii)). Take the closed subset $D = V(y, z)$ in X, then it is a prime divisor isomorphic to $\mathbb{A}_k^1$. But, as $\mathcal{O}_X(D)$ is not invertible, D is not a Cartier divisor. On the other hand, $\mathcal{O}_X(2D)$ is a free $\mathcal{O}_X$-Module generated by y^{-1}, therefore D is a $\mathbb{Q}$-Cartier divisor.

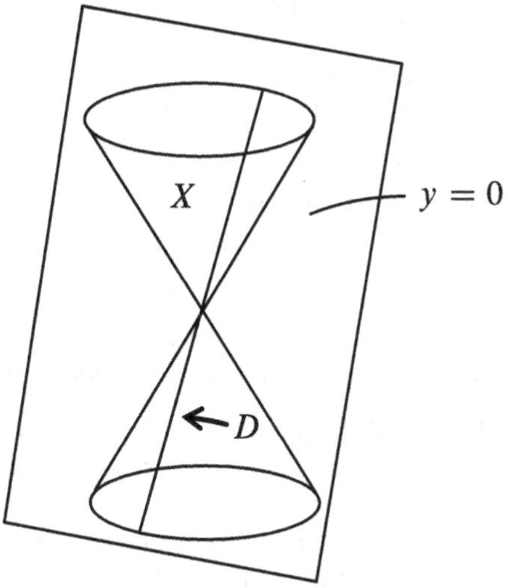

Definition 5.2.14. Let D be a Cartier divisor on a normal variety X. For a surjective morphism $f : Y \to X$ of varieties, $f^*\mathcal{O}(D)$ is an invertible subsheaf of $k(Y)$. The Cartier divisor f^*D on Y is defined by $\mathcal{O}_Y(f^*D) = f^*\mathcal{O}(D)$. Moreover, for a $\mathbb{Q}$-Cartier divisor D, let $r \in \mathbb{Z}$ such that rD is a Cartier divisor, then f^*D is defined by $\frac{1}{r}f^*(rD)$.

5.3 Canonical Sheaf and Canonical Divisor

First we define the canonical sheaf on a non-singular variety, then we introduce the canonical sheaf for general projective varieties, by using the canonical sheaf on the projective space. The canonical sheaf of a quasi-projective variety X is the restriction of the canonical sheaf of the compactification $\overline{X}$ in the projective space. The canonical sheaf is defined not necessarily for a normal variety, but we usually study only a normal variety. So, we may think that Corollary 5.3.9 is the definition of the canonical sheaf.

Definition 5.3.1. Let X be an n-dimensional non-singular variety. The *canonical sheaf* $\widetilde{\omega}_X$ on X is defined by using the sheaf of differentials Ω_X as follows:

$$\widetilde{\omega}_X = \bigwedge^n \Omega_X.$$

Note 5.3.2. Note that for an n-dimensional non-singular algebraic variety X the sheaf of differentials Ω_X is a locally free $\mathcal{O}_X$-Module of rank n. In fact, by the proof of Theorem 4.1.7 (vi), for the maximal ideal $\mathfrak{m} \subset \mathcal{O}_{X,x}$, the canonical map $\mathfrak{m}/\mathfrak{m}^2 \xrightarrow{\sim} \Omega_{X,x}/\mathfrak{m}\Omega_{X,x}$, $f \mapsto df$ is isomorphic, therefore for every point $x \in X$ we have $\dim \Omega_{X,x}/\mathfrak{m}\Omega_{X,x} = n$. Then by Nakayama's lemma, the sheaf Ω_X is a locally free $\mathcal{O}_X$-Module of rank n (see the proof of Proposition 5.1.4). Let n-elements $x_1, \cdots, x_n \in \mathcal{O}_{X,x}$ be generators of $\mathfrak{m}$ (such a system is called a *regular system of parameters*). Then Ω_X is generated by $dx_1, \cdots, dx_n$ on a neighborhood of x. Hence $\widetilde{\omega}_X$ is generated by $dx_1 \wedge \cdots \wedge dx_n$. The following are known about a regular system of parameters:

Proposition 5.3.3 ([Ma2, Th. 14.2 p. 105]). *Let R be an n-dimensional regular local ring and let $\mathfrak{m}$ the maximal ideal. Take i elements $x_1, \cdots, x_i \in \mathfrak{m}$, then the following are equivalent:*

(i) $x_1, \cdots, x_i$ *is a part of a regular system of parameters.*
(ii) $R/(x_1, \cdots, x_i)$ *is an $(n-i)$-dimensional regular local ring.*

Proposition 5.3.4 ([Ma1, Th. 36]). *Let R and $\mathfrak{m}$ be as in Proposition 5.3.3. If R/I is again regular for an ideal $I \subset \mathfrak{m}$, then there exists a regular parameter system $x_1, \cdots, x_n$ of R satisfying the following:*
$(x_1, \cdots, x_i) = I$, $\overline{x}_{i+1}, \cdots, \overline{x}_n \in R/I$ *is a regular parameter system of R/I and I/I^2 is a free R/I-module generated by $\overline{x}_1, \cdots, \overline{x}_i$.*

Definition 5.3.5. Let X be a quasi-projective integral variety, i.e., X is an open subvariety of a projective integral variety $\overline{X} \subset \mathbb{P}^N$. (Of course, $X = \overline{X}$ is also possible.) If $\dim X = n$, we define the *canonical sheaf* ω_X of X as follows:

$$\omega_X := \mathcal{E}xt^{N-n}_{\mathcal{O}_{\mathbb{P}^N}}(\mathcal{O}_{\overline{X}}, \widetilde{\omega}_{\mathbb{P}^N})|_X.$$

Theorem 5.3.6. *For an n-dimensional projective integral variety $X \subset \mathbb{P}^N$ the following hold:*

(i) There is a homomorphism $t : H^n(X, \omega_X) \to k$ such that for every coherent $\mathcal{O}_X$-Module $\mathcal{F}$ the following holds: There exists a canonical bilinear map

$$\mathrm{Hom}_{\mathcal{O}_X}(\mathcal{F}, \omega_X) \times H^n(X, \mathcal{F}) \longrightarrow H^n(X, \omega_X)$$

which gives an isomorphism:

$$\mathrm{Hom}(\mathcal{F}, \omega_X) \xrightarrow{\sim} H^n(X, \mathcal{F})^*$$

by composing with t. Here, $*$ means the dual vector space over k.
(ii) The sheaf ω_X is independent of the embedding $X \subset \mathbb{P}^N$.
(iii) For every integer $i \geq 0$ and every coherent $\mathcal{O}_X$-Module $\mathcal{F}$, there exists a canonical isomorphism: $\mathrm{Ext}^i_{\mathcal{O}_X}(\mathcal{F}, \omega_X) \simeq H^{n-i}(X, \mathcal{F})^*$ *if and only if X is Cohen–Macaulay.*

For the proof see, for example, [AK, IV.5].

We will study properties of ω_X. The following is the first step.

Lemma 5.3.7. *Let P be an N-dimensional non-singular variety, $X \subset P$ an n-dimensional non-singular closed subvariety, and $\mathscr{J}$ the defining ideal. For every coherent $\mathscr{O}_P$-Module $\mathscr{F}$, there is the following canonical isomorphism:*

$$\mathscr{E}xt^{N-n}_{\mathscr{O}_P}(\mathscr{O}_X, \mathscr{F}) \simeq \mathscr{H}om_{\mathscr{O}_X}\left(\bigwedge^{N-n} \mathscr{J}/\mathscr{J}^2, \mathscr{F}/\mathscr{J}\mathscr{F}\right).$$

Proof. On a sufficiently small affine open neighborhood of each point of X $U \subset P$, $\mathscr{J}$ is generated by $r = N - n$ elements $x_1, \cdots, x_r$. Let $F = \Gamma(U, \mathscr{F})$, $J = \Gamma(U, \mathscr{J})$, $A = \Gamma(U, \mathscr{O}_P)$. We will prove the following:

$$\mathrm{Ext}^r_A(A/J, F) \simeq \mathrm{Hom}_{A/J}\left(\bigwedge^r J/J^2, F/JF\right). \tag{5.1}$$

For a system $\mathfrak{X} = (x_1, \cdots, x_r)$ we define the complex $K.(\mathfrak{X})$ as follows (this is called the *Koszul complex*):

$$K_p(\mathfrak{X}) = \begin{cases} \bigwedge^p \left(\bigoplus_{i=1}^r Ae_i\right) & 0 \le p \le r \\ 0 & p < 0, \quad p \ge r+1 \end{cases}$$

$$\begin{aligned} d_p : K_p(\mathfrak{X}) &\longrightarrow K_{p-1}(\mathfrak{X}); \\ e_{i_1} \wedge \cdots \wedge e_{i_p} &\longmapsto \sum (-1)^j x_{i_j} e_{i_1} \wedge \cdots \wedge \widehat{e_{i_j}} \wedge \cdots \wedge e_{i_p}. \end{aligned}$$

Here e_i $(i = 1, \cdots, r)$ are free generators over A. Then it is known that $K.(\mathfrak{X}) \to A/J$ is a projective resolution of A/J ([Ma1, Theorem 43, p. 135]). Therefore, by Corollary 3.3.10, the cohomology of $\mathrm{Hom}_A(K.(\mathfrak{X}), F)$ gives $\mathrm{Ext}^p_A(A/J, F)$:

$$H^p(\mathrm{Hom}\,(K.(\mathfrak{X}), F)) = \mathrm{Ext}^p_A(A/J, F).$$

On the other hand, in particular if $p = r$, we define the map:

$$\varphi_{\mathfrak{X}} : \mathrm{Ext}^r_A(A/J, F) = H^r\big(\mathrm{Hom}\,(K.(\mathfrak{X}), F)\big) \longrightarrow \mathrm{Hom}_{A/J}\left(\bigwedge^r J/J^2, F/JF\right)$$

as follows:

For $f : K_r(\mathfrak{X}) \to F$, define

$$\varphi_{\mathfrak{X}}(f) : \bigwedge^r J/J^2 = (A/J)x_1 \wedge \cdots \wedge x_r \longrightarrow F/JF$$

by

$$x_1 \wedge \cdots \wedge x_r \longmapsto \overline{f(e_1 \wedge \cdots \wedge e_r)}.$$

Then $\varphi_{\mathfrak{X}}$ is an isomorphism. Indeed, the subjectivity is obvious and the injectivity is proved as follows:

Assume $\varphi_{\mathfrak{X}}(f) = 0$. As $f(e_1 \wedge \cdots \wedge e_r) = \sum_{i=1}^{r} x_i a_i$ $(\exists a_i \in F)$, by using these a_i we define $g : K_{r-1}(\mathfrak{X}) = \bigwedge^{r-1}(\bigoplus Ae_i) \to F$ by

$$g(e_1 \wedge \cdots \wedge \widehat{e_j} \wedge \cdots \wedge e_r) = (-1)^j a_j.$$

Then, clearly it follows that $d(g) = f$. Therefore $\varphi_{\mathfrak{X}}$ is injective. By this, (1) is proved.

Now it is sufficient to prove the compatibility of the patching of $\varphi_{\mathfrak{X}}$. For two systems of generators $\mathfrak{X} = (x_1, \cdots, x_r)$, $\mathscr{Y} = (y_1, \cdots, y_r)$ of J, let $y_i = \sum c_{ij} x_j$. Then, we have

$$\varphi_{\mathfrak{X}}(f)(y_1 \wedge \cdots \wedge y_r) = \varphi_{\mathfrak{X}}(f)(\det(c_{ij}) x_1 \wedge \cdots \wedge x_r) = \det(c_{ij}) \overline{f(e_1 \wedge \cdots \wedge e_r)}.$$

On the other hand, the isomorphism $K.(\mathfrak{X}) \cong K.(\mathscr{Y})$ is given by $e \mapsto \sum d_{ij} e'_j$, where $(d_{ij}) = (c_{ij})^{-1}$ and $(e'_1, \cdots, e'_r)$ is a basis of $K.(\mathscr{Y})$. Therefore by the diagram:

$$\begin{array}{ccc} K_r(\mathfrak{X}) & \xrightarrow{f} & F/JF \\ \wr\downarrow & \nearrow_{f'} & \\ K_r(\mathscr{Y}) & & \end{array} \qquad \begin{array}{ccc} e_1 \wedge \cdots \wedge e_r & \xrightarrow{f} & \overline{f(e_1 \wedge \cdots \wedge e_r)} \\ \updownarrow & \nearrow_{f'} & \\ \det(d_{ij}) e'_1 \wedge \cdots \wedge e'_r & & \end{array}$$

we have

$$\begin{aligned} \det(c_{ij}) \overline{f(e_1 \wedge \cdots \wedge e_r)} &= \det(c_{ij}) \overline{f'(\det(d_{ij}) e'_1 \wedge \cdots \wedge e'_r)} \\ &= \overline{f'(e'_1 \wedge \cdots \wedge e'_r)} = \varphi_{\mathscr{Y}}(f')(y_1 \wedge \cdots \wedge y_r). \end{aligned}$$

By this we obtain $\varphi_{\mathfrak{X}}(f) = \varphi_{\mathscr{Y}}(f')$.

Let us see some basic properties of ω_X.

Theorem 5.3.8. *For an n-dimensional projective variety $X \subset \mathbb{P}^N$ the following hold:*

(i) $\omega_X|_{X_{\mathrm{reg}}} \simeq \widetilde{\omega}_{X_{\mathrm{reg}}}$.
(ii) ω_X *is a torsion free* $\mathscr{O}_X$*-Module.*
(iii) ω_X *satisfies the condition of Theorem 5.1.11 (ii), i.e., if a torsion free* $\mathscr{O}_X$*-Module* $\mathscr{M}$ *satisfies* $\omega_X \subset \mathscr{M}$, $\operatorname{codim} \operatorname{Supp}(\mathscr{M}/\omega_X) \geq 2$ *then* $\mathscr{M} = \omega_X$.

Proof. (i) Apply Lemma 5.3.7 for non-singular variety X_{reg}, its open neighborhood P in $\mathbb{P}^N$ and $\mathscr{F} = \widetilde{\omega}_P$ and we obtain:

$$\mathscr{E}xt^{N-n}_{\mathscr{O}_P}(\mathscr{O}_{X_{\mathrm{reg}}}, \widetilde{\omega}_P) \simeq \mathscr{H}om_{\mathscr{O}_{X_{\mathrm{reg}}}}\left(\bigwedge^{N-n} \mathscr{I}/\mathscr{I}^2, \widetilde{\omega}_P/\mathscr{I}\widetilde{\omega}_P\right).$$

By Proposition 5.3.4, $\mathscr{I}/\mathscr{I}^2$ is a locally free $\mathscr{O}_{X_{\mathrm{reg}}}$-Module of rank $N-n$. Then we have that $\bigwedge^{N-n} \mathscr{I}/\mathscr{I}^2$ is an invertible sheaf on X_{reg}, which shows that the right-hand side is

$$\cong \left(\bigwedge^{N-n} \mathscr{I}/\mathscr{I}^2\right)^{-1} \otimes_{\mathscr{O}_{X_{\mathrm{reg}}}} (\widetilde{\omega}_P \otimes \mathscr{O}_{X_{\mathrm{reg}}}).$$

On the other hand, from the exact sequence:

$$0 \longrightarrow \mathscr{I}/\mathscr{I}^2 \longrightarrow \Omega_P \otimes \mathscr{O}_{X_{\mathrm{reg}}} \longrightarrow \Omega_{X_{\mathrm{reg}}} \longrightarrow 0$$

we have

$$\bigwedge^{N} \Omega_P \otimes \mathscr{O}_{X_{\mathrm{reg}}} \cong \left(\bigwedge^{N-n} \mathscr{I}/\mathscr{I}^2\right) \otimes \left(\bigwedge^{n} \Omega_{X_{\mathrm{reg}}}\right).$$

This shows the required isomorphism:

$$\bigwedge^{n} \Omega_{X_{\mathrm{reg}}} \cong \left(\bigwedge^{N-n} \mathscr{I}/\mathscr{I}^2\right)^{-1} \otimes (\widetilde{\omega}_P \otimes \mathscr{O}_{X_{\mathrm{reg}}}) \cong \mathscr{E}xt^{N-n}_{\mathscr{O}_P}(\mathscr{O}_{X_{\mathrm{reg}}}, \widetilde{\omega}_P).$$

(ii) Let $T = Tor_{\mathscr{O}_X}\omega_X$, then by Proposition 5.1.4, we have $\dim \operatorname{Supp} T \leqq n-1$. Then we obtain $H^n(X, T) = 0$. By Theorem 5.3.6 (i), it follows that $\operatorname{Hom}(T, \omega_X) = 0$, which implies the inclusion map $i : T \hookrightarrow \omega_X$ is zero map, i.e., $T = 0$.

(iii) By $\operatorname{codim}(\operatorname{Supp} \mathscr{M}/\omega_X) \geq 2$, for $i = n, n-1$ we have $H^i(\mathscr{M}/\omega_X) = 0$. Therefore,

$$H^n(X, \omega_X) \simeq H^n(X, \mathscr{M}).$$

By taking the dual of this isomorphism we obtain

$$\operatorname{Hom}_{\mathscr{O}_X}(\omega_X, \omega_X) \simeq \operatorname{Hom}(\mathscr{M}, \omega_X).$$

Under this isomorphism the map $\varphi : \mathscr{M} \to \omega_X$ corresponding to id_{ω_X} corresponds to a cross-section of the map $\omega_X \hookrightarrow \mathscr{M}$. Therefore the exact sequence $0 \to \omega_X \to \mathscr{M} \to \mathscr{M}/\omega_X \to 0$ splits and $\mathscr{M} = \omega_X \oplus \mathscr{M}/\omega_X$. Here, if $\mathscr{M}/\omega_X \neq 0$ then it contradicts the fact that $\mathscr{M}$ is torsion free.

Corollary 5.3.9. *If X is an n-dimensional normal quasi-projective variety, the sheaf ω_X is divisorial and*

$$\omega_X \simeq j_*\Big(\bigwedge^n \Omega_{X_{\mathrm{reg}}}\Big),$$

where $j : X_{\mathrm{reg}} \hookrightarrow X$ is the inclusion map.

Proof. As the sheaf ω_X satisfies the condition of Theorem 5.1.11 (ii), the corollary follows immediately from the definition.

Definition 5.3.10. Let X be a normal quasi-projective variety. A divisor K_X satisfying $\omega_X \cong \mathcal{O}(K_X)$ is called a *canonical divisor*. By Theorem 5.2.11 a canonical divisor is unique up to linearly equivalence.

If a subvariety is Cartier divisor, i.e., it is defined by a principal ideal, we have the following relation between the canonical sheaf of X and the canonical sheaf of the subvariety.

Proposition 5.3.11. *Let X be a quasi-projective integral variety and $D \subset X$ a Cartier divisor. If X is a Cohen–Macaulay variety,*

$$\omega_D \cong \omega_X(D) \otimes_{\mathcal{O}_X} \mathcal{O}_D,$$

where $\omega_X(D) = \omega_X \otimes_{\mathcal{O}_X} \mathcal{O}_X(D)$. This formula is called the adjunction formula.

Proof. We may assume that there is an embedding $D \subset X \subset \mathbb{P}^N$. Let $\dim X = n$. From the exact sequence

$$0 \longrightarrow \mathcal{O}_X(-D) \longrightarrow \mathcal{O}_X \longrightarrow \mathcal{O}_D \longrightarrow 0$$

we obtain the following exact sequence:

$$\begin{aligned}\longrightarrow\ & \mathcal{E}xt^{N-n}_{\mathcal{O}_{\mathbb{P}^N}}(\mathcal{O}_D, \omega_{\mathbb{P}^N}) \longrightarrow \mathcal{E}xt^{N-n}_{\mathcal{O}_{\mathbb{P}^N}}(\mathcal{O}_X, \omega_{\mathbb{P}^N}) \longrightarrow \mathcal{E}xt^{N-n}_{\mathcal{O}_{\mathbb{P}^N}}(\mathcal{O}(-D), \omega_{\mathbb{P}^N})\\ \longrightarrow\ & \mathcal{E}xt^{N-n+1}_{\mathcal{O}_{\mathbb{P}^N}}(\mathcal{O}_D, \omega_{\mathbb{P}^N}) \longrightarrow \mathcal{E}xt^{N-n+1}_{\mathcal{O}_{\mathbb{P}^N}}(\mathcal{O}_X, \omega_{\mathbb{P}^N}).\end{aligned}$$

By Theorem 3.5.14, $\mathcal{E}xt^{N-n}_{\mathcal{O}_{\mathbb{P}^N}}(\mathcal{O}_D, \omega_{\mathbb{P}^N})_x = H^n_{\{x\}}(\mathcal{O}_D) = 0$. On the other hand, as X is n-dimensional and Cohen–Macaulay, by Proposition 3.5.13 we have

$$\mathcal{E}xt^{N-n+1}_{\mathcal{O}_{\mathbb{P}^N}}(\mathcal{O}_X, \omega_{\mathbb{P}^N})_x = H^{n-1}_{\{x\}}(\mathcal{O}_X) = 0.$$

Therefore we obtain

$$\omega_D = \mathcal{E}xt^{N-n}_{\mathcal{O}_{\mathbb{P}^N}}(\mathcal{O}_X(-D), \omega_{\mathbb{P}^N})/\omega_X.$$

Here, since for every point of X there are an open neighborhood $U \subset \mathbb{P}^N$ and a divisor $\widetilde{D} \subset U$ on it such that $\widetilde{D}|_{U\cap X} = D|_U$, it follows that

$$\mathscr{E}xt^{N-n}_{\mathscr{O}_{\mathbb{P}^N}}(\mathscr{O}_X(-D), \omega_{\mathbb{P}^N}) = \mathscr{E}xt^{N-n}_{\mathscr{O}_{\mathbb{P}^N}}(\mathscr{O}_{\mathbb{P}^N}(-\widetilde{D}) \otimes \mathscr{O}_X, \omega_{\mathbb{P}^N})$$
$$\cong \mathscr{E}xt^{N-n}_{\mathscr{O}_{\mathbb{P}^N}}(\mathscr{O}_X, \omega_{\mathbb{P}^N}) \otimes_{\mathscr{O}_{\mathbb{P}^N}} \mathscr{O}_{\mathbb{P}^N}(\widetilde{D}) = \omega_X \otimes_{\mathscr{O}_X} \mathscr{O}_X(D).$$

Hence we have

$$\omega_D = \omega_X(D)/\omega_X \cong \omega_X \otimes_{\mathscr{O}_X} \left(\mathscr{O}_X(D)/\mathscr{O}_X\right)$$
$$\cong \omega_X(D) \otimes_{\mathscr{O}_X} \mathscr{O}_X/\mathscr{O}_X(-D) \simeq \omega_X(D) \otimes_{\mathscr{O}_X} \mathscr{O}_D.$$

Now we compare properties of X and properties of a Cartier divisor on X.

Proposition 5.3.12. *(i) A singularity (X, x) of an integral variety X is a Cohen–Macaulay singularity if and only if for a Cartier divisor $D \subset X$ containing x, the singularity (D, x) is a Cohen–Macaulay singularity.*
(ii) A singularity (X, x) of an integral variety X is a Gorenstein singularity if and only if for a Cartier divisor $D \subset X$ containing x, the singularity (D, x) is a Gorenstein singularity.

Proof. (i) Let h be a generator of $\mathscr{O}_X(-D)$ at the point x, then as h is not a torsion element, we have the following exact sequence:

$$0 \longrightarrow \mathscr{O}_{X,x} \xrightarrow{\times h} \mathscr{O}_{X,x} \longrightarrow \mathscr{O}_{D,x} \longrightarrow 0.$$

By this sequence we have the following exact sequence:

$$\mathrm{Ext}^i_{\mathscr{O}_{X,x}}(k, \mathscr{O}_{X,x}) \longrightarrow \mathrm{Ext}^i_{\mathscr{O}_{X,x}}(k, \mathscr{O}_{D,x}) \longrightarrow \mathrm{Ext}^{i+1}_{\mathscr{O}_{X,x}}(k, \mathscr{O}_{X,x}).$$

Here, if (X, x) is a Cohen–Macaulay singularity, we have depth $\mathscr{O}_{X,x} = \dim X$. Then, by Proposition 3.5.13, we obtain $\mathrm{Ext}^i_{\mathscr{O}_{X,x}}(k, \mathscr{O}_{X,x}) = 0$, $i < \dim X$. By the above exact sequence, we have

$$\mathrm{Ext}^i_{\mathscr{O}_{X,x}}(k, \mathscr{O}_{D,x}) = \mathrm{Ext}^i_{\mathscr{O}_{D,x}}(k, \mathscr{O}_{D,x}) = 0, \quad i < \dim X - 1.$$

Again by Proposition 3.5.13 it follows that depth $\mathscr{O}_{D,x} \geq \dim X - 1 = \dim D$, which shows that (D, x) is Cohen–Macaulay. Conversely, if (D, x) is Cohen–Macaulay. Let $h_1, \cdots, h_{n-1}$ $(n-1 = \dim D)$ be a regular sequence of $\mathscr{O}_{D,x}$. As h is not a torsion element, the sequence $h, h_1, h_2, \cdots, h_{n-1}$ is a regular sequence of $\mathscr{O}_{X,x}$. Hence (X, x) is a Cohen–Macaulay singularity.

(ii) By (i) we may assume that (X, x) and (D, x) are both Cohen–Macaulay. By Proposition 5.3.11, we have $\omega_D \cong \omega_X(D) \otimes \mathscr{O}_D$, which shows that ω_D is invertible around x if and only if ω_X is invertible around x.

5.4 Intersection Number of Divisors

In this section we introduce the intersection number of Cartier divisors.

Theorem 5.4.1 (Snapper [Sn]). *Let X be an algebraic variety, $D_1, \cdots, D_r$ Cartier divisors on X, $\mathscr{F}$ a coherent $\mathscr{O}_X$-Module such that* Supp $\mathscr{F}$ *is complete of n-dimension. Then there exists a polynomial $P(z_1, \cdots, z_r)$ of degree $\leqq n$ with coefficients in $\mathbb{Q}$ such that*

$$\begin{aligned} & P(m_1, \cdots, m_r) \\ =& \chi\left(X, \mathscr{F} \otimes_{\mathscr{O}_X} \mathscr{O}(m_1 D_1 + \cdots + m_r D_r)\right) \\ :=& \sum_i (-1)^i \dim H^i\left(X, \mathscr{F} \otimes \mathscr{O}(m_1 D_1, + \cdots + m_r D_r)\right), \end{aligned}$$

for every $m_1, \cdots, m_r \in \mathbb{Z}$.

Definition 5.4.2. In the above theorem, when $r \geq n$, the coefficient of the monomial $z_1 \cdots z_r$ in $P(z_1, \cdots, z_r)$ is called the *intersection number* of $D_1, \cdots, D_r$ with respect to $\mathscr{F}$ and denoted by $(D_1, \cdots, D_r; \mathscr{F})$.

In particular, if $\mathscr{F} = \mathscr{O}_Y$ for a complete closed subvariety $Y \subset X$, we denote $(D_1, \cdots, D_r; \mathscr{F})$ by $(D_1, \cdots, D_r; Y)$. If, moreover, $Y = X$, we denote it simply by $(D_1, \cdots, D_r)$. If D_i all coincide with D, then we denote it by D^r.

By the definition, the intersection number is determined by $\mathscr{F}, \mathscr{O}_X(D_i)$, $i = 1, \cdots, r$, therefore if $D_i \sim D_i'$, $i = 1, \cdots, r$ we have:

$$(D_1, \cdots, D_r; \mathscr{F}) = (D_1', \cdots, D_r'; \mathscr{F}).$$

If X is a complete algebraic variety of dimension n and the closed subvariety Y is a Cartier divisor on X, we have

$$(D_1, \cdots, D_{n-1}; \mathscr{O}_Y) = (D_1, \cdots, D_{n-1}, Y).$$

The following is a basic property on the intersection number.

Proposition 5.4.3 (Kleiman [Klm]). *The following hold on the intersection number:*

(i) $(D_1, \cdots, D_r; \mathscr{F})$ *is an integer.*
(ii) The map $\mathrm{CDiv}(X) \times \cdots \times \mathrm{CDiv}(X) \to \mathbb{Z}$, $(D_1, \cdots, D_r) \mapsto (D_1, \cdots, D_r; \mathscr{F})$ *is a symmetric multilinear map.*

Proposition 5.4.4. *If C is a non-singular complete curve on an algebraic variety X, for a Cartier divisor D on X we have:*

$$(D; C) = \deg D|_C.$$

Here, $D|_C$ means the divisor on C such that $\mathcal{O}_C(D|_C) = \mathcal{O}_X(D) \otimes \mathcal{O}_C$.

Proof. First by the definition of $(D; C)$, we have $\chi(C, \mathcal{O}_X(mD) \otimes \mathcal{O}_C) = (D; C)$ $m + \chi(C, \mathcal{O}_C)$. In particular, for $m = 1$ we have:

$$\chi(C, \mathcal{O}_X(D) \otimes \mathcal{O}_C) = (D; C) + \chi(C, \mathcal{O}_C).$$

Here we put $D|_C = D_1 - D_2$, $D_1 = \sum n_i P_i \geq 0$, $D_2 = \sum m_j P_j \geq 0$. From the exact sequence:

$$0 \longrightarrow \mathcal{O}_C(-D_2) \longrightarrow \mathcal{O}_C(D_1 - D_2) \longrightarrow \mathcal{O}_{D_1}(D_1 - D_2) \cong \mathcal{O}_{D_1} \longrightarrow 0,$$
$$0 \longrightarrow \mathcal{O}_C(-D_2) \longrightarrow \mathcal{O}_C \longrightarrow \mathcal{O}_{D_2} \longrightarrow 0$$

we have

$$\chi\big(C, \mathcal{O}_C(D_1 - D_2)\big) = \chi\big(C, \mathcal{O}_C(-D_2)\big) + \sum n_i,$$
$$\chi(C, \mathcal{O}_C) = \chi\big(C, \mathcal{O}_C(-D_2)\big) + \sum m_j.$$

Therefore,

$$\chi(C, \mathcal{O}_X(D) \otimes \mathcal{O}_C) = \sum n_i - \sum m_j + \chi(C, \mathcal{O}_C).$$

Thus we conclude that $(D; C) = \sum n_i - \sum m_j$.

The following is important.

Theorem 5.4.5 (Riemann–Roch [Ha2, IV, 1.3 & V, 1.6]).

(i) Let X be a non-singular complete curve and D a Cartier divisor on X. Then, the following holds:

$$\chi(X, \mathcal{O}_X(D)) = \deg D + \chi(X, \mathcal{O}_X).$$

(ii) Let X be a non-singular complete surface and D a Cartier divisor on X. Then the following holds:

$$\chi(X, \mathcal{O}_X(D)) = \frac{D^2 - K_X \cdot D}{2} + \chi(X, \mathcal{O}_X).$$

Here D^2 means $D \cdot D$.

Definition 5.4.6. Let X be an algebraic variety and C a complete curve on X. For a Cartier divisor D on X, the intersection number $(D; C)$ is denoted simply by $D \cdot C$. For $D \in \mathrm{CDiv}\,(X) \otimes \mathbb{Q}$, assume $rD \in \mathrm{CDiv}\,(X)$, then define

$$D \cdot C := \frac{1}{r}(rD) \cdot C.$$

For a proper morphism $f : X \to Y$ of algebraic varieties, if any complete curve C such that $f(C) =$ one point satisfies $D \cdot C \geq 0$, then D is called *f-nef*. In particular, when Y is a one point we call f-nef just *nef*.

Chapter 6
Differential Forms Around a Singularity

An attractive conjecture cannot be proved.
A big theorem's proof is wrong.
If the proof is correct, the statement is trivial.

In this chapter, we study the singularity by looking at the poles of the pull-back of differential forms around the singularity onto the resolved space. By this consideration, plurigenera of isolated singularities are defined and the order of growth of the plurigenera gives a rough classification of isolated singularities. Here, a variety is always integral and defined over the complex number field $\mathbb{C}$.

6.1 Ramification Formula

Definition 6.1.1. For a normal n-dimensional integral variety X, its rational function field is denoted by $\mathbb{C}(X)$. For a rational m-form

$$\theta \in \Gamma\big(X, \omega_X^{[m]} \otimes_{\mathcal{O}_X} \mathbb{C}(X)\big)$$

on X, define the valuation $\nu_D(\theta)$ (D is a prime divisor on X). If x is a general point of D, then X is non-singular at x, therefore for a regular system of parameters $x_1, \cdots, x_n$ the form $(dx_1 \wedge \cdots \wedge dx_n)^{\otimes m}$ is a generator of $\omega_X^{[m]}$ at x. Let θ be denoted as $\theta = f(dx_1 \wedge \cdots \wedge dx_n)^{\otimes m}$ $\big(f \in \mathbb{C}(X)\big)$, then we define $\nu_D(\theta) := \nu_D(f)$. In particular

$$\theta \in \Gamma\Big(X, \omega_X^{[m]}\Big(\sum_{i=1}^{r} m_i D_i\Big)\Big) = \Gamma\Big(X, \mathcal{O}_X\Big(mK_X + \sum_{i=1}^{r} m_i D_i\Big)\Big)$$

if and only if $\nu_{D_i}(\theta) \geq -m_i$ $(i = 1, \cdots, r)$, $\nu_D(\theta) \geq 0$ $(D \neq D_i)$.

S. Ishii, *Introduction to Singularities*,
https://doi.org/10.1007/978-4-431-56837-7_6

If $\nu_D(\theta) > 0$, we say that θ has a *zero* at D and if $\nu_D(\theta) < 0$, we say that θ has a *pole* at D.

Definition 6.1.2. A proper morphism $f : Y \to X$ of integral algebraic varieties is called a *finite morphism* if for every point $x \in X$ the inverse image $f^{-1}(x)$ consists of finite points. A proper morphism $f : Y \to X$ is called a *generically finite morphism* if there exists a non-empty open subset $U \subset X$ such that the restriction $f|_{f^{-1}(U)} : f^{-1}(U) \to U$ is a finite morphism. Let U be the maximal open subset such that $f|_{f^{-1}(U)}$ is finite and the union of irreducible components E of the closed subset $Y \setminus f^{-1}(U)$ such that $\dim E > \dim f(E)$ is called the *exceptional set*. A divisor contained in the exceptional set is called an *exceptional divisor*.

The sum $\sum_{i=1}^{r} E_i$ of all irreducible exceptional divisors $E_1, \cdots, E_r$ is called the *total exceptional divisor*.

In general, let P be a property P of a point of a topological space X. If the subset $U := \{x \in X \mid \text{a point } x \text{ has property } P\}$ is a dense open subset of X, then we say that a *general point* of X has the property P. For example if X is an integral algebraic variety, a general point of X is non-singular.

As the intersection of finite number of open dense subsets is again open dense, for a finite number of properties $P_1, \cdots, P_r$ if a general point of X has property P_i for each $i = 1, \cdots, r$, then a general point of X has all properties $P_1, \cdots, P_r$.

Let $f : Y \to X$ be a generically finite morphism of normal varieties. Let D_1 and D_2 be prime divisors on Y and X, respectively and assume $f(D_1) = D_2$. As on a neighborhood of a general point x of D_2 both D_2 and X are non-singular, by Proposition 5.3.4, there exists a regular system of parameters $x_1, \cdots, x_n$ such that D_2 is defined by $x_1 = 0$. In the same way at a neighborhood of $y \in f^{-1}(x)$ there exists a regular system of parameters $y_1, \cdots, y_n$ such that D_1 is defined by $y_1 = 0$. As $f(D_1) = D_2$, we have an expression $f^*x_1 = y_1^a u$, $a \geq 1$ (u is a regular function which is non-zero at a general point of D_1). Note that $a = \nu_{D_1}(f^*x_1)$.

Definition 6.1.3. If $a > 1$, we say that f ramifies at D_1 and call D_1 a *ramification divisor*. In this case D_2 is called a *branch divisor*. The number a is called the *ramification index* of f at D_1.

Remark 6.1.4. For a generically finite morphism $f : Y \to X$, there are at most a finite number of ramification divisors. (It is proved in the same way as in the case of dimension 1 in [Ha2, IV, 2.2].)

In particular, if $f : Y \to X$ is a finite morphism of non-singular varieties over $\mathbb{C}$, it has no ramification divisor if and only if it is an étale morphism [SGA, XI,3, 13, i]. And it is equivalent to the fact that for every $x \in X$ the inverse image $f^{-1}(x)$ is reduced [Mu, III, Sect. 5, Theorem 4]. In this case, we note that $f^*\omega_X = \omega_Y$.

Definition 6.1.5. Let X be a non-singular variety and D a divisor on X. We say that a divisor D is of *normal crossings* at a point $x \in X$ if there exists a regular system of parameters $x_1, \cdots, x_n$ at x such that D is defined by $x_1 \cdot \cdots \cdot x_k = 0$, $(k \leq n)$.

Here, for a generically finite morphism $f : Y \to X$ we describe the difference between K_Y and K_X.

Theorem 6.1.6 (Log ramification formula). *Let $f : Y \to X$ be a generically finite morphism of non-singular varieties. Let D and $E = f^{-1}(D)_{\text{red}}$ be divisors of normal crossings on X and Y, respectively. Here, the notation A_{red} means the divisor with the support on A and the coefficients one. Assume that f is not ramified outside E. Then, we have:*

$$K_Y + E = f^*(K_X + D) + R,$$

where R is an effective divisor supported on the exceptional divisor of f.

Proof. Let E_i be an irreducible component of E. On a neighborhood of a general point x of $f(E_i)$ the closed subset $f(E_i)$ is considered as a non-singular closed subvariety in X. Then, by Proposition 5.3.4, there exists a regular system of parameters $x_1, \cdots, x_n$ of X such that $f(E_i)$ is defined by $x_1 = \cdots = x_r = 0$ $(r \geq 1)$ and D is defined by $x_1 \cdot \cdots \cdot x_s = 0$ $(s \leq r)$. Let E_i be defined by $y_1 = 0$ on a neighborhood of a general point y of $f^{-1}(x)$.

Then, we have

$$f^* x_i = y_1^{a_i} u_i.$$

Here u_i is a regular function on a neighborhood of y such that it is not divided by y_1 and for $i = 1, \cdots, r$, $a_i \geq 1$ and for $i = r+1, \cdots, n$, $a_i = 0$. It is sufficient to prove that for a rational form

$$\theta := \frac{dx_1 \wedge \cdots \wedge dx_n}{x_1 \cdot \cdots \cdot x_s}$$

the inequality $\nu_{E_i}(f^*\theta) \geq -1$ holds and in particular, if E_i is not exceptional the equality $= -1$ holds.

$$\begin{aligned} f^*\theta &= \frac{dy_1^{a_1} u_1 \wedge \cdots \wedge dy_1^{a_r} u_r \wedge du_{r+1} \wedge \cdots \wedge du_n}{y_1^{a_1 + \cdots + a_s} u_1 \cdots u_s} \\ &= y_1^{\sum_{i=1}^r a_i - \sum_{i=1}^s a_i - 1} \theta', \end{aligned}$$

where θ' is a regular form at a general point of E_i.

Therefore it follows that

$$\nu_{E_i}(f^*\theta) \geq \sum_{i=1}^{r} a_i - \sum_{i=1}^{s} a_i - 1 \geq -1.$$

In particular, if E_i is not exceptional, we may assume that $r = s = 1$ and $y_1, u_2, \cdots, u_n$ is a regular system of parameters around y. Hence, it follows that $\nu_{E_i}(\theta') = 0$ and therefore $\nu_{E_i}(f^*\theta) = -1$. □

Theorem 6.1.7 (Ramification formula). *Let $f : Y \to X$ be a generically finite morphism of non-singular varieties. Let E_i $(i = 1, \cdots, t)$ be irreducible exceptional divisors of f and irreducible ramification divisors. Let $r_i := \dim E_i - \dim f(E_i)$. Then, we have*

$$K_Y = f^*K_X + \sum_{i=1}^{t} m_i E_i, \ (m_i \geq r_i).$$

Proof. Take a general point x of $f(E_i)$. We use the same notation as in the proof of Theorem 6.1.6. The image set $f(E_i)$ is defined by $x_1 = \cdots = x_r = 0$ and E_i is defined by $y_1 = 0$, then we have $r_i = r - 1$. By $f^*(dx_1 \wedge \cdots \wedge dx_n) = y_1^{a_1+\cdots+a_r-1}\theta'$ $(a_i \geq 1, \theta'$ is regular), we obtain

$$\nu_{E_i}\big(f^*(dx_1 \wedge \cdots \wedge dx_n)\big) \geq a_1 + \cdots + a_r - 1 \geq r - 1.$$

Example 6.1.8. Let X be an n-dimensional non-singular variety and $Z \subset X$ an irreducible s-dimensional non-singular closed subvariety. Let $b : Y := B_Z(X) \to X$ be the blow-up of X with the center Z and E the exceptional divisor. Then E is irreducible and

$$K_Y = b^*K_X + (n - s - 1)E.$$

In fact it is sufficient to prove the statement for a sufficiently small neighborhood of a point $x \in Z$. So we may assume that X is an affine variety. By choosing an appropriate regular system of parameters $x_1, \cdots, x_n$ of X at x, by Proposition 5.3.4 Z is defined by $x_{s+1} = \cdots = x_n = 0$. Then Y has the affine covering $\{V_j\}_{j=s+1}^n$ such that the affine coordinate ring of V_j is the polynomial ring

$$R\left[\frac{x_{s+1}}{x_j}, \cdots, \frac{x_n}{x_j}\right]$$

over the affine coordinate ring R of X. As on each V_j, E is defined by $I(Z)$, the affine coordinate ring of $E \cap V_j$ is

$$R/I(Z)\left[\frac{x_{s+1}}{x_j}, \cdots, \frac{x_n}{x_j}\right]$$

and by looking at the patching of them we have

$$E \cong Z \times \mathbb{P}^{n-s-1},$$

which is irreducible.

Next, for a point $y \in b^{-1}(x)$, functions $x_1, \cdots, x_s, x_j, \frac{x_{s+1}}{x_j}, \cdots, \frac{x_{j-1}}{x_j}, \frac{x_{j+1}}{x_j}, \cdots, \frac{x_n}{x_j}$ are a regular system of parameters of Y at y. Then we have

$$
\begin{aligned}
&b^*(dx_1 \wedge \cdots \wedge dx_n)\\
&= dx_1 \wedge \cdots \wedge dx_s \wedge dx_j \wedge d\left(x_j \frac{x_{s+1}}{x_j}\right) \wedge \cdots \wedge d\left(x_j \frac{x_n}{x_j}\right)\\
&= x_j^{n-s-1} dx_1 \wedge \cdots \wedge dx_s \wedge dx_j \wedge d\left(\frac{x_{s+1}}{x_j}\right) \wedge \cdots \wedge d\left(\frac{x_n}{x_j}\right),
\end{aligned}
$$

which show the required equality.

Corollary 6.1.9. *Let $f : Y \to X$ be a generically finite morphism of non-singular varieties. Then, an irreducible component of the exceptional set which is not contained in the ramification divisor is of codimension one, i.e., it is a prime divisor on Y.*

Proof. Assume the contrary. Let $C \subset Y$ be an irreducible component of the exceptional set with codim ≥ 2. As it is an irreducible component of the exceptional set, we have $\dim f(C) < \dim C$. Now, let $\sigma : \widetilde{Y} \to Y$ be the blow-up with the center C and E_C the prime exceptional divisor with respect to σ such that $\sigma(E_C) = C$. As C is non-singular at a general point, by Example 6.1.8 we have

$$
\begin{aligned}
K_{\widetilde{Y}} &= \sigma^* K_Y + (n - \dim C - 1)E_C + (\text{other components})\\
&= \sigma^* f^* K_X + (n - \dim C - 1)E_C + (\text{other components}).
\end{aligned}
$$

Then, the coefficient of the exceptional divisor E_C with respect to $f \circ \sigma$ in $K_{\widetilde{Y}}$ is $n - \dim C - 1 < n - \dim f(C) - 1$, which is a contradiction to Theorem 6.1.7.

This statement does not hold for general X, Y. In fact if X has a singular point, there is an example with the exceptional set of codimension ≥ 2 (cf. Chap. 8).

Next we define a "good" resolution of singularities. By Theorem 4.3.4 there is always a "good" resolution over $\mathbb{C}$.

Definition 6.1.10. For an integral variety X, a resolution $f : Y \to X$ of the singularities of X is called a *good resolution* if the exceptional set is a divisor of normal crossings. A weak resolution is called a *good weak resolution* if the exceptional set is a divisor of normal crossings.

By the ramification formula and the log ramification formula we obtain the following invariance property. This is useful to define log canonical singularities, log terminal singularities and also plurigenera.

Corollary 6.1.11. *(i) Let X be a normal variety and $f : Y \to X$ a weak resolution of the singularities of X. Then for every positive integer m the sheaf $f_*\mathcal{O}_Y(mK_Y)$ is a coherent $\mathcal{O}_X$-Module and independent of choices of f.*

(ii) *Let X be a normal variety and $f : Y \to X$ a weak good resolution of the singularities of X and E the total exceptional divisor. Then for every positive integer m the sheaves $f_*\mathcal{O}_Y(mK_Y + mE)$ and $f_*\mathcal{O}_Y(mK_Y + (m-1)E)$ are coherent $\mathcal{O}_X$-Modules and independent of choices of f.*

Proof. The coherence property in (i), (ii) follows from Grauert's direct image theorem [Gr1], [EGA III, 3.2.1]. For the uniqueness, any two weak resolutions (or weak good resolutions) $f : Y \to X$, $f' : Y' \to X$ have a new weak resolution (or weak good resolution) $g : \widetilde{Y} \to X$ such that the following diagram is commutative:

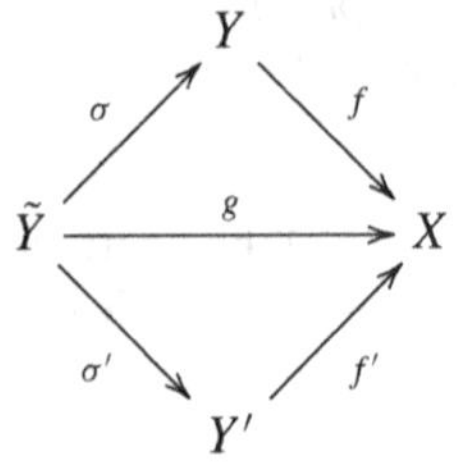

To see this, it is sufficient to take as $\widetilde{Y}$ a good resolution of the irreducible component of fiber product $Y' \times_X Y$ that dominates X.

Therefore, it is sufficient to prove that for two weak resolutions (or weak good resolutions) f, f' with $f' = f \circ \sigma : Y' \to Y \to X$ it follows that $f_*\mathcal{O}_Y(mK_Y) = f'_*\mathcal{O}_{Y'}(mK_{Y'})$ (resp. $f_*\mathcal{O}_Y(mK_Y + (m-1)E) = f'_*\mathcal{O}_{Y'}(mK_{Y'} + (m-1)E')$, etc.).

(i) Let $f' = f \circ \sigma : Y' \to X$ and $f : Y \to X$ be weak resolutions of the singularities of X. As $\sigma : Y' \to Y$ satisfies the condition in Theorem 6.1.7, for an exceptional prime divisors $\{E_i\}$ with respect to σ we have:

$$\begin{aligned}\sigma_*\mathcal{O}_{Y'}(mK_{Y'}) &= \sigma_*\Big(\sigma^*\mathcal{O}_Y(mK_Y) \otimes_{\mathcal{O}_{Y'}} \mathcal{O}_{Y'}\Big(\sum m_i E_i\Big)\Big)\\ &= \mathcal{O}_Y(mK_Y) \otimes_{\mathcal{O}_Y} \sigma_*\mathcal{O}_{Y'}\Big(\sum m_i E_i\Big).\end{aligned}$$

Here we obtain $\sigma_*\mathcal{O}_{Y'}(\sum m_i E_i) = \mathcal{O}_Y$. Indeed by Theorem 6.1.7 it follows that $m_i > 0$ ($\forall i$), therefore the inclusion $\supset$ is obvious. On the other hand, since an element h in the left-hand side is holomorphic outside the exceptional divisors, by normality of Y and by Theorem 4.1.11, (v), h is holomorphic on Y, which yields the inclusion $\subset$. Hence, it follows that $f_*\sigma_*\mathcal{O}(mK_{Y'}) = f_*\mathcal{O}_Y(mK_Y)$.

(ii) Let $f' = f \circ \sigma, Y' \to X, f : Y \to X$ be weak good resolutions of the singularities of X, $E \subset Y$ be the total exceptional divisor with respect to f and $E' \subset Y'$ the total exceptional divisor with respect to f'. Let $E'_0 = \sigma^{-1}(E)_{\mathrm{red}}$, then by Theorem 6.1.6, we have $mK_{Y'} + mE'_0 = \sigma^*(mK_Y + mE) + mR$.
On the other hand, the coefficient of each prime divisor E_i in σ^*E is 1 if E_i is not an exceptional divisor with respect to σ and it is ≥ 1 if E_i is an exceptional divisor with respect to σ. Therefore, by subtracting $\sigma^*(E)$ from both sides of the above equality, we have

$$mK_{Y'} + (m-1)E'_0 - R' = \sigma^*\bigl(mK_Y + (m-1)E\bigr) + mR,$$

where R, R' are effective divisors with the support on the exceptional set with respect to σ. Noting that $E' \geq E'_0$, we add $(m-1)(E' - E'_0)$ on both sides of the equality and transpose R', then we obtain

$$mK_{Y'} + (m-1)E' = \sigma^*\bigl(mK_Y + (m-1)E\bigr) + R''.$$

Here, as R'' is an effective divisor with the support on the exceptional set with respect to σ, in the same way as in (i) we obtain

$$f_*\sigma_*\mathcal{O}_{Y'}\bigl(mK_{Y'} + (m-1)E'\bigr) = f_*\mathcal{O}_Y\bigl(mK_Y + (m-1)E\bigr).$$

The proof for $mK + mE$ is similar.

Here, we introduce a useful vanishing theorem which will be used often.

Theorem 6.1.12. (Grauert–Riemenschneider vanishing theorem [GR]). *Let $f : Y \to X$ be a weak resolution of a normal singularity (X, x), then it follows that*

$$R^i f_*\mathcal{O}_Y(K_Y) = R^i f_*\omega_Y = 0, \quad i > 0.$$

6.2 Canonical, Terminal, Rational Singularities

In this section a singularity is not necessarily isolated. The singularities treated here, canonical singularities, terminal singularities, and rational singularities are considered as mild singularities.

Definition 6.2.1. A normal singularity (X, x) is called a $\mathbb{Q}$-*Gorenstein singularity* if on a neighborhood X of x, there exists a positive integer r such that $\omega_X^{[r]}$ is invertible. We call the minimum such r the *index* of (X, x) and call (X, x) an r-*Gorenstein singularity*. A singularity which is 1-Gorenstein and Cohen–Macaulay is called a *Gorenstein singularity*. This is equivalent to the fact that $\mathcal{O}_{X,x}$ is a Gorenstein ring ([Ma2, p. 142]).

For a $\mathbb{Q}$-Gorenstein singularity we define the canonical cover.

Definition 6.2.2. For an r-Gorenstein singularity (X, x), if we take a sufficiently small affine neighborhood X of x, then there exists $\theta \in \Gamma(X, \omega_X^{[r]})$ such that

$$\omega_X^{[r]} = \mathcal{O}_X \cdot \theta \cong \mathcal{O}_X.$$

Let an $\mathcal{O}_X$-Module $\mathcal{A}$ be $\mathcal{O}_X \oplus \omega_X \oplus \cdots \oplus \omega_X^{[r-1]}$ and put $\mathcal{O}_X$-algebra structure on $\mathcal{A}$.

For two elements $\lambda \in \omega^{[i]}$, $\mu \in \omega^{[j]}$ in $\mathscr{A}$ the multiplication $\lambda \cdot \mu$ is defined as follows:

$$\begin{aligned} &\text{If } i+j \le r-1, \ \lambda \cdot \mu = \text{image}(\lambda \otimes \mu) \in \omega_X^{[i+j]}. \\ &\text{If } i+j \ge r, \qquad \lambda \cdot \mu = \theta^{-1} \cdot \text{image}(\lambda \otimes \mu) \in \omega_X^{[i+j-r]}. \end{aligned}$$

Here, image$(\lambda \otimes \mu)$ means the image of $\lambda \otimes \mu$ by the map $\omega_X^{[i]} \otimes \omega_X^{[j]} \to \omega_X^{[i+j]}$.

The ring $A = \Gamma(X, \mathscr{A})$ is a finitely generated $\Gamma(X, \mathscr{O}_X)$-algebra, therefore it is also a finitely generated $\mathbb{C}$-algebra. Let $\overline{X}$ be the affine variety with the affine coordinate ring A and the natural morphism $\rho : \overline{X} \to X$ is called the *canonical cover*. In this case we have a relation $\rho_* \mathscr{O}_{\overline{X}} = \mathscr{A}$.

Proposition 6.2.3. *The canonical cover $\rho : \overline{X} \to X$ of an r-Gorenstein singularity (X, x) has the following properties:*

(i) ρ is a finite morphism without ramification divisor.
(ii) ρ is the quotient morphism by the action of the group $\mathbb{Z}/r\mathbb{Z}$.
(iii) $\overline{X}$ is normal and has at worst 1-Gorenstein singularities.

Proof. (i) In general a morphism $f : Y \to X$ of algebraic varieties is a finite morphism if and only if there is an affine open covering $\{U_i\}_i$ of X such that $\{f^{-1}(U_i)\}$ is also an affine covering of Y and $\Gamma(f^{-1}(U_i), \mathscr{O}_Y)$ is a finitely generated $\Gamma(U_i, \mathscr{O}_X)$-module. Hence, by the construction of $\overline{X}$ and ρ, it follows that ρ is a finite morphism. On the other hand, for any sufficiently small open subset $U \subset X_{\text{reg}}$ we have $\mathscr{A}|_U \cong \mathscr{O}_U[T]/(T^r - u)$ (u is a unit of $\mathscr{O}_U$), therefore it is unramified over U.

(ii) Let σ be a generator of the group $G = \mathbb{Z}/r\mathbb{Z}$. We define the action of G on $\mathscr{A}$ as follows:

$$\text{For } \forall \eta = \eta_0 + \eta_1 + \cdots + \eta_{r-1} \in \mathscr{A}, \quad \eta_i \in \omega_X^{[i]}$$

define

$$\eta^\sigma = \eta_0 + \varepsilon \eta_1 + \varepsilon^2 \eta_2 + \cdots + \varepsilon^{r-1} \eta_{r-1},$$

where ε is a primitive r-th root of unity. Then the invariant subring of this action is $\mathscr{A}^G = \mathscr{O}_X$ (cf. Proposition 7.4.2).

(iii) Note that $V = \rho^{-1}(X_{\text{reg}})$ is non-singular open subset of $\overline{X}$ and $\text{codim}_{\overline{X}} \overline{X} \setminus V \ge 2$. We also have

$$\Gamma(V, \mathscr{O}_{\overline{X}}) = \Gamma(X_{\text{reg}}, \mathscr{A}) = \Gamma(X_{\text{reg}}, \mathscr{O}_X) \oplus \cdots \oplus \Gamma(X_{\text{reg}}, \omega_X^{[r-1]}).$$

Now as $\omega_X^{[i]}$ $(i = 0, \cdots, r-1)$ are divisorial, we have

$$\Gamma(X_{\text{reg}}, \omega_X^{[i]}) = \Gamma(X, \omega_X^{[i]}).$$

Then we obtain $\Gamma(V, \mathscr{O}_{\overline{X}}) = \Gamma(\overline{X}, \mathscr{O}_{\overline{X}})$, which yields that $\overline{X}$ is normal by Theorem 4.1.11, (v).

As ρ is unramified over X_{reg}, we have $\omega_{\overline{X}}|_V = \rho^*(\omega_X|_{X_{\text{reg}}})$, therefore it follows that

$$\Gamma(\overline{X}, \omega_{\overline{X}}) = \Gamma(V, \omega_{\overline{X}}|_V) = \Gamma(X_{\text{reg}}, \omega_{X_{\text{reg}}} \otimes \mathscr{A}) = \Gamma(X_{\text{reg}}, \omega_{X_{\text{reg}}} \oplus \cdots \oplus \omega_{X_{\text{reg}}}^{[r]})$$

$$\cong \Gamma(X, \mathscr{A}) = \Gamma(\overline{X}, \mathscr{O}_{\overline{X}}).$$

Hence $\omega_{\overline{X}}$ is invertible.

Definition 6.2.4. A normal singularity (X, x) is called a *canonical singularity* (resp. *terminal singularity*) if the following hold:

(1) (X, x) is a $\mathbb{Q}$-Gorenstein singularity.
Let $f : Y \to X$ be a weak resolution of the singularities of X and $E_1, \cdots, E_r$ prime exceptional divisors with respect to f.
(2) If we express a canonical divisor on Y in $\operatorname{Div}(Y) \otimes \mathbb{Q}$ as follows:

$$K_Y = f^*K_X + \sum_{i=1}^{r} m_i E_i,$$

then for every i the inequality $m_i \geq 0$ (resp. $m_i > 0$) holds.

Example 6.2.5. By the ramification formula Theorem 6.1.7, a non-singular point is a terminal singularity.

Proposition 6.2.6. *For a normal singularity (X, x), if the condition in Definition 6.2.4, (2) holds for a weak resolution $f : Y \to X$, then it holds for every weak resolution.*

Proof. It is sufficient to prove the statement for a weak resolution $f' : Y' \to X$ factoring as $f' = f \circ \pi$, $\pi : Y' \to Y$. As Y is non-singular, by Example 6.2.5, we have

$$K_{Y'} = \pi^*K_Y + \sum_{j=1}^{s} n_j E_j, \quad n_j > 0.$$

Here, if condition (2) holds for f,

$$K_Y = f^*K_X + \sum_{i=1}^{r} m_i E_i, \quad m_i \geq 0 \ \ (\text{resp. } m_i > 0).$$

Substitute this into the expression of $K_{Y'}$, then we have that every prime exceptional divisor has a coefficient ≥ 0 (resp. > 0).

Definition 6.2.7. A normal singularity (X, x) is called a *log canonical singularity* (resp. *log terminal singularity*) if the following hold:

(1) (X, x) is a $\mathbb{Q}$-Gorenstein singularity.
Let $f : Y \to X$ be a weak good resolution and $E_1, \cdots, E_r$ the prime exceptional divisors with respect to f.
(2) Express a canonical divisor of Y by

$$K_Y = f^* K_X + \sum m_i E_i$$

in Div $(Y) \otimes \mathbb{Q}$, then for every i the inequality $m_i \geq -1$ (resp. $m_i > -1$) holds.

Proposition 6.2.8. *For a normal singularity (X, x), if the condition in Definition 6.2.7, (2) holds for a weak good resolution $f : Y \to X$, then it holds for every weak good resolution.*

Proof. It is sufficient to prove the statement for a weak good resolution $f' : Y' \to X$ factoring as $f' = f \circ \pi$, $\pi : Y' \to Y$. Let $E_1, \cdots, E_r$ be the exceptional prime divisors with respect to f and $E_{r+1}, \cdots, E_{r+s}$ be the exceptional divisors with respect to π. The strict transform of E_i $(i = 1, \cdots, r)$ on Y' is also denoted by E_i. Noting that $\pi^* \left(\sum_{i=1}^r E_i\right)_{\mathrm{red}} \leq \sum_{i=1}^{r+s} E_i$, by log ramification formula Theorem 6.1.6 we obtain

$$K_{Y'} + \sum_{i=1}^{r+s} E_i = \pi^*\Big(K_Y + \sum_{i=1}^{r} E_i\Big) + \sum_{j=1}^{s} n_j E_{r+j}, \quad n_j \geq 0.$$

By Theorem 6.1.7 if in particular $\pi(E_{r+j}) \not\subset \bigcup_{i=1}^r E_i$ holds, then $n_j \geq 1$. If f satisfies condition (2), then

$$K_Y = f^* K_X + \sum_{i=1}^{r} m_i E_i, \quad m_i \geq -1 \text{ (resp. } m_i > -1).$$

Therefore we obtain

$$K_Y + \sum_{i=1}^{r} E_i = f^* K_X + \sum_{i=1}^{r} r_i E_i, \quad r_i \geq 0 \text{ (resp. } r_i > 0).$$

Substituting this into the equality above we obtain

$$K_{Y'} = \pi^* f^* K_X + \sum_{i=1}^{r+s} m_i E_i, \quad m_i \geq -1 \text{ (resp. } m_i > -1).$$

The four kinds of singularities defined above have the following relation:

terminal $\Longrightarrow$ canonical $\Longrightarrow$ log terminal $\Longrightarrow$ log canonical.

If there is a finite morphism between two singularities, how does the property of one singularity affect the other? When there is no ramification divisor, we have the following theorem.

Theorem 6.2.9. *Let X, Y be normal varieties and $\pi : Y \to X$ a finite morphism without a ramification divisor. Then the following hold:*

(i) If X has at worst canonical singularities (resp. terminal singularities), then Y has also canonical singularities (resp. terminal singularities).
(ii) Y has log canonical singularities (resp. log terminal singularities) if and only if X has log canonical singularities (resp. log terminal singularities).

Proof. As there is no ramification divisor, it follows that $\omega_Y \simeq \pi^*\omega_X$ outside a closed subset of codim 2. First we prove that X is $\mathbb{Q}$-Gorenstein if and only if so is Y. If X is $\mathbb{Q}$-Gorenstein, then it is clear that Y is also $\mathbb{Q}$-Gorenstein. Let us show the converse. It is sufficient to prove that there exists r such that

$$\omega_X^{[mr]} \simeq \mathscr{O}_X$$

for affine varieties X, Y with $\omega_Y^{[m]} \simeq \mathscr{O}_Y$. Let $\pi \circ \sigma : \overline{Y} \to X$ be the Galois closure of $\pi : Y \to X$. The Galois closure is constructed as follows: Let K be the Galois closure of $k(Y)$ over $k(X)$. Let $\overline{A(Y)}$ be the integral closure of the affine coordinate ring $A(Y)$ of Y in K. We should note that $\overline{A(Y)}$ is a finite $A(Y)$-module ([ZS, I, p.267]). Let $\overline{Y}$ be the affine variety with the affine coordinate ring $\overline{A(Y)}$. Then there is a natural finite morphism $\sigma : \overline{Y} \to Y$. We call

$$\pi \circ \sigma : \overline{Y} \to X$$

the *Galois closure of* π. Here the Galois group $G = \mathrm{Gal}\big(K/k(X)\big)$ acts on $\overline{A(Y)}$, then it naturally acts on $\overline{Y}$. As it acts on X trivially, the morphism $\pi \circ \sigma : \overline{Y} \to X$ factors as $\pi \circ \sigma = \tau \circ \lambda$ through the quotient $\lambda : \overline{Y} \to \overline{Y}/G$ by G. As the function field of $\overline{Y}/G$ is $K^G = k(X)$, the morphism $\tau : \overline{Y}/G \to X$ is birational. But the morphism $\pi \circ \sigma = \tau \circ \lambda$ is finite, and the morphism τ is also finite; then by the Zariski Main Theorem [Ha2, V, 5.2] the morphism τ is an isomorphism.

Now by $\omega_Y^{[m]} \simeq \mathscr{O}_Y$ there is an open subset Y_0 of Y such that $\mathrm{codim}(Y \setminus Y_0) \geq 2$ and

$$\pi^*\omega_X^{\otimes m} \mid_{Y_0} \simeq \omega_{Y_0}^{\otimes m} \simeq \mathscr{O}_{Y_0}.$$

Then for $\overline{Y}_0 := \sigma^{-1}(Y_0)$ we have

$$\sigma^*\pi^*\omega_X^{\otimes m} \mid_{\overline{Y}_0} \simeq \mathscr{O}_{\overline{Y}_0}.$$

Therefore there is an element θ of $\Gamma(\overline{Y}_0, \sigma^*\pi^*\omega_X^{\otimes m})$ that does not vanish on $\overline{Y}_0$. Take the element

$$\overline{\theta} := \otimes_{g \in G}\theta^g \in \Gamma(\overline{Y}_0, \sigma^*\pi^*\omega_X^{\otimes mr}) = \Gamma(\overline{Y}_0, \lambda^*\tau^*\omega_X^{\otimes mr}), \quad (r = \#G);$$

then, $\overline{\theta}$ is invariant under the action of G, therefore it comes from an element of $\Gamma(\overline{Y}_0/G, \tau^*\omega_X^{\otimes mr})$. As $\overline{\theta}$ has no zeros on $\overline{Y}_0/G$, we obtain

$$\tau^*\omega_X^{\otimes mr}\,|_{\widetilde{Y}_0/G} \simeq \mathcal{O}_{\widetilde{Y}_0/G},$$

which implies that on an open subset X_0 of X such that $\operatorname{codim} X \setminus X_0 \geq 2$ we have $\omega_X^{\otimes mr}\,|_{X_0} \simeq \mathcal{O}_{X_0}$, which implies $\omega_X^{[mr]} \simeq \mathcal{O}_X$. This completes the proof of the equivalence of the $\mathbb{Q}$-Gorenstein property on X and that on Y.

Let r be a common multiple of the indices of X and Y, then $\omega_Y^{[r]} = \pi^*\omega_X^{[r]}$. Therefore as $\mathbb{Q}$-divisors we have $K_Y = \pi^* K_X$. Now consider the commutative diagram:

$$\begin{array}{ccc} \widetilde{Y} & \xrightarrow{\widetilde{\pi}} & \widetilde{X} \\ {\scriptstyle g}\downarrow & & \downarrow{\scriptstyle f} \\ Y & \xrightarrow{\pi} & X \end{array}$$

Here, f and g are both weak good resolutions of X and Y, respectively. Let $D \subset \widetilde{Y}$ be a prime exceptional divisor with respect to g and express a canonical divisor of $\widetilde{Y}$ as

$$K_{\widetilde{Y}} = g^* K_Y + aD + (\text{other divisors}).$$

If $\widetilde{\pi}(D)$ is a divisor on $\widetilde{X}$, we can write

$$K_{\widetilde{X}} = f^* K_X + b\widetilde{\pi}(D) + (\text{other divisors}).$$

By the equality $\pi^* K_X = K_Y$ obtained above and by the log ramification formula Theorem 6.1.6 we obtain

$$a + 1 = e(b + 1),$$

where e is the ramification index at D. Hence, we have the following relations:

$$a \geq -1 \Longleftrightarrow b \geq -1,$$

$$a > -1 \Longleftrightarrow b > -1,$$

$$b \geq 0 \Longrightarrow a \geq 0,$$

$$b > 0 \Longrightarrow a > 0.$$

Among these, from $a \geq -1 \Longrightarrow b \geq -1$ and $a > -1 \Longrightarrow b > -1$, it follows that if Y is log canonical (resp. log terminal) then so is X.

Next, if $\widetilde{\pi}(D)$ is not a divisor on $\widetilde{X}$: Express $K_{\widetilde{X}} = f^* K_X + \sum b_i E_i$ and assume $b_i \geq 0$ for every E_i. Then by the ramification formula Theorem 6.1.7 we obtain $a > 0$. On the other hand, if $b_i \geq -1$ (resp. $b_i > -1$), then by the log ramification formula Theorem 6.1.6 we obtain $a \geq -1$ (resp. $a > -1$).

Now we give the definition of a rational singularity.

Definition 6.2.10. A normal singularity (X, x) is called a *rational singularity* if for every weak resolution $f : Y \to X$ the following holds in a neighborhood of x:

$$R^i f_* \mathcal{O}_Y = 0, \quad (i > 0).$$

Elkik proved in [El2] that a canonical singularity is a rational singularity. Here, we will prove it by using Fujita's vanishing theorem, which follows from the Kawamata–Viehweg vanishing theorem.

Theorem 6.2.11 (Fujita [Fu], [KMM, Theorem 1-2-3]). *Let $f : Y \to X$ be a weak resolution of the singularities. For a Cartier divisor L on Y we assume that there exist divisors D, D^*, E on Y satisfying the following, where for a $\mathbb{Q}$-divisor $D = \sum m_i E_i$ (E_i is irreducible for i) we define $[D] = \sum [m_i] E_i$ ($[m_i]$ is the greatest integer which does not exceed m_i):*

(i) $[D] = [D^*] = 0$.
(ii) *E is an effective exceptional divisor.*
(iii) *$L + E - K_Y - D$, $-L - D^*$ are both f-nef.*

Then it follows that $R^i f_* \mathcal{O}_Y(L) = 0$, $i > 0$.

Theorem 6.2.12. *A log terminal singularity is rational, in particular a canonical singularity is rational.*

Proof. Let $f : Y \to X$ be a weak resolution of the singularities. Then, by the assumption we have $K_Y = f^* K_X + \sum m_i E_i, m_i > -1$. Now we define for a rational number m_i, $m_i' := \lceil m_i \rceil - m_i$, where $\lceil m_i \rceil$ means the round up of m_i. We put $E = \sum \lceil m_i \rceil E_i$, $D = \sum m_i' E_i$, $D^* = 0$, $L = 0$, then these satisfy the conditions of Theorem 6.2.11, therefore we obtain $R^i f_* \mathcal{O}_Y = 0$ $(i > 0)$.

By the following theorem, we can see that the vanishing $R^i f_* \mathcal{O}_Y = 0$ $(i > 0)$ for one weak resolution $f : Y \to X$ yields the rationality of the singularity.

Proposition 6.2.13. *For two weak resolutions $f : Y \to X$, $f' : Y' \to X$ of a normal singularity (X, x), the isomorphism*

$$R^p f_* \mathcal{O}_Y \simeq R^p f'_* \mathcal{O}_{Y'}, \quad (\forall p \in \mathbb{Z})$$

holds. In particular, (X, x) is a rational singularity if and only if there is a weak resolution $f : Y \to X$ satisfying $R^i f_* \mathcal{O}_Y = 0$ $(i > 0)$.

Proof. It is sufficient to prove the statement for f, f' such that $f' = f \circ \sigma, \sigma : Y' \to Y$. Consider Leray's spectral sequence (Example 3.6.9):

$$E_2^{p,q} = R^p f_*(R^q \sigma_* \mathcal{O}_{Y'}) \Longrightarrow E^{p+q} = R^{p+q} f'_* \mathcal{O}_{Y'}.$$

As Y is non-singular, by Example 6.2.5 and Theorem 6.2.12, it follows that $R^q \sigma_* \mathcal{O}_{Y'} = 0$ for $q > 0$. Then we have $E_2^{p,q} = 0$ $(q \neq 0)$, and by Proposition 3.6.3, it follows that $E_2^{p,0} \cong E^p$, which means $R^p f_* \mathcal{O}_Y \cong R^p f'_* \mathcal{O}_{Y'}$.

According to the definition, a rational singularity seems independent of differential forms. But actually they are related as follows:

Theorem 6.2.14. *For a normal singularity (X, x) the following are equivalent:*

(i) (X, x) is a rational singularity.
(ii) (X, x) is a Cohen–Macaulay singularity and for a weak resolution $f : Y \to X$ the canonical injection $f_*\omega_Y \hookrightarrow \omega_X$ becomes an isomorphism.

Proof. Applying the duality theorem (Theorem 3.5.10) on $\mathcal{O}_Y$ we have the following isomorphism:

$$\mathbb{R}f_*\big(\mathbb{R}\mathcal{H}om_{\mathcal{O}_Y}(\mathcal{O}_Y, D_Y^{\bullet})\big) \cong \mathbb{R}\mathcal{H}om_{\mathcal{O}_X}(\mathbb{R}f_*\mathcal{O}_Y, D_X^{\bullet}). \tag{6.1}$$

As Y is non-singular, it is a Cohen–Macaulay variety. Then by Proposition 3.5.12, it follows that$D_{\dot{Y}} \cong \omega_Y[n]$, where $n = \dim_x X$. By the Grauert–Riemenschneider vanishing theorem $R^i f_*\omega_Y = 0$, $i > 0$, we obtain:

$$f_*\omega_Y[n] \cong \mathbb{R}\mathcal{H}om_{\mathcal{O}_X}(\mathbb{R}f_*\mathcal{O}_Y, D_X^{\bullet}). \tag{6.2}$$

Now assume (i), then $\mathbb{R}f_*\mathcal{O}_Y = \mathcal{O}_X$. Then (2) turns out to be $f_*\omega_Y[n] \cong D_{\dot{X}}$ and therefore by Proposition 3.5.12, X is a Cohen–Macaulay variety with $f_*\omega_Y = \omega_X$.

Conversely, assume (ii), apply $\mathbb{R}\mathcal{H}om_{\mathcal{O}_X}(\ , D_{\dot{X}})$ on both sides of (2), and we obtain

$$\mathbb{R}\mathcal{H}om_{\mathcal{O}_X}(f_*\omega_Y[n], D_X^{\bullet}) \cong \mathbb{R}f_*\mathcal{O}_Y. \tag{6.3}$$

Here, by condition (ii) it follows that $f_*\omega_Y[n] \cong D_{\dot{X}}$. Hence the left-hand side of (3) turns out to be $\mathcal{O}_X$, which gives (i).

Corollary 6.2.15. *If a normal singularity (X, x) is 1-Gorenstein, then the following are equivalent:*

(i) (X, x) is a rational singularity.
(ii) (X, x) is a canonical singularity.

Proof. (ii)$\Rightarrow$(i) follows from Theorem 6.2.12. Assume (i), then by Theorem 6.2.14, for a weak resolution $f : Y \to X$ the equality $f_*\omega_Y = \omega_X$ holds. Therefore if we express $K_Y = f^*K_X + \sum m_i E_i$, we obtain $m_i \geq 0$, which implies that (X, x) is a canonical singularity.

Next we introduce a special version of the Kawamata–Viehweg vanishing theorem.

Theorem 6.2.16 ([KMM, 1.2.6]). *Let $f : Y \to X$ a proper morphism and Y with at worst log terminal singularities. Let D be a Weil divisor on Y such that it is $\mathbb{Q}$-Cartier, $D - K_Y$ is f-big and f-nef. Then the following holds:*

$$R^i f_*\mathcal{O}_Y(D) = 0 \quad (i > 0).$$

In particular, if $X = \{point\}$, $D - K_Y$ is ample and Y is non-singular, then the theorem coincides with Kodaira's vanishing theorem.

6.3 Classification of Normal Isolated Singularities

In this section (X, x) is a germ of an algebraic variety over $\mathbb{C}$ at an isolated singularity. We denote a small affine neighborhood of x also by X. Here, we introduce three kinds of plurigenera γ_m, δ_m, d_m for a normal isolated singularity and according to the growth order we classify roughly the isolated singularities.

Definition 6.3.1 (Knöller [Kn]). For a normal isolated singularity (X, x) we define

$$\gamma_m(X, x) := \dim_{\mathbb{C}} \omega_X^{[m]} / f_*(\omega_Y^{\otimes m}) \quad (m \in \mathbb{N}),$$

where $f : Y \to X$ is a resolution of the singularity (X, x) and $\dim_{\mathbb{C}}$ means the dimension of the $\mathbb{C}$-vector space.

Here, by Corollary 6.1.11 (i) the sheaf $\omega_X^{[m]}/f_*(\omega_Y^{\otimes m})$ is independent of choices of resolutions f. As the sheaves $\omega_X^{[m]}$, $f_*(\omega_Y^{\otimes m})$ are both coherent $\mathcal{O}_X$-Modules, so is the quotient. Then, $\omega_X^{[m]}/f_*(\omega_Y^{\otimes m})$ is a finitely generated $\mathcal{O}_V := \mathcal{O}_X/Ann\left(\omega_X^{[m]}/f_*(\omega_Y^{\otimes m})\right)$-Module. But $\mathcal{O}_V$ is a finite-dimensional $\mathbb{C}$-vector space with the support on x, therefore $\gamma_m(X, x)$ is finite. Here, the $\mathcal{O}_X$-ideal sheaf $Ann(\mathcal{F})$ is defined by $Ann(\mathcal{F})(U) = \{f \in \Gamma(U, \mathcal{O}_X) \mid f\mathcal{F} = 0\}$.

Definition 6.3.2 (Watanabe [W1], Ishii [I7]). For a normal isolated singularity (X, x) and $m \in \mathbb{N}$ we define

$$\delta_m(X, x) := \dim_{\mathbb{C}} \omega_X^{[m]} / f_*\big(\omega_Y^{\otimes m}((m-1)E)\big),$$
$$d_m(X, x) := \dim_{\mathbb{C}} f_*\omega_Y^{\otimes m}(mE) / f_*\big(\omega_Y^{\otimes m}((m-1)E)\big),$$

where $f : Y \to X$ is a good resolution of the singularity (X, x) and E is the total exceptional divisor with respect to f. We should note that $\omega_Y^{\otimes m}\big((m-1)E\big) = \mathcal{O}_Y\big(mK_Y + (m-1)E\big)$.

Here also as in Definition 6.3.1, by using Corollary 6.1.11 (ii), δ_m, d_m are independent of choices of good resolutions and also the values of $\delta_m(X, x)$, $d_m(X, x)$ are finite.

Now we obtain three kinds of number series $\{\gamma_m(X, x)\}_{m=1}^{\infty}$, $\{\delta_m(X, x)\}_{m=1}^{\infty}$, $\{d_m(X, x)\}_{m=1}^{\infty}$ for an isolated singularity (X, x). These are called *plurigenera*. Next we introduce the growth order of the number series.

Definition 6.3.3. For a real number series $a = \{a_m\}_{m=1}^{\infty}$ if there exists the following real number k, we say that a_m *grows in order* k and write it as $a_m \sim m^k$.

$$0 < \limsup_{m\to\infty} \frac{a_m}{m^k} < \infty.$$

Remark 6.3.4. Such a value k as above does not necessarily exist. For example, let $a_m = \log m$, then for $k \leq 0$ we have $\limsup_{m\to\infty} \frac{a_m}{m^k} = \infty$, while for $k > 0$ we have $\limsup_{m\to\infty} \frac{a_m}{m^k} = 0$.

First consider the series $\{\gamma_m(X,x)\}_{m=1}^{\infty}$. The following is proved for the 2-dimensional case by [Kn] and for the general case by [I7].

Theorem 6.3.5. *For an n-dimensional normal isolated singularity (X,x), one of the following holds:*

(i) *For all $m \in \mathbb{N}$ we have $\gamma_m(X,x) = 0$,*
(ii) *$\gamma_m(X,x) \sim m^n$.*

For other plurigenera $\delta_m(X,x)$, $d_m(X,x)$ the following is proved in [I7].

Theorem 6.3.6. *For an n-dimensional normal isolated singularity (X,x) one of the following holds:*

(i) *For all $m \in \mathbb{N}$ we have $\delta_m(X,x) = d_m(X,x) = 0$.*
(ii) *$\delta_m(X,x) = d_m(X,x) \sim m^k$, where $k = 0, 1, 2, \cdots, n-2$.*
(iii) *$\delta_m(X,x) \sim m^n$ and $d_m(X,x) \sim m^{n-1}$.*

Definition 6.3.7. In the cases of Theorem 6.3.5 (i) and 6.3.6 (i), we define $\kappa_\gamma(X,x) = -\infty$ and $\kappa_d(X,x) = \kappa_\delta(X,x) = -\infty$, respectively. In the cases of Theorem 6.3.5 (ii) and 6.3.6 (ii) (iii), we denote the orders of the growth of $\gamma_m(X,x)$, $\delta_m(X,x)$ and $d_m(X,x)$ by $\kappa_\gamma(X,x)$, $\kappa_\delta(X,x)$ and $\kappa_d(X,x)$, respectively.

By Theorem 6.3.5, n-dimensional normal isolated singularities are classified into two classes by the value of κ_γ; on the other hand, by Theorem 6.3.6 the singularities are classified into $n+1$-classed by the value of κ_δ. The next proposition shows that if the singularities are isolated $\mathbb{Q}$-Gorenstein, then only three classes are possible.

Proposition 6.3.8 (Tomari–Watanabe [ToW]). *Let (X,x) be an n-dimensional isolated $\mathbb{Q}$-Gorenstein singularity, then $\kappa_\delta(X,x) = -\infty$ or 0 or n.*

Remark 6.3.9. For every good resolution $f : Y \to X$ of the singularities, there is the inclusion

$$f_*\omega_Y^{\otimes m} \subset f_*\omega_Y^{\otimes m}\big((m-1)E\big),$$

therefore it follows that $\gamma_m(X,x) \geq \delta_m(X,x)$. In particular, for $m = 1$ we have $\gamma_1(X,x) = \delta_1(X,x)$.
And this value is related with the following invariant:

Definition 6.3.10. For an n-dimensional normal isolated singularity (X,x) we define

$$p_g(X, x) := \dim_{\mathbb{C}} R^{n-1} f_* \mathcal{O}_Y$$

and call it the *geometric genus* of (X, x). Here $f : Y \to X$ is a weak resolution of the singularity (X, x).

This is independent of choices of weak resolutions f (cf. Proposition 6.2.13).

Proposition 6.3.11. *For a normal isolated singularity (X, x) the following holds:*

$$p_g(X, x) = \gamma_1(X, x) = \delta_1(X, x).$$

Proof. Let $f : Y \to X$ be a resolution of the singularity (X, x) and E the exceptional set on Y with respect to f. For a neighborhood V of $x \in X$ take $U := f^{-1}(V)$. Then we have an exact sequence as follows:

$$0 = \Gamma_E(\omega_Y) \longrightarrow \Gamma(U, \omega_Y) \longrightarrow \Gamma(U - E, \omega_Y) \longrightarrow H^1_E(\omega_Y) \longrightarrow H^1(U, \omega_Y),$$

therefore we have the following exact sequence:

$$0 \longrightarrow f_*\omega_Y \longrightarrow \omega_X \longrightarrow H^1_E(\omega_Y) \longrightarrow R^1 f_*\omega_Y.$$

By the Grauert–Riemenschneider vanishing theorem we have $R^1 f_*\omega_Y = 0$, which yields $H^1_E(\omega_Y) = \omega_X / f_*\omega_Y$. Now by the local duality theorem (Corollary 3.5.15) $\dim R^{n-1} f_*\mathcal{O}_Y = \dim H^1_E(\omega_Y)$, and it follows that $p_g(X, x) = \gamma_1(X, x)$.

The relation of the singularities defined in the previous section and the plurigenera is as follows:

Proposition 6.3.12. *For a $\mathbb{Q}$-Gorenstein singularity (X, x) the following hold:*

(i) (X, x) *is a canonical singularity* $\Longleftrightarrow \kappa_\gamma(X, x) = -\infty$.
(ii) (X, x) *is a log terminal singularity* $\Longleftrightarrow \kappa_\delta(X, x) = -\infty$.
(iii) (X, x) *is a log canonical and not a log terminal singularity* $\Longleftrightarrow \kappa_\delta(X, x) = 0$. *More precisely, let r be the index, then $\delta_m(X, x) = 1$ $(r | m)$, $\delta_m(X, x) = 0$ $(r \nmid m)$.*

Proof. Let r be the index of (X, x) and $f : Y \to X$ a resolution of the singularity (X, x). Express a canonical divisor as $K_Y = f^*K_X + \sum m_i E_i$.

(i) If (X, x) is a canonical singularity, then for an integer m such that $r | m$ we have $\omega_Y^{\otimes m} = f^*\omega_X^{[m]}\left(m \sum m_i E_i\right)$, $m_i \geq 0$, therefore we have

$$f_*\omega_Y^{\otimes m} \supset f_* f^* \omega_X^{[m]} = \omega_X^{[m]}.$$

Therefore for such an m it follows that $\gamma_m(X, x) = 0$. If there is an integer $s \in \mathbb{N}$ such that $\gamma_s(X, x) \neq 0$, then there exists $\theta \in \omega_X^{[s]}$ such that $\theta \notin f_*\omega_Y^s$. Then there is a prime divisor E_i such that $\nu_{E_i}(\theta) < 0$, which implies $\nu_{E_i}(\theta^r) < 0$. Hence by $\theta^r \notin f_*\omega_Y^{rs}$ we obtain $\gamma_{rs}(X, x) \neq 0$, a contradiction. This completes the proof that $\kappa_\gamma(X, x) = -\infty$. The converse is trivial.

(ii) Let $f : Y \to X$ be a good resolution of the singularity (X, x). If (X, x) is a log terminal singularity, then for all m such that $r|m$ we have

$$\omega_Y^{\otimes m} = f^*\omega_X^{[m]}(m \sum m_i E_i).$$

Here, as $m_i > -1$ we have $m - 1 + mm_i \geq 0$. Hence

$$f_*(\omega_Y^{\otimes m}((m-1)E)) = f_*(f^*\omega_X^{[m]}(\sum_i (m - 1 + mm_i)E_i)) = \omega_X^{[m]},$$

where E is as in Definition 6.3.2. This yields $\delta_m(X, x) = 0$ for such an m. For general m, we can prove the statement as in the proof of (i). The converse is trivial.

(iii) Let (X, x) be a log canonical and not log terminal singularity. For m such that $r|m$ take a generator θ_m of $\omega_X^{[m]}$ over $\mathcal{O}_X$, then for each prime divisor E_i the equality $\nu_{E_i}(\theta_m) = mm_i$ holds. By the assumption, this value is $\geq -m$ for every i and is $= -m$ for some i. Now if an element $h \in \mathcal{O}_X$ satisfies

$$h\theta_m \in f_*\omega_Y^{\otimes m}\big((m-1)E\big),$$

then f^*h has zero on some E_i, therefore it has zero whole on E which is a connected compact set. This implies $h(x) = 0$. Hence, $h \in \mathfrak{m}_{X,x}$. Conversely, if $h \in \mathfrak{m}_{X,x}$, then it is clear that

$$h\theta_m \in f_*\omega_Y^m\big((m-1)E\big).$$

Then, we have

$$\omega_X^{[m]}/f_*\omega_Y^{\otimes m}\big((m-1)E\big) \simeq \mathcal{O}_X/\mathfrak{m}_{X,x},$$

which yields $\delta_m(X, x) = 1$. Assume $\delta_m(X, x) \neq 0$ for m such that $r \not| m$. Then there is an element $\eta \in \omega_X^{[m]}$ such that $\eta \notin f_*\omega_Y^{\otimes m}\big((m-1)E\big)$. As $\eta^r \notin f_*\omega_Y^{\otimes mr}\big((mr-1)E\big)$, we obtain $\eta^r = h\theta_{rm}$, $h \notin m_{X,x}$. Then η^r has no zero in a neighborhood of x, therefore η does not. This implies that η is a generator of $\omega_X^{[m]}$. Let $d = (r, m)$, then $\omega_X^{[d]}$ is invertible and this contradicts the minimality of r. In conclusion we have $\delta_m(X, x) = 0$ $(r \not| m)$.

Assume conversely $\kappa_\delta(X, x) = 0$. If (X, x) is not a log canonical singularity, there exists a prime divisor E_i such that $-t := \nu_{E_i}(\theta_r) < -r$. Take a regular function h on Y such that $h(E_i) = 0$ and let $\nu_{E_i}(h) = s$. As $\theta_r^m, h\theta_r^m, \cdots, h^{[\frac{m(t-r)}{s}]}\theta_r^m$ are linearly independent elements in $\omega_X^{[mr]}/f_*\omega_Y^{\otimes mr}\big((mr-1)E\big)$, we have that $\delta_{mr}(X, x) \underset{m\to\infty}{\longrightarrow} \infty$, which yields a contradiction to the assumption $\kappa_\delta(X, x) = 0$.

The following is a typical example which attains the given value for κ_d or κ_δ.

Example 6.3.13. Let S be an $(n-1)$-dimensional non-singular projective variety and $\mathscr{L}$ an ample sheaf on S. Let $\widetilde{X}$ be the line bundle $\mathbb{V}(\mathscr{L})$ defined by $\mathscr{L}$ and E the zero section. Then, by [EGA, II 8.9.1], there exists a proper morphism $f : \widetilde{X} \to X$ such that $f(E) = \{x\}$ one point and $\widetilde{X} - E \xrightarrow{\sim} X - \{x\}$ isomorphic. Such a singularity (X, x) is called a *cone over S with respect to $\mathscr{L}$*.

In this case we have $d_m(X, x) = P_m(E)$. Therefore, in particular the equality $\kappa_d(X, x) = \kappa(E)$ holds. Here, $P_m(E)$ means $\dim \Gamma(E, \mathscr{O}_E(mK_E))$ and κ is the order of the growth of P_m and is called *Kodaira dimension*.

In fact let $p : \widetilde{X} \to S$ be the natural projection, then $K_{\widetilde{X}} = p^*K_S - E$. Then the restriction map

$$\Gamma\big(\widetilde{X}, \mathscr{O}_{\widetilde{X}}(m(K_{\widetilde{X}} + E))\big) \longrightarrow \Gamma\big(E, \mathscr{O}_E(m(K_{\widetilde{X}} + E))\big) = \Gamma\big(E, \mathscr{O}_E(mK_E)\big)$$

is surjective. Therefore by

$$f_*\omega_{\widetilde{X}}^{\otimes m}(mE)/f_*\omega_{\widetilde{X}}^{\otimes m}((m-1)E) \simeq \Gamma(E, \mathscr{O}_E(mK_E))$$

we obtain $d_m(X, x) = P_m(E)$.

Chapter 7
Normal 2-Dimensional Singularities

If an experiment works, something has gone wrong [Mur]

In this chapter we consider normal singularities of 2-dimensional varieties over $\mathbb{C}$. A 2-dimensional integral algebraic variety is called a *surface*. A normal singularity on a surface is an isolated singularity and by Corollary 3.5.17 it is a Cohen–Macauley singularity.

7.1 Resolutions of Singularities on a Surface

The exceptional set of a normal singularity on a surface is a union of curves. We will see what kinds of properties it has. Looking at the exceptional set, we can somehow see the properties of the singularity.

Proposition 7.1.1. *For divisors D_1, D_2 on the projective line $\mathbb{P}^1_{\mathbb{C}}$, the following are equivalent:*

(i) $\deg D_1 = \deg D_2$,
(ii) $D_1 \sim D_2$,
(iii) $\mathscr{O}_{\mathbb{P}^1}(D_1) \simeq \mathscr{O}_{\mathbb{P}^1}(D_2)$.

Proof. The implication (ii) $\Leftrightarrow$ (iii) follows from Theorem 5.2.11. For the proof of (i) $\Rightarrow$ (ii), let the homogeneous coordinates of $\mathbb{P}^1_{\mathbb{C}}$ be Y_0, Y_1, then the rational function field of $\mathbb{P}^1_{\mathbb{C}}$ is $\mathbb{C}(Y_1/Y_0)$. Now express the divisors as $D_1 = \sum_P n_P P$, $D_2 = \sum_Q m_Q Q$ and the homogeneous coordinates of P and Q as $(a_P; b_P)$ and $(c_Q; d_Q)$, respectively. Define a function as

S. Ishii, *Introduction to Singularities*,
https://doi.org/10.1007/978-4-431-56837-7_7

$$f = \prod_P (b_P Y_0 - a_P Y_1)^{n_P} / \prod_Q (d_Q Y_0 - c_Q Y_1)^{m_Q},$$

then by $\sum n_P = \sum m_Q$ we obtain $f \in \mathbb{C}(Y_1/Y_0)$, and, moreover, $(f) = D_1 - D_2$. Conversely, assume (ii), then there exists $f \in \mathbb{C}(Y_1/Y_0)$ such that $(f) = D_1 - D_2$. Here, by $\deg(f) = 0$ we obtain (i).

Definition 7.1.2. An invertible sheaf $\mathcal{O}_{\mathbb{P}^1}(D)$ with $\deg D = m$ on $\mathbb{P}^1$ is written as $\mathcal{O}_{\mathbb{P}^1}(m)$.

Example 7.1.3. There is an isomorphism $\omega_{\mathbb{P}^1} \cong \mathcal{O}_{\mathbb{P}^1}(-2)$. Indeed on $\mathbb{P}^1$ there is the following exact sequence ([Ha2], II, 8.13):

$$0 \longrightarrow \Omega_{\mathbb{P}^1} \longrightarrow \mathcal{O}_{\mathbb{P}^1}(-1) \oplus \mathcal{O}_{\mathbb{P}^1}(-1) \longrightarrow \mathcal{O}_{\mathbb{P}^1} \longrightarrow 0.$$

$\bigwedge^2 \mathcal{O}(-1)^{\oplus 2} = \Omega_{\mathbb{P}^1} = \omega_{\mathbb{P}^1}$.

The following example is simple but important.

Example 7.1.4. Let $b : X = B_0(\mathbb{A}^2) \to \mathbb{A}^2$ be the blow-up at the origin. Then the fact that $E = b^{-1}(0) \cong \mathbb{P}^1$ is observed in Example 4.3.9. We will show that $E^2 = -1$. By Example 6.1.8 we have $K_X = b^*K_{\mathbb{A}^2} + E$. Applying the adjunction formula (Proposition 5.3.11), we have $K_E = K_X + E|_E = b^*K_{\mathbb{A}^2} + 2E|_E \sim 2E|_E$. Here, by Example 7.1.3 we obtain that $E^2 = \deg E|_E = -1$.

Definition 7.1.5. An irreducible curve C on a non-singular surface X is called an *exceptional curve of the first kind* or just a (-1)-*curve* if the following conditions are satisfied:

(1) $C \cong \mathbb{P}^1$,
(2) $C^2 = -1$.

The following gives the condition for an exceptional curve to satisfy.

Theorem 7.1.6 (Mumford). *Let $f : Y \to X$ be a weak resolution of 2-dimensional normal singularity (X, x). Let $f^{-1}(x)_{\text{red}} = \sum_{i=1}^r E_i$ be the exceptional set, then the intersection matrix (i.e., the square matrix whose (i, j)-entry is the intersection number $E_i \cdot E_j$) is negative definite.*

Proof. Let h be a regular function on X such that $h(x) = 0$. Consider the principal divisor $(f^*h) = D + \sum_{i=1}^r m_i E_i$ (D does not contain E_i) on Y, then f^*h vanishes on E_i, therefore $m_i > 0$ ($\forall i = 1, \cdots, r$). To prove that the matrix $(E_i \cdot E_j)_{ij}$ is negative definite, it is sufficient to prove that $(m_i m_j E_i \cdot E_j)_{ij}$ is negative definite. Define $e_{ij} := m_i m_j E_i \cdot E_j$. As $(f^*h) \cdot E_j = 0$, it follows that

$$0 = \Big(D + \sum_i m_i E_i\Big) \cdot E_j = D \cdot E_j + \Big(\sum m_i E_i\Big) \cdot E_j.$$

Here, by $D \cdot E_j \geq 0$ we obtain $m_j(\sum m_i E_i) \cdot E_j \leq 0$ and therefore we have $\sum_{i=1}^r e_{ij} \leq 0$ $(\forall j)$. In particular, for j such that $D \cdot E_j > 0$ we have $\sum_{i=1}^r e_{ij} < 0$.

Now for every $\boldsymbol{x} = \begin{pmatrix} x_1 \\ \vdots \\ x_r \end{pmatrix}$, the quadratic form

$$\begin{aligned}
{}^t\boldsymbol{x}(e_{ij})\boldsymbol{x} &= \sum_{\substack{i=1,\cdots,r \\ j=1,\cdots,r}} x_i x_j e_{ij} = \sum_{i=1}^r e_{ii} x_i^2 + 2\sum_{i<j} e_{ij} x_i x_j \\
&= \sum_{j=1}^r e_{jj} x_j^2 + \left(\sum_{\substack{ij \\ i \neq j}} e_{ij} x_j^2 - \sum_{\substack{ij \\ i \neq j}} e_{ij} x_j^2 \right) + 2\sum_{i<j} e_{ij} x_i x_j \\
&= \sum_{j=1}^r \left(\sum_{i=1}^r e_{ij} \right) x_j^2 - \sum_{i<j} e_{ij}(x_i - x_j)^2.
\end{aligned}$$

Here, by $\sum_{i=1}^r e_{ij} \leq 0$ and by $e_{ij} \geq 0$ $(i \neq j)$ the expression above is ≤ 0. If the equality $= 0$ holds, then for k such that $D \cap E_k \neq \emptyset$ it follows that $\sum_{i=1}^r e_{ik} < 0$, which requires $x_k = 0$. For k' $(k' \neq k)$ such that $E_{k'} \cap E_k \neq \emptyset$, we have $e_{kk'} > 0$, therefore by $(x_k - x_{k'})^2 = 0$ we obtain $x_{k'} = 0$. As the exceptional set $\cup E_i$ is connected by the Zariski Main Theorem (see for Example [Ha2, V, 5.2]), by performing this procedure successively, we finally obtain that $x_i = 0$ $(\forall i)$, which implies the negative definiteness of the matrix.

The following is the converse of this theorem. We skip the proof, because it needs many notions which cannot be treated in this book.

Theorem 7.1.7 (Grauert [Gr2]). *If a compact divisor $E = \sum_{i=1}^r E_i$ on a non-singular surface Y has the negative definite intersection matrix $(E_i \cdot E_j)_{ij}$, then there exist a unique normal singularity (X, x) and a proper morphism $f : Y \to X$ such that $f|_{Y \setminus E} : Y \setminus E \to X \setminus \{x\}$ is isomorphic.*

The following theorem also by Grauert gives a sufficient condition for the singularities obtained by contractions of divisors to be the same.

Theorem 7.1.8 (Grauert [Gr2, Satz 7, Corollary]). *Assume that a compact non-singular divisor E on a non-singular variety Y satisfies the following:*

(i) $H^1(E, \mathscr{I}_E^m/\mathscr{I}_E^{m+1}) = 0 \quad \forall m \geq 1.$
(ii) $H^1(E, \Omega_E^* \otimes \mathscr{I}_E^m/\mathscr{I}_E^{m+1}) = 0 \quad \forall m \geq 1.$

Then there exist an analytic neighborhood $U \subset Y$ of E and an analytic neighborhood $U_0 \subset \mathbb{V}(\mathscr{I}_E/\mathscr{I}_E^2)$ of the zero section E_0 of the line bundle $\mathbb{V}(\mathscr{I}_E/\mathscr{I}_E^2)$ such that $(U, E) \simeq (U_0, E_0)$.

Corollary 7.1.9. *Assume compact non-singular prime divisors E, E' on non-singular surfaces Y, Y', respectively satisfy $E \simeq E'$, $\mathcal{I}_E/\mathcal{I}_E^2 \simeq \mathcal{I}_{E'}/\mathcal{I}_{E'}^2$ and $E^2 < 0$ and also the conditions (i) and (ii) of Theorem 7.1.8. Let $f : (U, E) \to (X, x)$, $f' : (U', E') \to (X', x')$ be the contractions of E, E' obtained by Theorem 7.1.7. Then we have a commutative diagram:*

$$\begin{array}{ccc} (U, E) & \longrightarrow & (X, x) \\ \wr\downarrow & & \wr\downarrow \\ (U', E') & \longrightarrow & (X', x'). \end{array}$$

Example 7.1.10. Let a prime divisor E on a non-singular surface Y be a (-1)-curve. By Theorem 7.1.7, there exists a morphism $f : Y \to X$ with $f(E) = \{x\}$ and $f|_{Y-E} : Y - E \xrightarrow{\sim} X - \{x\}$. Here (X, x) is non-singular and the morphism f is the blow-up of X with the center at x. Indeed, let $b : Y' \to \mathbb{A}^2$ be the blow-up at the origin, then the exceptional divisor E' of b is also a (-1)-curve (Example 7.1.4). Therefore by Proposition 7.1.1, we have the isomorphisms

$$E \simeq E' \simeq \mathbb{P}^1, \quad \mathcal{I}_E/\mathcal{I}_E^2 \simeq \mathcal{I}_{E'}/\mathcal{I}_{E'}^2 \simeq \mathcal{O}_{\mathbb{P}^1}(1).$$

On the other hand, by isomorphisms

$$\mathcal{I}_E^m/\mathcal{I}_E^{m+1} \simeq \mathcal{I}_{E'}^m/\mathcal{I}_{E'}^{m+1} \simeq \mathcal{O}_{\mathbb{P}^1}(m)$$

and the exact sequence of $\Omega_{\mathbb{P}^1}$ (in Example 7.1.3) and by the Serre vanishing theorem ([Se2, III, Sect. 3, Proposition 8]), we obtain $H^1\big(\mathbb{P}^1, \mathcal{O}_{\mathbb{P}^1}(m)\big) = 0$ $(m \geq 0)$ and thus conditions (i) and (ii) of Theorem 7.1.8. Then, by Corollary 7.1.9, the two germs (X, x) and $(\mathbb{A}^2_{\mathbb{C}}, 0)$ are isomorphic.

The following gives a criterion for the point obtained by contracting a prime divisor to be non-singular.

Theorem 7.1.11 (Castelnuovo's criterion). *For a non-singular compact prime divisor E on a non-singular surface Y, there exist a non-singular surface X and a proper morphism $f : Y \to X$ satisfying $f(E) = x$ and $f|_{Y\setminus E} : Y \setminus E \xrightarrow{\sim} X \setminus \{x\}$ if and only if E is a (-1)-curve.*

Proof. If E is a (-1)-curve, then there exists a required morphism (Example 7.1.10). Conversely, assume there exists a proper morphism $f : Y \to X$ satisfying the conditions. By the exact sequence:

$$0 \longrightarrow \mathcal{I}_E \longrightarrow \mathcal{O}_Y \longrightarrow \mathcal{O}_E \longrightarrow 0,$$

we obtain the exact sequence:

$$R^1 f_* \mathcal{O}_Y \longrightarrow R^1 f_* \mathcal{O}_E = H^1(E, \mathcal{O}_E) \longrightarrow R^2 f_* \mathcal{I}_E = 0.$$

Here, as (X, x) is a rational singularity, we have $R^1 f_* \mathcal{O}_Y = 0$. Then, $H^1(E, \mathcal{O}_E) = 0$, which implies $E \cong \mathbb{P}^1$. Now let $E^2 = -n$ $(n > 0)$. If we show

$$\dim \Gamma(E, \mathcal{I}_E/\mathcal{I}_E^2) \leq \dim_{\mathbb{C}} \mathfrak{m}_{X,x}/\mathfrak{m}_{X,x}^2, \tag{7.1}$$

then, by $\dim \Gamma(E, \mathcal{I}_E/\mathcal{I}_E^2) = \dim \Gamma(\mathbb{P}^1, \mathcal{O}_{\mathbb{P}^1}(n)) = n + 1$ ([Se2, III, Sect. 3, Proposition 8]) and the fact that (X, x) is a 2-dimensional non-singular point, it follows that $n = 1$.

We will prove (1). As $\mathcal{I}$ is f-ample, by Serre's vanishing theorem ([EGA, III, 2.6.1]), we have $R^1 f_* \mathcal{I}_E^m = 0$ for $m \gg 0$. On the other hand, for every $i \geq 1$, the equality $H^1(E, \mathcal{I}_E^i/\mathcal{I}_E^{i+1}) = 0$ holds, therefore inductively we obtain

$$R^1 f_* \mathcal{I}_E^i = 0 \text{ for every } i \geq 1.$$

In particular, we have $R^1 f_* \mathcal{I}_E^2 = 0$, and we obtain the following exact sequence:

$$0 \longrightarrow f_* \mathcal{I}_E^2 \longrightarrow f_* \mathcal{I}_E \longrightarrow \Gamma(E, \mathcal{I}_E/\mathcal{I}_E^2) \longrightarrow 0.$$

Here, by $f_* \mathcal{I}_E = \mathfrak{m}_{X,x}$ and $f_* \mathcal{I}_E^2 \supset \mathfrak{m}_{X,x}^2$ we obtain the surjection $\mathfrak{m}_{X,x}/\mathfrak{m}_{X,x}^2 \longrightarrow \to \Gamma(E, \mathcal{I}_E/\mathcal{I}_E^2)$.

The next problem is how do we judge a curve to be a (-1)-curve?

Proposition 7.1.12. *A compact curve E on a non-singular surface Y is a (-1)-curve if and only if $K_Y \cdot E < 0$ and $E^2 < 0$.*

Proof. If E is a (-1)-curve, then it satisfies $-2 = \deg K_E = E^2 + K_Y \cdot E$, therefore it is clear that $K_Y \cdot E < 0$, $E^2 < 0$. We will prove the converse. By Nagata's theorem, Y is embedded into a complete algebraic variety as an open subvariety, therefore we may assume that Y is a complete non-singular integral variety. By the exact sequence $0 \to \mathcal{O}_Y(-E) \to \mathcal{O}_Y \to \mathcal{O}_E \to 0$ on Y we have $\chi(E, \mathcal{O}_E) = \chi(Y, \mathcal{O}_Y) - \chi(Y, \mathcal{O}(-E))$. Here, by the Riemann–Roch Theorem (Theorem 5.4.5) we have

$$\chi(E, \mathcal{O}_E) = -\frac{1}{2}(K_Y \cdot E + E^2).$$

As $h^0(E, \mathcal{O}_E) = 1$, it follows that $h^1(E, \mathcal{O}_E) = \frac{1}{2}(K_Y \cdot E + E^2) + 1$. Now by the assumption, we obtain that $h^1(E, \mathcal{O}_E) = 0$, $K_Y \cdot E + E^2 = -2$. Hence, we have $E^2 = K_Y \cdot E = -1$. In order to show that $E \simeq \mathbb{P}^1$, take the normalization $\pi : \overline{E} \to E$, then by the exact sequence:

$$0 \longrightarrow \mathcal{O}_E \longrightarrow \pi_* \mathcal{O}_{\overline{E}} \longrightarrow \pi_* \mathcal{O}_{\overline{E}}/\mathcal{O}_E \longrightarrow 0$$

we obtain the exact sequence:

$$0 \longrightarrow \Gamma(E, \mathscr{O}_E) \longrightarrow \Gamma(E, \pi_* \mathscr{O}_{\overline{E}}) \longrightarrow \Gamma(E, \pi_* \mathscr{O}_{\overline{E}}/\mathscr{O}_E) \longrightarrow H^1(E, \mathscr{O}_E) = 0.$$

As the first and the second terms are both isomorphic to $\mathbb{C}$, we obtain $\pi_* \mathscr{O}_{\overline{E}}/\mathscr{O}_E = 0$, therefore π is isomorphic. Then E is non-singular and $H^1(E, \mathscr{O}_E) = 0$, this means that E is nothing but $\mathbb{P}^1$.

The blow-up with the center at a point plays an important role on a proper morphism of non-singular varieties.

Proposition 7.1.13. *Let $f : Y \to X$ be a proper morphism of non-singular algebraic surfaces such that it is isomorphic outside finite point set of X. Then there exist blow-ups with the center at a point $\sigma_i : X_i \to X_{i-1}, i = 1, \ldots, k$ $(X_k = Y, X_0 = X)$ such that $f = \sigma_1 \circ \sigma_2 \circ \cdots \circ \sigma_k$.*

Proof. Let $E_1, \cdots, E_k$ be the prime exceptional divisors with respect to f, then as X is non-singular, by Example 6.2.5 we obtain

$$K_Y = f^* K_X + \sum_{i=1}^{k} m_i E_i, \quad m_i > 0 \ (\forall i = 1, \cdots, k).$$

Here, since the matrix $(E_i \cdot E_j)_{ij}$ is negative definite, we have $(\sum m_i E_i)^2 < 0, E_i^2 < 0$ $(\forall i)$. Then, in particular there exists E_j such that $K_Y \cdot E_j = (\sum m_i E_i) \cdot E_j < 0$, $E_j^2 < 0$. Therefore E_j is a (-1)-curve and there exists a non-singular surface X_{k-1} and a morphism $\sigma_k : Y \to X_{k-1}$, where σ_k is the blow-up at a point with the exceptional divisor E_j. The morphism $X_{k-1} \setminus \sigma_k(E_j) \to X \setminus f(E_j)$ is canonically extended to a morphism $f_{k-1} : X_{k-1} \to X$, which yields a factorization $f = f_{k-1} \circ \sigma_k$. Next perform the same procedure for f_{k-1}. By successive procedures we obtain $\sigma_{k-1}, \cdots, \sigma_1$, as required.

Given a resolution of the singularities of X, by blowing up at a point on an exceptional divisor, we have a new resolution of the singularities of X. Therefore for one singularity, there exist an infinite number of resolutions of singularity. We would like to avoid unnecessary procedures and get the most economical resolution.

Definition 7.1.14. A resolution $f : Y \to X$ of the singularities on X is called a *minimal resolution* if for every resolution $g : Y' \to X$ of X there exists a unique morphism $\varphi : Y' \to Y$ such that g factors as $g = f \circ \varphi$.

By the definition, if there exists a minimal resolution, then it is unique up to the isomorphism over X. The following shows that there exists a minimal resolution for a 2-dimensional resolution.

Theorem 7.1.15. *Assume* $\dim X = 2$. *A resolution $f : Y \to X$ of a singularity (X, x) is the minimal resolution if and only if $f^{-1}(x)$ does not contain a (-1)-curve. In particular there exists a minimal resolution.*

Proof. It is clear that the minimal resolution does not contain a (-1)-curve. Conversely, let $f : Y \to X$ be a resolution whose exceptional divisor does not contain a (-1)-curve. Let $g : Y' \to X$ be an arbitrary resolution. By taking a resolution of the irreducible component of $Y \times_X Y'$ that dominates X, we obtain a resolution $\widetilde{f} : \widetilde{Y} \to X$ such that

$$f \circ \varphi = \widetilde{f} = g \circ \psi$$

for some $\varphi : \widetilde{Y} \to Y$, $\psi : \widetilde{Y} \to Y'$. Now by Proposition 7.1.13, there is a decomposition

$$\psi = \sigma_1 \circ \cdots \circ \sigma_k$$

($\sigma_i (i = 1, \cdots, k)$, is a point blow-up). We may assume that we take $\widetilde{f}$ such that k is minimal. If $k > 0$, let E be the exceptional divisor on $\widetilde{Y}$ with respect to σ_k. In the case of a curve $\varphi(E) = C$, as $C^2 = (\varphi^* C)^2$, we have $E^2 \leq C^2$ and the equality holds if and only if φ is isomorphic on a neighborhood of C. Since C is an exceptional divisor with respect to f, by $-1 = E^2 \leq C^2 < 0$ the equality $C^2 = E^2$ holds and φ is isomorphic in a neighborhood of C. Then C is a (-1)-curve as well as E, which contradicts the definition of $f : Y \to X$. Therefore, $\varphi(E)$ is one point and φ factors through σ_k which contracts E. This contradicts the minimality of k. Hence $k = 0$. In conclusion, we obtain $\varphi : Y' = \widetilde{Y} \to Y$. The uniqueness of φ follows from that φ with the property $g = f \circ \varphi$ is unique on an open dense subset. On the other hand, a resolution $f : Y \to X$ without a (-1)-curve in its exceptional divisors is obtained by contracting (-1)-curves of any resolution, which implies the existence of the minimal resolution.

For a $\mathbb{Q}$-Cartier divisor D the pull-back by a morphism f is defined in Definition 5.2.14. This can be generalized to every Weil divisor on a normal surface.

Definition 7.1.16. Let X be a normal surface, $f : Y \to X$ a resolution of the singularities and $E_i (i = 1, \cdots, r)$ the exceptional prime divisors. For a divisor D on X, we define the *pull-back* $f^* D \in \mathrm{Div}\,(Y) \otimes \mathbb{Q}$ as follows:

Let $[D]$ be the strict transform of D and let $f^* D := [D] + \sum_{i=1}^r m_i E_i$. where $m_i \in \mathbb{Q}$'s are the solution of the r equations $([D] + \sum m_i E_i) \cdot E_j = 0\,(j = 1, \cdots, r)$. Here, we note that the solution exists uniquely because the matrix $(E_i \cdot E_j)_{ij}$ is negative definite. It is clear that this definition is a generalization of the pull-back of the $\mathbb{Q}$-Cartier divisor.

A resolution $f : Y \to X$ of the singularities of X is isomorphic outside the total exceptional divisor; the difference between K_Y and $f^* K_X$ appears on the exceptional divisor. What is this difference? We can see that it has a distinguished property if f is the minimal resolution.

Proposition 7.1.17. *Let $f : Y \to X$ be the minimal resolution of a normal 2-dimensional singularity (X, x) and E_i $(i = 1, \cdots, r)$ the prime exceptional divisors. Then the following holds:*

$$K_Y = f^* K_X + \sum_{i=1}^{r} m_i E_i, \quad m_i \le 0 \quad (\forall i = 1, \cdots, r).$$

Proof. Assume that there exists i with $m_i > 0$, then we may assume that $m_i > 0$ $(i = 1, \cdots, s)$ and $m_i \le 0$ $(i = s+1, \cdots, r)$. As the matrix $(E_i \cdot E_j)_{ij}$ is negative definite, we have $\left(\sum_{i=1}^{s} m_i E_i\right)^2 < 0$. Then, there exists j $(1 \le j \le s)$ such that $\left(\sum_{i=1}^{s} m_i E_i\right) \cdot E_j < 0$. For this E_j we have:

$$K_Y \cdot E_j = \left(\sum_{i=1}^{s} m_i E_i\right) \cdot E_j + \left(\sum_{i=s+1}^{r} m_i E_i\right) \cdot E_j < 0,$$

which implies that E_j is a (-1)-curve and therefore a contradiction to the minimality of f.

By using this theorem we obtain the characterization of 2-dimensional terminal singularities.

Corollary 7.1.18. *A singularity (X, x) is terminal if and only if it is non-singular.*

Proof. If a terminal singularity (X, x) is singular, then the minimal resolution $f : Y \to X$ of the singularity (X, x) is not isomorphic. Then by Proposition 7.1.17 it follows that $m_i \le 0$ $(\forall i = 1, \cdots, r)$, a contradiction to the fact that (X, x) is terminal.

In order to catch the shape of the total exceptional divisor, we introduce a graph of the exceptional divisor.

Definition 7.1.19. On a good resolution $f : Y \to X$ of a 2-dimensional normal singularity (X, x) assume that every exceptional prime divisor E_i is non-singular. We define the *weighted dual graph* of the exceptional set $\bigcup_{i=1}^{r} E_i$ as follows:

each prime divisor	$\Longrightarrow$	vertex ○
E_i and E_j intersect	$\Longrightarrow$	connect the vertices corresponding to E_i and E_j by an edge: ○—○
$E_i^2 = -a$	$\Longrightarrow$	write $-a$ in the vertex corresponding to E_i: ($-a$) in particular, if $a = 2$, do not write anything

Example 7.1.20. If the intersection matrix is

$$\begin{pmatrix} -2 & 1 & 0 \\ 1 & -2 & 1 \\ 0 & 1 & -2 \end{pmatrix},$$

then the corresponding dual weighted graph is 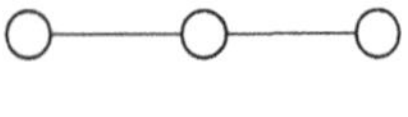

If the intersection matrix is

$$\begin{pmatrix} -3 & 0 & 1 & 0 \\ 0 & -4 & 1 & 0 \\ 1 & 1 & -5 & 1 \\ 0 & 0 & 1 & -6 \end{pmatrix},$$

then the corresponding dual weighted graph is

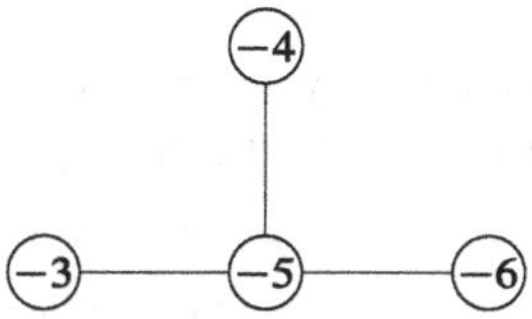

7.2 The Fundamental Cycle

In this section we introduce the fundamental cycle supported on the exceptional set on a resolution of the singularities. This cycle will be used for deciding the rationality of the singularity.

Lemma 7.2.1. *Let S be a non-singular surface, E_i $(i = 1, \cdots, r)$ complete prime divisors on S. If the intersection matrix $(E_i \cdot E_j)_{ij}$ is negative definite, there exists an effective divisor $Z = \sum m_i E_i$ such that $Z \cdot E_i < 0$ $(i = 1, \cdots, r)$.*

Proof. Let the intersection matrix $(E_i \cdot E_j)_{ij}$ be A. Since A is a regular integer matrix, there exist integers $m_1, \cdots, m_r$ such that $A \begin{pmatrix} m_1 \\ \vdots \\ m_r \end{pmatrix} = \begin{pmatrix} \alpha_1 \\ \vdots \\ \alpha_r \end{pmatrix}$ with α_i $(i = 1, \cdots, r)$ negative integers.

Let $Z := \sum m_i E_i$, then we have $Z \cdot E_i = \alpha_i < 0$ for each i. It is sufficient to show that $Z > 0$. Express $Z = Z_1 - Z_2$ with Z_1, Z_2 effective divisors without common components. By the properties of Z we have

$$0 \geq Z \cdot Z_2 = (Z_1 - Z_2) \cdot Z_2 = Z_1 \cdot Z_2 - Z_2^2 \geq 0.$$

Therefore it follows that $Z_1 \cdot Z_2 = Z_2^2 = 0$, which yields $Z_2 = 0$.

Lemma 7.2.2. *Under the same conditions in Lemma 7.2.1, assume that $E = \sum_{i=1}^r E_i$ is connected. Then the set $\{Z > 0 \mid Z \cdot E_i \leq 0,\ i = 1, \cdots, r\}$ of divisors on S has a minimum element.*

Proof. By Lemma 7.2.1, this set is not empty. Note that every element of this set has the support whole on E. Take any two elements $Z = \sum_{i=1}^r m_i E_i$, $Z' = \sum_{i=1}^r m_i' E_i$

and define $n_i := \min\{m_i, m'_i\}$ and put $Z'' := \sum_{i=1}^{r} n_i E_i$, then the support of Z'' is also whole on E. We will prove that $Z'' \cdot E_i \leq 0$ for every i. Without loss of generality, we may assume $m_i \leq m'_i$. Then, $Z'' \cdot E_i = m_i E_i^2 + \sum_{j \neq i} n_j E_j \cdot E_i \leq m_i E_i^2 + \sum_{j \neq i} m_j E_j \cdot E_i = Z \cdot E_i \leq 0$.

Definition 7.2.3. Let (X, x) be a 2-dimensional normal singularity, $f : Y \to X$ a resolution of the singularity and $E = \sum_{i=1}^{r} E_i$ the exceptional divisor. We call the minimum element of the set $\{Z > 0 \mid Z \cdot E_i \leq 0,\ i = 1, \cdots, r\}$ the *fundamental cycle* of Y and denote it by Z_f.

It is difficult to find the fundamental cycle only by using the definition. How can we find it? One practical way is by a computation sequence.

Proposition 7.2.4. *Under the notation in Definition 7.2.3, choose any E_{i_1} and define $Z_1 := E_{i_1}$. Assume that $Z_1, \cdots, Z_{\nu-1}$ are defined, if there exists E_{i_ν} such that $Z_{\nu-1} \cdot E_{i_\nu} > 0$, define $Z_\nu := Z_{\nu-1} + E_{i_\nu}$. Then, this procedure will stop at a finite stage. The last Z_{ν_0} is the fundamental cycle. The sequence $\{Z_1, Z_2, \cdots, Z_{\nu_0}\}$ reaching Z_f is called a computation sequence.*

Proof. By the construction, we have $Z_1 < Z_2 < \cdots$, therefore it is sufficient to prove $Z_\nu \leq Z_f$ for every Z_ν. We will prove the statement by induction on ν. It is clear that $Z_1 = E_{i_1} \leq Z_f$. Assume $\nu \geq 2$ and $Z_{\nu-1} \leq Z_f$. When $Z_{\nu-1} \neq Z_f$, then there exists E_{i_ν} such that $Z_{\nu-1} E_{i_\nu} > 0$. By the definition we have $Z_\nu := Z_{\nu-1} + E_{i_\nu}$. Here, if we write $Z_{\nu-1} = \sum_i n_i E_i$, $Z_f = \sum_i m_i E_i$, then for every i, we have $n_i \leq m_i$. If $n_{i_\nu} = m_{i_\nu}$, then

$$Z_{\nu-1} \cdot E_{i_\nu} = n_{i_\nu} E_{i_\nu}^2 + \sum_{i \neq i_\nu} n_i E_i \cdot E_{i_\nu} \leq m_{i_\nu} E_{i_\nu}^2 + \sum_{i \neq i_\nu} m_i E_i \cdot E_{i_\nu} = Z_f \cdot E_{i_\nu} \leq 0,$$

which is a contradiction. Therefore it follows that $n_{i_\nu} < m_{i_\nu}$, which implies that $Z_\nu = Z_{\nu-1} + E_{i_\nu} \leq Z_f$.

Example 7.2.5. The following graphs consisting of (-2)-curves give the following fundamental cycles, where each number expresses the coefficient of the fundamental cycle at the component:

(i) A_n :
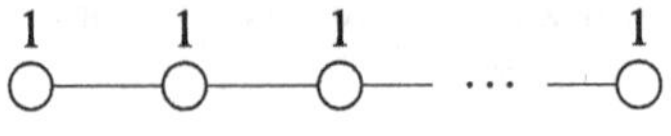

(ii) D_4 :
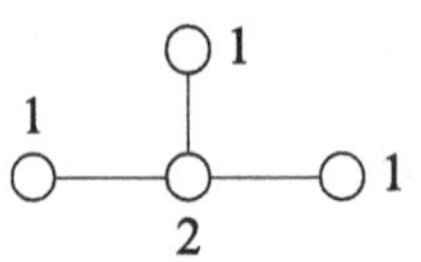

D_5 :
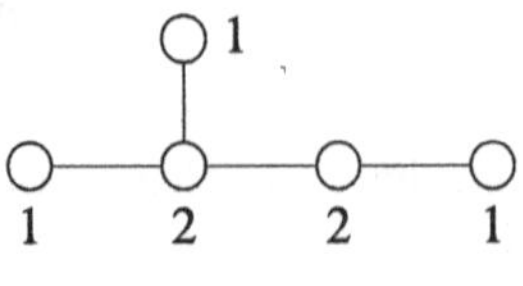

D_6 :

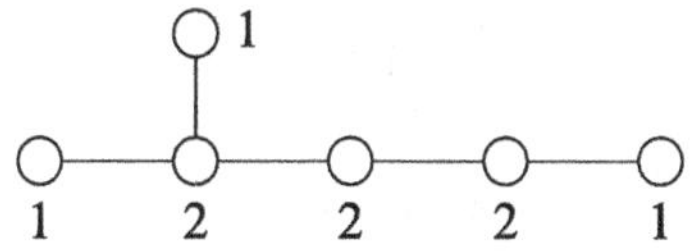

(iii) E_6 :

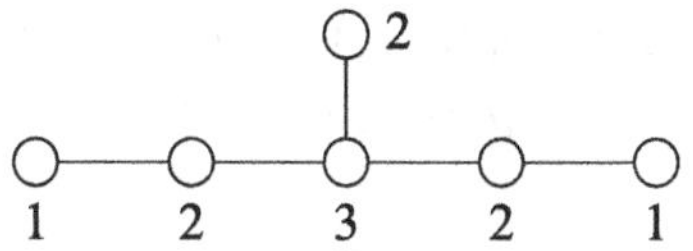

E_7 :

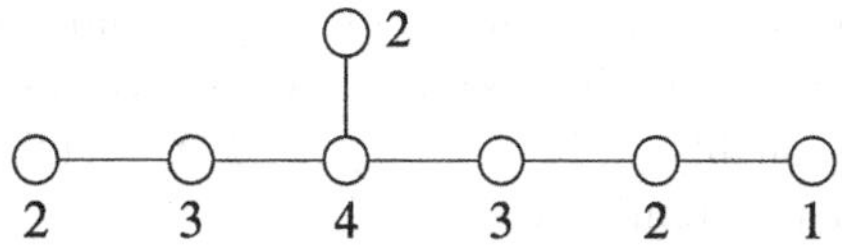

E_8 :

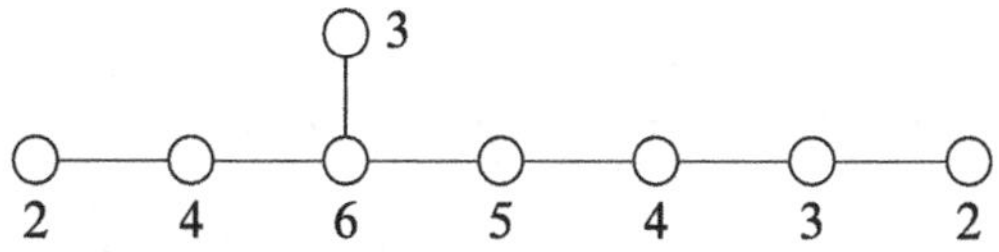

The fundamental cycles on two resolutions of a singularity satisfy the following relation:

Proposition 7.2.6. *If two resolutions $f : Y \to X$, $g : Y' \to X$ of a normal surface singularity (X, x) are connected by $\sigma : Y' \to Y$ such that $g = f \circ \sigma$, then the fundamental cycles Z_1 of Y and Z_2 of Y' have the relation: $Z_2 = \sigma^* Z_1$.*

Proof. By Proposition 7.1.13, it is sufficient to prove the statement for the case that σ is the blow-up at a point. Let $E_1, \cdots, E_r$ be the prime divisors of $f^{-1}(x)$. Let E_i' be the strict transform of E_i at Y' and F the exceptional divisor with respect to σ. Here, by $\sigma^* Z_1 \cdot E_i' = Z_1 \cdot E_i \leq 0$ and $\sigma^* Z_1 \cdot F = 0$, it follows that $Z_2 \leq \sigma^* Z_1$ by the minimality of the fundamental cycle. Now put $Z_2 = \sum_{i=1}^r a_i E_i' + bF$, $\sigma^* Z_1 = \sum_{i=1}^r \alpha_i E_i' + \beta F$, then it it sufficient to prove that $a_i = \alpha_i$, $b = \beta$. If there is an i such that $a_i < \alpha_i$, then it follows that $\sigma_* Z_2 < Z_1$. Therefore for some i_0

$$0 < \sigma_* Z_2 \cdot E_{i_0} = \sigma^* \sigma_* Z_2 \cdot E_{i_0}'.$$

This yields $E_{i_0}' \cap F \neq \emptyset$. Here if we put $\sigma^* \sigma_* Z_2 = \sum a_i E_i' + b' F$, then by the above inequality we obtain $b' > b$. But

$$0 = \sigma^* \sigma_* Z_2 \cdot F = \sum a_i E_i' \cdot F + b' F^2 < \sum a_i E_i' \cdot F + b F^2 = Z_2 \cdot F \leq 0,$$

which gives a contradiction. Hence, we obtain that $a_i = \alpha_i$ $(i = 1, \cdots, r)$.

Next, if $b < \beta$, then it produces an inequality $0 \geq Z_2 \cdot F = \sum a_i E_i' \cdot F + bF^2 > \sum \alpha_i E_i' \cdot F + \beta F^2 = 0$, a contradiction.

Here, we introduce an invariant for a divisor.

Definition 7.2.7. Let S be a non-singular surface and D be a complete divisor on S. We define the *virtual genus* of D denoted by $p(D)$ as follows:

$$p(D) := \frac{D^2 + K_S \cdot D}{2} + 1.$$

Let us look at the basic properties on the virtual genus.

Proposition 7.2.8. *(i) If $D > 0$, the equality $p(D) = 1 - \chi(D, \mathcal{O}_D)$ holds.*
(ii) If D is a prime divisor, then the inequality $p(D) \geq 0$ holds. Here, the equality holds if and only if $D \cong \mathbb{P}^1$.

Proof. (i) By Nagata's compactification theorem, we may assume that S is a complete non-singular surface. By the exact sequence $0 \to \mathcal{O}_S(-D) \to \mathcal{O}_S \to \mathcal{O}_D \to 0$ on S, we have the equality $\chi(D, \mathcal{O}_D) = \chi(S, \mathcal{O}_S) - \chi(S, \mathcal{O}_S(-D))$. Now, by the Riemann–Roch Theorem (Theorem 5.4.5)

$$\chi(D, \mathcal{O}_D) = -\frac{1}{2}(D^2 + K_S \cdot D).$$

(ii) As D is irreducible and reduced, the equality $h^0(D, \mathcal{O}_D) = 1$ holds. Hence, we have $p(D) = 1 - \chi(D, \mathcal{O}_D) = h^1(D, \mathcal{O}_D) \geq 0$. If $D \cong \mathbb{P}^1$, obviously the equality in (ii) holds. Conversely, if the equality $p(D) = 0$, then taking the normalization $\overline{D} \to D$ in the same way as in the proof of Proposition 7.1.12 we obtain $D \cong \mathbb{P}^1$.

Proposition 7.2.9. *For an arbitrary resolution of a normal surface singularity (X, x), the virtual genus $p(Z_f)$ of the fundamental cycle Z_f is constant.*

Proof. For the morphism $\sigma : Y' \to Y$ of two resolutions, we have $p(\sigma^* Z) = p(Z)$ for every cycle Z, therefore by Proposition 7.2.6 the required statement follows.

Definition 7.2.10. For a normal surface singularity (X, x), the virtual genus $p(Z_f)$ of the fundamental cycle Z_f of a resolution $f : Y \to X$ is called the *fundamental genus* and is denoted by $p_f(X, x)$.

Proposition 7.2.11. *The inequality $p_f(X, x) \geq 0$ holds for every normal surface singularity (X, x).*

Proof. Let Z_f be the fundamental cycle of the resolution $f : Y \to X$ and let the computation sequence be $Z_1, \cdots, Z_{\nu_0} = Z_f$. By the construction of the sequence, for every ν we have $Z_\nu = Z_{\nu-1} + E_{i_\nu}$ and $Z_{\nu-1} \cdot E_{i_\nu} > 0$. On the other hand, by Proposition 7.2.8, we have $p(E_{i_\nu}) \geq 0$. Then, we obtain

$$p(Z_\nu) = p(Z_{\nu-1} + E_{i_\nu}) = p(Z_{\nu-1}) + p(E_{i_\nu}) + Z_{\nu-1} \cdot E_{i_\nu} - 1 \geq p(Z_{\nu-1}).$$

In particular, for $\nu = 1$, as Z_1 is a prime divisor, $p(Z_1) \geq 0$. Therefore, for arbitrary ν we obtain $p(Z_\nu) \geq 0$.

Next we introduce a new invariant for singularities, called the *arithmetic genus*. To this end we prepare some notions.

For a resolution $f : Y \to X$ of the normal surface singularity (X, x), we define:

$$p_a(f) = \sup\{p(Z) \mid Z > 0,\ |Z| \subset f^{-1}(x)\},$$

where $|Z|$ is the support of the divisor Z.

Lemma 7.2.12. *For two resolutions $f : Y \to X$, $g : Y' \to X$ of the singularity (X, x), the equality $p_a(f) = p_a(g)$ holds.*

Proof. It is sufficient to prove for f, g such that there is a morphism $\sigma : Y' \to Y$ with $g = f \circ \sigma$. By Proposition 7.1.13, it is sufficient to prove the statement in the case where σ is a blow-up at one point. For every cycle $Z > 0, |Z| \subset f^{-1}(x)$ on Y, we have $\sigma^* Z > 0$, $|\sigma^* Z| \subset g^{-1}(x)$ and $p(Z) = p(\sigma^* Z)$. Therefore the inequality $p_a(f) \leq p_a(g)$ is clear. For the converse it is sufficient to prove $p(\sigma^* \sigma_* Z) \geq p(Z)$ for every $Z > 0, |Z| \subset g^{-1}(x)$. Let E_i $(i = 1, \cdots, r)$ be the prime divisors of $f^{-1}(x)$, let E'_i $(i = 1, \cdots, r)$ be the strict transform of them and let F be the exceptional divisor of σ. Put $\sigma^* E_i = E'_i + m_i F$ for each i, then by $F^2 = -1$ we have

$$E'_i \cdot F = m_i.$$

Decompose Z as $Z = Z_1 + nF$, $Z_1 = \sum_{i=1}^r n_i E'_i$, then

$$\sigma_* Z = \sum n_i E_i \ \text{ and } \ \sigma^* \sigma_* Z = Z_1 + (\sum m_i n_i) F.$$

Here, $p(Z_1 + \nu F) = p(Z_1 + (\nu - 1)F) + p(F) + (Z_1 + (\nu - 1)F) \cdot F - 1$ for every ν, therefore it follows that $p(Z_1 + \nu F) = p(Z_1 + (\nu - 1)F) + \sum_{i=1}^r m_i n_i - \nu$. Then, we obtain:

$$\text{If } \nu \leq \sum m_i n_i, \text{ then } p(Z_1 + (\nu - 1)F) \leq p(Z_1 + \nu F).$$

$$\text{If } \nu > \sum m_i n_i, \text{ then } p(Z_1 + \nu F) < p(Z_1 + (\nu - 1)F).$$

Hence, the inequality $p(Z_1 + (\sum m_i n_i)F) \geq p(Z_1 + nF)$ holds.

By this lemma we can see that the value of $p_a(f)$ is independent of choices of resolutions f, therefore we can define a new invariant as follows:

Definition 7.2.13. For a normal surface singularity (X, x), define the *arithmetic genus* as follows:

$$p_a(X, x) = \sup\{p(Z) \mid Z > 0,\ |Z| \subset f^{-1}(x)\}$$

for a resolution $f : Y \to X$ of the singularity.

By this we get three kinds of genera. The relation of these is as follows:

Theorem 7.2.14. *For a normal surface singularity (X, x), the following holds:*

$$p_g(X, x) \geq p_a(X, x) \geq p_f(X, x) \geq 0.$$

Proof. The inequality $p_f(X,x) \geq 0$ was proved in Proposition 7.2.11. The inequality $p_a(X,x) \geq p_f(X,x)$ is clear from the definition. We will prove that $p_g(X,x) \geq p_a(X,x)$. Let $f: Y \to X$ be a resolution of the singularity. For a divisor $Z > 0$, $|Z| \subset f^{-1}(x)$ there exists an exact sequence $0 \to \mathcal{O}_Y(-Z) \to \mathcal{O}_Y \to \mathcal{O}_Z \to 0$. As the dimension of Y is 2, the equality $R^2 f_* \mathcal{O}_Y(-Z) = 0$ holds, which gives the surjection $R^1 f_* \mathcal{O}_Y \to R^1 f_* \mathcal{O}_Z = H^1(Z, \mathcal{O}_Z)$. On the other hand, by Proposition 7.2.8 it follows that

$$p(Z) = 1 - h^0(Z, \mathcal{O}_Z) + h^1(Z, \mathcal{O}_Z) \leq h^1(Z, \mathcal{O}_Z),$$

therefore we obtain $p(Z) \leq \dim R^1 f_* \mathcal{O}_Y = p_g(X,x)$.

7.3 Rational Singularities

Let (X,x) be a 2-dimensional normal singularity. By Definitions 6.2.10 and 6.3.10 a singularity (X,x) is rational if and only if $p_g(X,x) = 0$. In this section we introduce the results by Artin [A2, A1]. First we characterize a rational singularity.

Theorem 7.3.1. *For a normal surface singularity (X,x) the following are equivalent:*

(i) (X,x) is rational.
(ii) $p_a(X,x) = 0$.
(iii) $p_f(X,x) = 0$.

Proof. By Theorem 7.2.14: $p_g(X,x) \geq p_a(X,x) \geq p_f(X,x) \geq 0$. The implications (i) $\Rightarrow$ (ii) $\Rightarrow$ (iii) are obvious.

Proof of (iii) $\Rightarrow$ (ii): It is sufficient to prove $p(Z) \leq 0$ for every effective divisor Z and this is proved by constructing a computation sequence $Z_1 = Z, Z_2, \cdots, Z_{\nu_0} = Z_f$ such that $p(Z_1) \leq p(Z_2) \leq \cdots \leq p(Z_{\nu_0})$. Actually, define $\{Z_\nu\}$ as follows: First define $Z_1 := Z$, next for $\nu > 1$ assume that $Z_1, \cdots, Z_{\nu-1}$ are already defined. Then, we define Z_ν as follows:

(a) If $Z_{\nu-1} > Z_f$, then define $Z_\nu := Z_{\nu-1} - Z_f$.
(b) If $Z_{\nu-1} \not\geq Z_f$, then there exists E_i such that $Z_{\nu-1} \cdot E_i > 0$. Then define $Z_\nu := Z_{\nu-1} + E_i$.

In case (a), $p(Z_{\nu-1}) = p(Z_\nu + Z_f) = p(Z_\nu) + p(Z_f) + Z_\nu \cdot Z_f - 1$
$\leq p(Z_\nu) - 1.$

In case (b), we have $p(Z_\nu) = p(Z_{\nu-1} + E_i) \geq p(Z_{\nu-1})$.
By this we obtain $p(Z_1) \leq p(Z_2) \leq \cdots$. If this procedure terminates at a finite stage, then there exists ν_0 such that $Z_{\nu_0} = Z_f$. If this procedure continues infinitely, then by Theorem 7.2.14 $p(Z_\nu)$ is bounded; the procedure (a) appears only finitely many times. Therefore the procedures after ν_1 are all procedure (b). As $Z_{\nu_1} \leq mZ_f$ for a sufficiently big m, in the same way as in the proof of Proposition 7.2.4 we obtain

$$Z_{\nu_1} < Z_{\nu_1+1} < \cdots \leq mZ_f.$$

Hence, this procedure finishes at a finite stage.

Proof of (ii) $\Rightarrow$ (i). Take a resolution $f : Y \to X$ factoring through the blow-up at the point x. Then, there exists an effective divisor D with the support on $f^{-1}(x)$ such that

$$\mathfrak{m}_{X,x}\mathcal{O}_Y = \mathcal{O}(-D),$$

by [EGA II, 8.1.11]. Now by [EGA III 4.2.1] we have

$$(R^1 f_*\mathcal{O}_Y)\widehat{_x} = \varprojlim_n H^1(f^{-1}(x), \mathcal{O}_Y/\mathfrak{m}^n_{X,x}\mathcal{O}_Y) = \varprojlim_n H^1(f^{-1}(x), \mathcal{O}_{nD}).$$

Therefore it is sufficient to prove $H^1(Z, \mathcal{O}_Z) = 0$ for an effective divisor with the support on $f^{-1}(x)$. Let $Z = \sum_{i=1}^r r_i E_i$ and we will prove the statement by induction on $\sum r_i$.

If $\sum r_i = 1$, then Z is a prime divisor. By condition (ii) it follows that $p(Z) \leq 0$, but by Proposition 7.2.8 it turns out to be $p(Z) = 0$, which implies that $Z \cong \mathbb{P}^1$, therefore of course the equality $H^1(Z, \mathcal{O}_Z) = 0$ holds. For $\sum r_i > 1$ assume that the statement holds true up to $\sum r_i - 1$. Take any prime divisor $E_i < Z$ and define $Z_i := Z - E_i$. Then by the assumption of the induction, it follows that $H^1(Z_i, \mathcal{O}_{Z_i}) = 0$. Therefore by the exact sequence $0 \to \mathcal{O}_{E_i}(-Z_i) \to \mathcal{O}_Z \to \mathcal{O}_{Z_i} \to 0$, we obtain the surjection $H^1(E_i, \mathcal{O}_{E_i}(-Z_i)) \to H^1(Z, \mathcal{O}_Z)$. Here, by $E_i \cong \mathbb{P}^1$, if for some E_i, $E_i \cdot Z_i \leq 1$ holds, then

$$h^1(E_i, \mathcal{O}_{E_i}(-Z_i)) = h^0(\mathbb{P}^1, \mathcal{O}(-2 + E_i \cdot Z_i)) = 0$$

which implies $H^1(Z, \mathcal{O}_Z) = 0$ as required.

Now assume the contrary, i.e., assume $E_i \cdot Z_i > 1$ for every prime divisor $E_i < Z$. Then, $Z \cdot E_i \geq E_i^2 + 2$ $(i = 1, \cdots, r)$. Therefore

$$\begin{aligned} K_Y \cdot Z + Z^2 &= (K_Y + Z) \cdot \left(\sum r_i E_i\right) \\ &= \sum r_i (K_Y \cdot E_i + Z \cdot E_i) \\ &\geq \sum_{i=1}^r r_i (-E_i^2 - 2 + 2 + E_i^2) = 0, \end{aligned}$$

which is a contradiction to the fact that $p(Z) \leq 0$.

A 2-dimensional rational singularity has the following distinguished property, which cannot be expected for higher-dimensional singularities.

Theorem 7.3.2. *A 2-dimensional rational singularity (X, x) is analytically $\mathbb{Q}$-factorial, i.e., there exists an analytic neighborhood V of x such that every divisor* *on V is a $\mathbb{Q}$-Cartier divisor. In particular, a rational singularity is a $\mathbb{Q}$-Gorenstein singularity.*

Proof. Let $f : Y \to X$ be a resolution of the singularity (X, x) and V a sufficiently small Stein neighborhood of x and let $\widetilde{V} := f^{-1}(V)$. Let Z be an effective divisor with the support on $f^{-1}(x)_{\text{red}} = \sum_{i=1}^{r} E_i$. Consider the following commutative diagram:

$$\begin{array}{ccccccccc} 0 & \longrightarrow & \mathbb{Z} & \longrightarrow & \mathcal{O}_{\widetilde{V}}^{\text{hol}} & \xrightarrow{\exp 2\pi\sqrt{-1}} & \mathcal{O}_{\widetilde{V}}^{\text{hol}\,*} & \longrightarrow & 0 \\ & & \downarrow & \curvearrowright & \downarrow & \curvearrowright & \downarrow & & \\ 0 & \longrightarrow & \mathbb{Z} & \longrightarrow & \mathcal{O}_{Z}^{\text{hol}} & \xrightarrow{\exp 2\pi\sqrt{-1}} & \mathcal{O}_{Z}^{\text{hol}\,*} & \longrightarrow & 0. \end{array}$$

By Theorem 2.4.11, Z is a deformation retract of $\widetilde{V}$, we have $H^2(\widetilde{V}, \mathbb{Z}) \simeq H^2(Z, \mathbb{Z})$. On the other hand by rationality, we have $H^i(\widetilde{V}, \mathcal{O}_{\widetilde{V}}^{\text{hol}}) = 0$ $(i > 0)$. Moreover, by the proof of Theorem 7.3.1, we have $H^1(Z, \mathcal{O}_Z^{\text{hol}}) = 0$. Then we obtain the following commutative diagram:

$$\begin{array}{ccccccc} 0 = H^1(\widetilde{V}, \mathcal{O}_{\widetilde{V}}^{\text{hol}}) & \longrightarrow & H^1(\widetilde{V}, \mathcal{O}_{\widetilde{V}}^{\text{hol}\,*}) & \longrightarrow & H^2(\widetilde{V}, \mathbb{Z}) & \longrightarrow & H^2(\widetilde{V}, \mathcal{O}_{\widetilde{V}}^{\text{hol}}) = 0 \\ \wr\downarrow & & \downarrow & & \wr\downarrow & & \wr\downarrow \\ 0 = H^1(Z, \mathcal{O}_Z^{\text{hol}}) & \longrightarrow & H^1(Z, \mathcal{O}_Z^{\text{hol}\,*}) & \longrightarrow & H^2(Z, \mathbb{Z}) = \mathbb{Z}^r & \longrightarrow & H^2(Z, \mathcal{O}_Z^{\text{hol}}) = 0. \end{array} \tag{7.2}$$

Therefore the map $H^1(\widetilde{V}, \mathcal{O}_{\widetilde{V}}^{\text{hol}\,*}) \longrightarrow \mathbb{Z}^r$, $\mathcal{L} \mapsto (\deg \mathcal{L}|_{E_1}, \cdots, \deg \mathcal{L}|_{E_r})$ is isomorphic. Now for a Weil divisor D on V by Definition 7.1.16, the pull-back $f^*D \in \operatorname{Div}(\widetilde{V}) \otimes \mathbb{Q}$ is defined. By taking an appropriate $m \in \mathbb{N}$ it satisfies $mf^*D \in \operatorname{Div}(\widetilde{V})$. As $mf^*D \cdot E_i = 0$ $(i = 1, \cdots, r)$, by the isomorphism above we have $mf^*D \sim 0$ on $\widetilde{V}$. Therefore mD is a Cartier divisor.

In particular, for K_X there is an integer $r > 0$ such that rK_X is a Cartier divisor on an analytic neighborhood, therefore by Mori's theorem [Fo, 6.12, p. 35] it is a Cartier divisor as an algebraic divisor.

The following is a lemma for the forthcoming theorem.

Lemma 7.3.3. (Lichtenbaum [Li]). *For a complete effective divisor $Z = \sum_{i=1}^{r} r_i E_i$ on a non-singular surface, assume $H^1(Z, \mathcal{O}_Z) = 0$. Let $\mathcal{F}$ be an invertible sheaf on Z and let $d_i = \deg \mathcal{F}|_{E_i} \geq 0$ $(\forall i)$. Then the following holds:*

$$\dim_{\mathbb{C}} H^0(Z, \mathcal{F}) = \sum r_i d_i + 1,$$
$$H^1(Z, \mathcal{F}) = 0.$$

By using this lemma, we obtain the following theorem which shows that the fundamental cycle plays an important role for a rational singularity:

Theorem 7.3.4. *For a resolution $f : Y \to X$ of a rational surface singularity (X, x), the following hold, where Z_f is the fundamental cycle on Y.*

(i) $\mathscr{O}_Y(-nZ_f) = \mathfrak{m}_{X,x}^n \mathscr{O}_Y$,
(ii) $H^0(nZ_f, \mathscr{O}_{nZ_f}) = \mathscr{O}_{X,x}/\mathfrak{m}_{X,x}^n$,
(iii) $\dim_{\mathbb{C}} \mathfrak{m}_{X,x}^n/\mathfrak{m}_{X,x}^{n+1} = -nZ_f^2 + 1$.

Proof. Take a sufficiently small analytic neighborhood V of x and let $\widetilde{V} = f^{-1}(V)$. As we know that for a point $y \in Z_f$

$$\mathscr{O}_Y(-Z_f)_y = \mathscr{O}_{\widetilde{V}}^{\text{hol}}(-Z_f)_y \cap \mathscr{O}_{Y,y}, \ \mathfrak{m}_{X,x}\mathscr{O}_{Y,y} = (\mathfrak{m}_{V,x}^{\text{hol}})\mathscr{O}_{\widetilde{V},y}^{\text{hol}} \cap \mathscr{O}_{Y,y},$$
$$\mathscr{O}_{X,x}/\mathfrak{m}_{X,x}^n = \mathscr{O}_{V,x}^{\text{hol}}/(\mathfrak{m}_{V,x}^{\text{hol}})^n, \ \mathfrak{m}_{X,x}^n/\mathfrak{m}_{X,x}^{n+1} = (\mathfrak{m}_{V,x}^{\text{hol}})^n/(\mathfrak{m}_{V,x}^{\text{hol}})^{n+1},$$
$$H^0(nZ_f, \mathscr{O}_{nZ_f}) = H^0(nZ_f, \mathscr{O}_{nZ_f}^{\text{hol}}),$$

it is sufficient to prove the following:

(i) $\mathscr{O}_{\widetilde{V}}^{\text{hol}}(-nZ_f) = (\mathfrak{m}_{V,x}^{\text{hol}})^n \mathscr{O}_V$,
(ii) $H^0(nZ_f, \mathscr{O}_{nZ_f}) = \mathscr{O}_{V,x}^{\text{hol}}/(\mathfrak{m}_{V,x}^{\text{hol}})^n$,
(iii) $\dim_{\mathbb{C}}(\mathfrak{m}_{V,x}^{\text{hol}})^n/(\mathfrak{m}_{V,x}^{\text{hol}})^{n+1} = -nZ_f^2 + 1$.

From now on, we omit the symbol hol.

Proof of (i). It is sufficient to prove in the case $n = 1$. First we prove the inclusion $\mathfrak{m}_{V,x}\mathscr{O}_{\widetilde{V}} \subset \mathscr{O}_{\widetilde{V}}(-Z_f)$. For an arbitrary element $h \in \mathfrak{m}_{V,x}$ the divisor (h) on $\widetilde{V}$ is represented as $(h) = C + D$ (where C is an effective divisor with the support on $f^{-1}(x)_{\text{red}} = \sum_{i=1}^r E_i$ and D is an effective divisor which does not contain E_i). As $(h) \cdot E_i = 0$ $(\forall i)$ and $D \cdot E_i \geq 0$ $(\forall i)$, it follows that $C \cdot E_i \leq 0$ $(\forall i)$. Hence, by the minimality of the fundamental cycle, we have $Z_f \leq C$. This completes the proof of $\mathfrak{m}_{V,x}\mathscr{O}_{\widetilde{V}} \subset \mathscr{O}_{\widetilde{V}}(-Z_f)$.

Next we prove the opposite inclusion. It is sufficient to prove that for an arbitrary point $P \in Z_f$ there is $h \in \mathfrak{m}_{V,x}$ such that $(h) = Z_f$ in a neighborhood of P. If we take a neighborhood $\widetilde{V}$ of P sufficiently small, then there exists a positive divisor D such that $D \cdot E_i = -Z_f \cdot E_i$ $(i = 1, \cdots, r)$ and D does not contain P.

It is sufficient to prove that for such a D there is a rational function h such that $(h) = Z_f + D$. Note that it is equivalent to the fact that $\mathscr{L} := \mathscr{O}_{\widetilde{V}}(-(Z_f + D))$ has a section nowhere vanishing on $\widetilde{V}$ $\left(\text{i.e., } h \notin \mathfrak{m}_{\widetilde{V},y}\mathscr{L}_y (\forall y \in \widetilde{V})\right)$. In order to prove this it is sufficient to show that $H^0(Z_f, \mathscr{L} \otimes \mathscr{O}_{Z_f})$ has a section h_0 nowhere vanishing and there is $\widetilde{h}$ in $\varprojlim_n H^0(\mathscr{L} \otimes \mathscr{O}_{nZ_f})$ corresponding to h_0. In fact by [EGA III, 4.2.1] there exists an element $h \in H^0(\widetilde{V}, \mathscr{L})$ corresponding to h_0 such that h is nowhere vanishing, since h_0 is nowhere vanishing. By the diagram (7.2), we have $H^1(nZ_f, \mathscr{O}_{nZ_f}^*) \xrightarrow{\sim} \mathbb{Z}^r$, which yields $-(Z_f + D) \cdot E_i = 0$ $(\forall i)$. Therefore it follows that

$$\mathscr{L} \otimes \mathscr{O}_{nZ_f} \simeq \mathscr{O}_{nZ_f}.$$

Thus if $n = 1$, there exists h_0. By the exact sequence $0 \to \mathscr{O}_{Z_f}(-nZ_f) \to \mathscr{O}_{(n+1)Z_f} \to \mathscr{O}_{nZ_f} \to 0$ we have the exact sequence

$$H^0(\mathcal{O}_{(n+1)Z_f}) \xrightarrow{\beta_n} H^0(\mathcal{O}_{nZ_f}) \longrightarrow H^1(\mathcal{O}_{Z_f}(-nZ_f)).$$

Here, by $-nZ_f \cdot E_i \geq 0$ and Lemma 7.3.3, we obtain $H^1(\mathcal{O}_{Z_f}(-nZ_f)) = 0$. By this, for every n, the homomorphism, β_n is surjective, then we can construct $\widetilde{h} \in \varprojlim_n H^0(Z_f, \mathcal{L} \otimes \mathcal{O}_{nZ_f})$.

Proof of (ii). By the exact sequence $0 \to \mathcal{O}_{\widetilde{V}}(-nZ_f) \to \mathcal{O}_{\widetilde{V}} \to \mathcal{O}_{nZ_f} \to 0$, we obtain the following exact sequence:

$$\begin{array}{ccccccc} 0 \longrightarrow & f_*\mathcal{O}_{\widetilde{V}}(-nZ_f) & \longrightarrow & f_*\mathcal{O}_{\widetilde{V}} & \xrightarrow{\gamma} & H^0(\mathcal{O}_{nZ_f}) \longrightarrow & R^1 f_*\mathcal{O}_{\widetilde{V}}(-nZ_f) \\ & \cup & & \| & & & \\ & \mathfrak{m}^n_{V,x} & & \mathcal{O}_V & & & \end{array} . \tag{7.3}$$

By the previous discussion, we know that γ is surjective, therefore the homomorphism $\mathcal{O}_V/\mathfrak{m}^n_{V,x} \xrightarrow{\overline{\gamma}} H^0(\mathcal{O}_{nZ_f})$ is also surjective. In order to show that $\overline{\gamma}$ is injective, it is sufficient to prove that if $g \in \mathcal{O}_V$ satisfies $(g) = nZ_f + D$ $(D \geq 0)$ on $\widetilde{V}$, then $g \in \mathfrak{m}^n_{V,x}$. First assume that D intersects E_i transversally. Then we have a decomposition $D = D_1 + \cdots + D_n$ such that $D_\nu \cdot E_i = -Z_f \cdot E_i$ $(\forall \nu, \forall i)$. Therefore by the previous discussion we obtain functions f_ν $(\nu = 1, \cdots, n)$ such that $(f_\nu) = Z_f + D_\nu$. Clearly it holds that $f_\nu \in \mathfrak{m}_{V,x}$ and $g = f_1 \cdot \cdots \cdot f_n u$ (u is a unit of $\mathcal{O}_V$), therefore we obtain $g \in \mathfrak{m}^n_{V,x}$. For a general g, take g_1 with the transversality as above and let $g_2 = g + ag_1$. then for an appropriate a, g_2 also have transversality. Hence, we have $g_2 \in \mathfrak{m}^n_{V,x}$ and as a consequence it follows that $g \in \mathfrak{m}^n_{V,x}$.

Proof of (iii). By the isomorphism (7.3) proved in (ii), it turns out that $\mathfrak{m}^n_{V,x} = f_*\mathcal{O}(-nZ_f)$.

By the exact sequence

$$0 \longrightarrow \mathcal{O}_{\widetilde{V}}(-(n+1)Z_f) \longrightarrow \mathcal{O}_{\widetilde{V}}(-nZ_f) \longrightarrow \mathcal{O}_{Z_f}(-nZ_f) \longrightarrow 0$$

we obtain the following:

$$\begin{array}{ccccc} 0 \longrightarrow & f_*\mathcal{O}_{\widetilde{V}}(-(n+1)Z_f) & \longrightarrow & f_*\mathcal{O}_{\widetilde{V}}(-nZ_f) & \longrightarrow H^0(\mathcal{O}_{Z_f}(-nZ_f)) \\ & \| & & \| & \\ & \mathfrak{m}^{n+1} & & \mathfrak{m}^n & \\ \longrightarrow & R^1 f_*\mathcal{O}_{\widetilde{V}}(-(n+1)Z_f) & & & \\ & \| & & & \\ & 0 & & & \end{array}$$

Therefore it follows that $H^0\big(\mathcal{O}_{Z_f}(-nZ_f)\big) = \mathfrak{m}^n/\mathfrak{m}^{n+1}$.

By Lemma 7.3.3, we obtain

$$\dim_{\mathbb{C}} \mathfrak{m}^n/\mathfrak{m}^{n+1} = -\sum r_i(nZ_f \cdot E_i) + 1 = -nZ_f^2 + 1.$$

Corollary 7.3.5. *For a 2-dimensional rational singularity (X, x), the following hold:*

(i) $\operatorname{mult}(X, x) := \operatorname{mult}\mathscr{O}_{X,x} = -Z_f^2$,
(ii) $\operatorname{emb}(X, x) = -Z_f^2 + 1$,

where the multiplicity mult R *of a local* $\mathbb{C}$*-algebra* R *with* $R/\mathfrak{m} = \mathbb{C}$ *and* $\dim R = d$ *is given by*

$$\dim_{\mathbb{C}} R/\mathfrak{m}^n = \frac{\operatorname{mult} R}{d!} n^d + O(n^{d-1})$$

(see for Example [Ma2, Sect. 14]).

On the other hand, the embedding dimension $\operatorname{emb}(X, x)$ *is the minimal k such that there is a closed immersion* $(X, x) \simeq (Y, y)$*, where* $(Y, y) \subset (\mathbb{C}^k, 0)$*.*

Proof. By Theorem 7.3.4 (iii) we have

$$\dim_{\mathbb{C}} \mathscr{O}_{X,x}/\mathfrak{m}_{X,x}^n = -\frac{Z_f^2}{2} n^2 + (\text{terms of degree } \leq 1).$$

We also have that $\operatorname{emb}(X, x) = \dim_{\mathbb{C}} \mathfrak{m}_{X,x}/\mathfrak{m}_{X,x}^2 = -Z_f^2 + 1$ from the following lemma.

Lemma 7.3.6. $\operatorname{emb}(X, x) = \dim_{\mathbb{C}} \mathfrak{m}_{X,x}/\mathfrak{m}_{X,x}^2$.

Proof. Assume that $(X, x) \subset (\mathbb{C}^n, 0)$, then as the canonical map $\mathfrak{m}_{\mathbb{C}^n,0}/\mathfrak{m}_{\mathbb{C}^n,0}^2 \twoheadrightarrow \mathfrak{m}_{X,x}/\mathfrak{m}_{X,x}^2$ is surjective, we obtain

$$\operatorname{emb}(X, x) \geq \dim \mathfrak{m}_{X,x}/\mathfrak{m}_{X,x}^2.$$

In order to show the converse $\leq$, put $\dim \mathfrak{m}_{X,x}/\mathfrak{m}_{X,x}^2 = k$ and we will show that there is an embedding $(X, x) \subset (\mathbb{C}^k, 0)$. Assume $(X, x) \subset (\mathbb{C}^n, 0)$ and let n be the minimum such number.

Assume $n > k$ and induce a contradiction. Let $x_1, \cdots, x_n$ be coordinates of $\mathbb{C}^n$. By the assumption $n > k$ it follows that

$$\operatorname{Ker}\left(\mathfrak{m}_{\mathbb{C}^n,0}/\mathfrak{m}_{\mathbb{C}^n,0}^2 \longrightarrow \mathfrak{m}_{X,x}/\mathfrak{m}_{X,x}^2\right) \neq 0,$$

therefore there exists $(a_1, \cdots, a_n) \neq (0, \cdots, 0)$ such that $\sum a_i x_i \equiv z \pmod{\mathscr{I}_X}$, $z \in \mathfrak{m}_{\mathbb{C}^n,0}^2$. Here $\mathscr{I}_X$ is the defining ideal of X. Now let $z_1 := \sum a_i x_i - z$, then by the implicit function theorem there exists a coordinate system $(z_1, z_2, \cdots, z_n)$ (containing z_1) around the origin. By the definition of z_1, we have $z_1 \in \mathscr{I}_X$, then

$$\mathscr{O}_{X,x} = \left(\mathscr{O}_{\mathbb{C}^n,0}/(z_1)\right)/\left(\mathscr{I}_{X,0}/(z_1)\right),$$

therefore (X, x) is contained in a hyperplane $(\mathbb{C}^{n-1}, 0)$ defined by $z_1 = 0$. This is a contradiction to the minimality of n.

7.4 Quotient Singularities

In this section, we discuss 2-dimensional quotient singularities. But for a moment, from Definition 7.4.1 to Corollary 7.4.10, we study under a more general setting, so we assume that the dimension of a variety is arbitrary, n. First we clarify what "quotient" is.

Definition 7.4.1. Let an algebraic group G act on an algebraic variety X (cf. Definition 4.4.3). A morphism $\phi : X \to Y$ of algebraic varieties is called the *geometric quotient* by G, if the following holds:

1. The following diagram is commutative:
$$\begin{array}{ccc} G \times X & \xrightarrow{\sigma} & X \\ \downarrow{\scriptstyle p_2} & & \downarrow{\scriptstyle \phi} \\ X & \xrightarrow{\phi} & Y \end{array}$$
2. For every point $y \in Y$ the fiber $\phi^{-1}(y)$ consists of only one orbit.
3. $U \subset Y$ is an open subset $\Longleftrightarrow \phi^{-1}(U)$ is an open subset.
4. An element $f \in \Gamma(U, \phi_* \mathcal{O}_X)$ belongs to $\Gamma(U, \mathcal{O}_Y) \Longleftrightarrow$ The following diagram is commutative:
$$\begin{array}{ccc} G \times \phi^{-1}(U) & \xrightarrow{\sigma} & \phi^{-1}(U) \\ \downarrow{\scriptstyle p_2} & & \downarrow{\scriptstyle f} \\ \phi^{-1}(U) & \xrightarrow[f]{} & \mathbb{A}^1_{\mathbb{C}} \end{array}$$

When $\phi : X \to Y$ is the geometric quotient, we denote Y by X/G.

Proposition 7.4.2 ([GIT, 1, Sect. 2]). *Let G be a finite group and X an affine variety, then the geometric quotient $\phi : X \to Y$ exists and it is a finite morphism. The quotient space $Y = X/G$ is an affine variety with the affine coordinate ring $A(X)^G$, where $A(X)^G$ is the subalgebra consisting of the invariant elements by the automorphisms $A(X) \xrightarrow{g^*} A(X)$ induced from $X \to X$, $x \mapsto \sigma(g, x)$ $g \in G$. Here, if X is normal, then X/G is also normal.*

If X is an analytic space, we can also define an action of a finite group and the geometric quotient by the group. Then, by Holmann's results (see, for Example [Hol, Satz 11]) we have the following:

Proposition 7.4.3. *Let G be a finite group acting on an analytic space X, then the geometric quotient $\phi : X \to Y$ exists and satisfies $\mathcal{O}_Y = \phi_* {\mathcal{O}_X}^G$.*

Definition 7.4.4. A singularity (X, x) is called a *quotient singularity* if there exist a non-singular (V, v) and a finite group G acting on an analytic neighborhood V such that $(X, x) = (V/G, \phi(v))$.

A group G giving a quotient singularity can be a linear group.

Theorem 7.4.5. *For every quotient singularity (X, x) of dimension n, there exists a finite subgroup $G' \subset GL(n, \mathbb{C})$ such that $(X, x) \simeq (\mathbb{C}^n/G', 0)$ as germs. In particular, a quotient singularity is algebraic.*

Proof. Let $(X, x) = \big(V/G, \phi(v)\big)$. If the isotropy subgroup $G_v := \{g \in G \mid gv = v\}$ of v does not coincide with G, take an appropriate neighborhood V' of v, we can see that $X = V'/G_v$. Therefore we may assume that $G_v = G$. As v is fixed under the action of G, the maximal ideal $\mathfrak{m}$ of v is invariant under this action. Therefore every element $g \in G$ gives an isomorphism

$$g' : \mathfrak{m}/\mathfrak{m}^2 \simeq \mathfrak{m}/\mathfrak{m}^2$$

of vector spaces.

Let $x_1, \cdots, x_n$ be a coordinate system of V around v and let

$$y_i = \frac{1}{\#G} \sum_{g \in G} g'^{-1} \cdot g(x_i),$$

then, as $\det\big(\frac{\partial y_j}{\partial x_i}\big)(v) = 1$, $y_1, \cdots, y_n$ becomes a new coordinate system. By this coordinate system, $g \in G$ is linear. Indeed, if $g'(x_i) = \sum_{j=1}^{n} a_{ji} x_j$, we can write $g(y_i) = \sum_{j=1}^{n} a_{ji} y_j$.

Definition 7.4.6. An element $g \in GL(n, \mathbb{C})$ is called a *reflection* if its order is finite and the fixed point set $\mathrm{Fix}(g) = \{x \in \mathbb{C}^n \mid g(x) = x\}$ has codimension one. A subgroup generated by reflections is called a *reflection group*. A finite subgroup of $GL(n, \mathbb{C})$ is called *small* if the subgroup does not contain any reflections.

Lemma 7.4.7 (Chevalley [Ch]). *If a subgroup $H \subset GL(n, \mathbb{C})$ is a finite reflection group, then $\mathbb{C}^n/H \cong \mathbb{C}^n$.*

By this, a reflection group is not effective to create a singularity.

Theorem 7.4.8. *For every n-dimensional quotient singularity (X, x), there exists a small finite subgroup $G \subset GL(n, \mathbb{C})$ such that $(X, x) \cong (\mathbb{C}^n/G, 0)$ as germs.*

Proof. By Theorem 7.4.5, we have $(X, x) \cong (\mathbb{C}^n/G, 0)$, $G \subset GL(n, \mathbb{C})$. Let H be the subgroup of G consisting of the reflections. Then, for every element $g \in G$ and a reflection $h \in H$ we have

$$\mathrm{Fix}(ghg^{-1}) = g(\mathrm{Fix}(h)).$$

Therefore H is a normal subgroup. Then the quotient group $\overline{G} = G/H$ acts on $\mathbb{C}^n/H \cong \mathbb{C}^n$ and we have isomorphisms

$$\mathbb{C}^n/\overline{G} \cong (\mathbb{C}^n/H)/(G/H) \cong \mathbb{C}^n/G.$$

Now it is sufficient to prove that $\overline{G}$ does not contain a reflection.

For an element $\overline{g} = gH \in \overline{G}$, $g \notin H$ and an element $h \in H$ we have $hg \notin H$. Then, we obtain $\operatorname{codim} \operatorname{Fix}(hg) \geq 2$. On the other hand, let ϕ be the projection $\mathbb{C}^n \to \mathbb{C}^n/H$, then as

$$\overline{g}\phi(p) = \phi(p) \Longleftrightarrow gp = hp(\exists h \in H)$$

we obtain

$$\operatorname{Fix}(\overline{g}) = \bigcup_{h \in H} \phi\big(\operatorname{Fix}(hg)\big).$$

Here, H is a finite group and ϕ is a finite morphism, therefore $\operatorname{codim} \operatorname{Fix}(\overline{g}) \geq 2$, which completes the proof that $\overline{g}$ is not a reflection.

We will study the properties of a quotient singularity.

Theorem 7.4.9. *A quotient singularity is a log terminal singularity.*

Proof. By Theorem 7.4.8, there exists a small finite group $G \subset GL(n, \mathbb{C})$ such that $(X, x) \cong (\mathbb{C}^n/G, 0)$. Then the projection $\phi : \mathbb{C}^n \to X = \mathbb{C}^n/G$ does not have a ramification divisor. Then, by Theorem 6.2.9 (ii) X has log terminal singularities.

Corollary 7.4.10. *A quotient singularity is rational.*

Proof. By Theorem 6.2.12 a log terminal singularity is rational.

So far, we discussed on arbitrary-dimensional singularities. From now on, we study 2-dimensional normal singularities. For the 2-dimensional case, the converse of Theorem 7.4.9 also holds.

Theorem 7.4.11. *A 2-dimensional normal singularity is a quotient singularity if and only if it is a log terminal singularity.*

Proof. For the proof we will use some knowledge which will appear in the next section. Let (X, x) be a log terminal singularity and $\pi : (Y, y) \to (X, x)$ the canonical cover. As π has no ramification divisor by Proposition 6.2.3 (i), the covering space (Y, y) is also log terminal singularity by Theorem 6.2.9 (ii). Here, the index of (Y, y) is 1, then it is a canonical singularity and by the equivalence (iv) $\Leftrightarrow$ (x) in Theorem 7.5.1, it follows that (Y, y) is a quotient singularity. Since $\mathbb{C}^2 \setminus \{0\}$ is the universal covering of $Y \setminus \{y\}$, it is also the universal covering of $X \setminus \{x\}$ and $\pi_{X,x} = \varprojlim \pi_1(U \setminus \{x\})$ is finite. By Theorem 7.4.18 we have that (X, x) is a quotient singularity.

Definition 7.4.12. If G is a finite cyclic group, the quotient $(\mathbb{C}^2/G, x)$ is called a *cyclic quotient singularity* of a *Jung singularity*.

Let us study cyclic quotient singularities.

Lemma 7.4.13. *For 2-dimensional cyclic quotient singularity* $(\mathbb{C}^2/G, 0)$, *we can take a generator of* G *as* $\begin{pmatrix} \varepsilon & 0 \\ 0 & \varepsilon^q \end{pmatrix}$, *where* ε *is a primitive* n*-th root of unity and* $(q, n) = 1$. *In this case we denote the singularity by* $X_{n,q}$.

Proof. We may assume that G is small. Let n be the order of G, then a generator σ satisfies $\sigma^n = 1$. Therefore the minimal polynomial of σ is a factor of $X^n - 1$. Hence, the roots are mutually distinct, which implies that σ can be diagonalized by general theory in linear algebra. A generator is represented by using the primitive n-th root of unity ε as $\begin{pmatrix} \varepsilon^p & 0 \\ 0 & \varepsilon^q \end{pmatrix}$. Here, if $(p, n) = d > 1$, then we have $\begin{pmatrix} \varepsilon^p & 0 \\ 0 & \varepsilon^q \end{pmatrix}^{\frac{n}{d}} = \begin{pmatrix} 1 & 0 \\ 0 & \varepsilon^{q\frac{n}{d}} \end{pmatrix}$. Here, if $\varepsilon^{q\frac{n}{d}} = 1$ then the order of a generator is $< n$, a contradiction. On the other hand, if $\varepsilon^{q\frac{n}{d}} \neq 1$, then this matrix is reflection, which is also a contradiction. Therefore we obtain $(p, n) = 1$. In the same way, it follows that $(q, n) = 1$. For the statement of the lemma, it is sufficient to let ε^p be ε.

One of the mysterious observations about a quotient singularity is that it has a relation with a continued fraction.

Lemma 7.4.14. *A rational number* $\frac{n}{q}$ $(n > q)$ *is uniquely expanded by using a finite number of integers* $b_1, \cdots, b_r$ $(b_i \geq 2)$ *as follows:*

$$\frac{n}{q} = b_1 - \cfrac{1}{b_2 - \cfrac{1}{b_3 - \ddots \cfrac{}{b_{r-1} - \cfrac{1}{b_r}}}}$$

This is called the Hirzebruch–Jung continued fraction and the right-hand side is denoted by $b_1 - 1\rfloor b_2 - 1\rfloor b_3 - \cdots - 1\rfloor b_r$. *We should be careful as it is different from the well-known normal continued fraction* $a_1 + 1\rfloor a_2 + \cdots + 1\rfloor a_s$.

Proof. Let $q_1 := q$ and take a positive integer b_1 such that $n = b_1 q_1 - q_2$ $(0 \leq q_2 < q_1)$. As $n > q_1$, the inequality $b_1 \geq 2$ holds. In a similar way, decompose successively:

$$q_1 = b_2 q_2 - q_3, \quad (0 \leq q_3 < q_2,\ b_2 \geq 2)$$
$$q_2 = b_3 q_3 - q_4, \quad (0 \leq q_4 < q_3,\ b_3 \geq 2)$$
$$\cdots\cdots$$

Then, as q_i is an integer, there exists an integer $r > 0$ such that $q_{r+1} = 0$. In this case we have

$$\frac{n}{q} = b_1 - 1\rfloor\overline{b_2} - \cdots - 1\rfloor\overline{b_r}\,.$$

For the proof of the uniqueness, assume that there is another decomposition $\frac{n}{q} = b'_1 - 1\rfloor\overline{b'_2} - \cdots - 1\rfloor\overline{b'_s}$. Noting that

$$b'_2 - 1\rfloor\overline{b'_3} - \cdots - 1\rfloor\overline{b'_s} > 1,$$

if $n = b'_1 q'_1 - q'_2$, then $0 \leq q'_2 < q'_1 = q$. Hence it follows that $b_1 = b'_1$, $q'_2 = q_2$. In the same way, we obtain successively $b_i = b'_i$.

Lemma 7.4.15 (Riemenschneider [Ri1, Satz 1]). *The affine coordinate ring of $X_{n,q}$ is $\mathbb{C}[u^i v^j]$, $i + qj \equiv 0 \pmod n$, $0 \leq i \leq n$, $0 \leq j \leq n$, where u, v are indeterminates.*

Proof. It is sufficient to find i, j such that the equality $u^i v^j = (\varepsilon u)^i (\varepsilon^q v)^j$ holds.

Theorem 7.4.16. *The weighted dual graph of the exceptional divisor on the minimal resolution of the singularity $X_{n,q}$ is as follows, where $\frac{n}{q} = b_1 - 1\rfloor\overline{b_2} - \cdots - 1\rfloor\overline{b_r}$:*

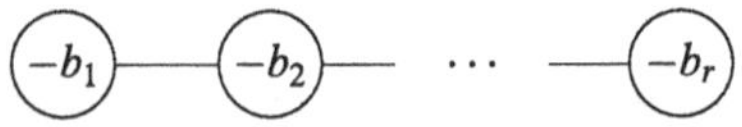

Proof. Let $Y_{n,q}$ be the affine variety with the following affine coordinate ring:

$$B_{n,q} := \mathbb{C}[u^n, u^{n-q}v, v^n] \cong \mathbb{C}[x_1, x_2, x_3]/(x_1^{n-q}x_3 - x_2^n).$$

Then the ring $\mathbb{C}[u^i v^j]$ that appeared in Lemma 7.4.15 is integral over $B_{n,q}$ and it is normal, therefore $X_{n,q}$ is the normalization of $Y_{n,q}$. Note that by Theorem 4.1.11 (iv), a resolution of the singularity of $Y_{n,q}$ factors through $X_{n,q}$. By using Hirzebruch's method we will construct an appropriate resolution of $Y_{n,q}$.

Take $b_1, \cdots, b_r$ such that $\frac{n}{q} = b_1 - 1\rfloor\overline{b_2} - \cdots - 1\rfloor\overline{b_r}$ and construct a non-singular variety $M = M(b_1, \cdots, b_r)$ by patching $U_i \cong \mathbb{C}^2$ $(i = 0, \cdots, r)$ together as follows, where we let (u_i, v_i) be a coordinate system of U_i:

$$U_0 \cap U_1 = \{u_0 \neq 0\}, \quad u_1 = \frac{1}{u_0} \quad v_1 = u_0^{b_1} v_0$$

$$U_1 \cap U_2 = \{v_1 \neq 0\}, \quad v_2 = \frac{1}{v_1} \quad u_2 = v_1^{b_2} u_1$$

$$U_2 \cap U_3 = \{u_2 \neq 0\}, \quad u_3 = \frac{1}{u_2} \quad v_3 = u_2^{b_3} v_2$$

$$\vdots$$

$$U_{2i} \cap U_{2i+1} = \{u_{2i} \neq 0\}, \quad u_{2i+1} = \frac{1}{u_{2i}} \quad v_{2i+1} = u_{2i}^{b_{2i+1}} v_{2i}$$

$$U_{2i+1} \cap U_{2i+2} = \{v_{2i+1} \neq 0\}, \quad v_{2i+2} = \frac{1}{v_{2i+1}} \quad u_{2i+2} = v_{2i+1}^{b_{2i+2}} u_{2i+1}$$

$$\vdots$$

Then, the closed subset $E = \{v_0 = v_1 = 0\} \cup \{u_1 = u_2 = 0\} \cup \{v_2 = v_3 = 0\} \cup \cdots$ is a complete closed subvariety isomorphic to $\mathbb{P}^1 \cup \cdots \cup \mathbb{P}^1$. By looking at the images of an open subset $(U_0 \cap U_1) \setminus E = \{u_0 v_0 \neq 0\}$ by the isomorphisms of patching, we obtain that if r is odd, then $M = \{u_0 v_0 \neq 0\} \cup E \cup \{u_0 = 0\} \cup \{v_r = 0\}$. On the other hand, if r is even, then $M = \{u_0 v_0 \neq 0\} \cup E \cup \{u_0 = 0\} \cup \{u_r = 0\}$. Let us assume that r is odd. For the case where r is even, it is sufficient to replace v_r by u_r in the following discussion.

Now we can see that $u_0 v_0$, v_0, v_r are regular functions on M. Indeed it is sufficient to show that these are written as $u_i^a v_i^b$ $(a, b \geq 0)$ on each U_i. Let us try on v_r:

$$\begin{aligned} v_r &= {u_{r-1}}^{b_r} v_{r-1} = {u_{r-2}}^{b_r} {v_{r-2}}^{b_{r-1}b_r - 1} \\ &= {u_{r-3}}^{b_{r-2}(b_{r-1}b_r - 1) - b_r} {v_{r-3}}^{b_{r-1}b_r - 1} \\ &= {u_{r-4}}^{b_{r-2}(b_{r-1}b_r - 1) - b_r} {v_{r-4}}^{b_{r-3}(b_{r-2}(b_{r-1}b_r - 1) - b_r) - (b_{r-1}b_r - 1)} \\ &= \cdots \end{aligned}$$

In this way we can see that on each U_i the indexes of u_i, v_i are non-negative. Moreover, the index of u and the index of v are prime to each other. Writing (index of u_i)/(index of v_i) if i is even, and writing (index of v_i)/(index of u_i) if i is odd, we obtain the following according to the order $i = r-1, r-2, r-3, \ldots$:

$$b_r, \; b_{r-1} - 1\rfloor b_r, \quad b_{r-2} - 1\rfloor b_{r-1} - 1\rfloor b_r, \cdots$$

Therefore let $v_r = u_0^\alpha v_0^\beta$ on U_0, then

$$\alpha/\beta = b_1 - 1\rfloor b_2 - \cdots - 1\rfloor b_r.$$

Also by $(\alpha, \beta) = 1$ we obtain that $\alpha = n$, $\beta = q$.

Now we obtain regular functions v_0, $u_0 v_0$, $v_r = u_0^n v_0^q$ on M and we can define a morphism $\Phi : M \longrightarrow \mathbb{C}^3$ by $(u_i, v_i) \longmapsto (v_0, u_0 v_0, u_0^n v_0^q)$. Then $\operatorname{Im} \Phi = Y_{n,q}$. Obviously Φ gives an isomorphism between $\{u_0 v_0 \neq 0\}$ and $\{(x_1, x_2, x_3) \in Y_{n,q} \mid x_1 x_3 x_3 \neq 0\}$. We can also see that on $Y_{n,q} \setminus \{0\}$ the morphism Φ is a finite morphism. By Theorem 4.1.11 (iv) the morphism Φ factors

$$M \xrightarrow{\phi} X_{n,q} \xrightarrow{\pi} Y_{n,q}$$

through the normalization $X_{n,q}$. As ϕ is isomorphic outside the singular point, it is a resolution of the singularity $X_{n,q}$. The exceptional divisor of ϕ is $E = \mathbb{P}^1 \cup \cdots \cup \mathbb{P}^1$ (r prime divisors with normal crossings) and the weighted dual graph is

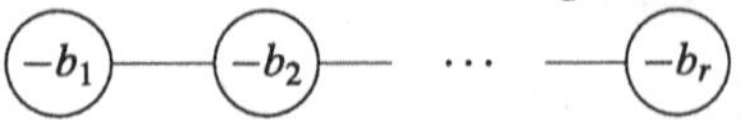

Indeed, let E_1 be the closed subset $\{v_0 = v_1 = 0\}$ in $U_0 \cup U_1$, then $E_1 \cong \mathbb{P}^1$. As we have $\mathscr{O}_M(E_1)|_{U_0} = v_0^{-1}\mathscr{O}_{U_0}$, $\mathscr{O}_M(E_1)|_{U_1} = v_1^{-1}\mathscr{O}_{U_1}$, $v_1 = u_0^{b_1} v_0$, it follows that

$$\mathscr{O}_M(E_1) \otimes \mathscr{O}_{E_1} \cong \mathscr{O}_{\mathbb{P}^1}(-b_1).$$

We can check the self-intersection number of other prime divisors in a similar way. Moreover, by $b_i \geq 2$ for every i, the resolution ϕ is the minimal resolution.

Next we show characterizations of cyclic quotient singularities.

Theorem 7.4.17. *For a 2-dimensional normal singularity (X, x), the following are equivalent:*

(i) (X, x) is a cyclic quotient singularity.
(ii) (X, x) is a singularity on a toric variety.
(iii) Every irreducible component of the exceptional divisor on the minimal resolution is $\mathbb{P}^1$ and the dual graph of the exceptional divisor on the minimal resolution without weights is in the form of a chain, i.e.,

Proof. (i) $\Rightarrow$ (ii). Note that the affine coordinate ring of $X_{n,q}$ is $\mathbb{C}[u^i v^j]$, $i + qj \equiv 0 \pmod{n}$, $0 \leq i \leq n$, $0 \leq j \leq n$. Here, define $S := \{(i, j) \in \mathbb{Z}^2 \mid i + qj \equiv 0 \pmod{n}$ $i \geq 0, j \geq 0\}$, $M := \{(i, j) \in \mathbb{Z}^2 \mid i + qj \equiv 0 \pmod{n}\}$, then we have $M \cong \mathbb{Z}^2$ and M is generated as a group by the saturated subgroup S. As the affine coordinate ring of $X_{n,q}$ is $\mathbb{C}[S]$, by Theorem 4.4.12, $X_{n,q}$ is a toric variety.

(ii) $\Rightarrow$ (iii). A 2-dimensional affine toric variety is defined by a cone σ in $N_{\mathbb{R}} \cong \mathbb{R}^2$. If the dimension of σ is 0 or 1, then $X_{n,q}$ is $(\mathbb{A}^1_{\mathbb{C}} - \{0\})^2$ or $(A^1_{\mathbb{C}} - \{0\}) \times \mathbb{A}^1_{\mathbb{C}}$ and both non-singular. So it is sufficient to consider the case $\dim \sigma = 2$.

A 2-dimensional cone is as in Fig. 7.1 and by subdividing this appropriately we obtain a resolution of the singularities (Fig. 7.2). By Fig. 7.2 and Theorem 4.4.12 (iii) the exceptional divisor is as follows:

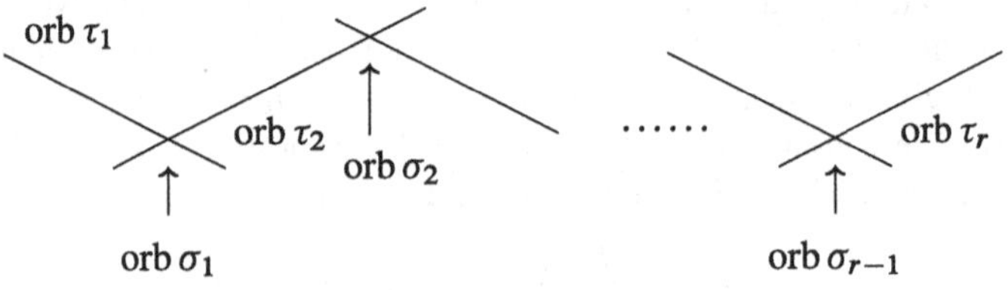

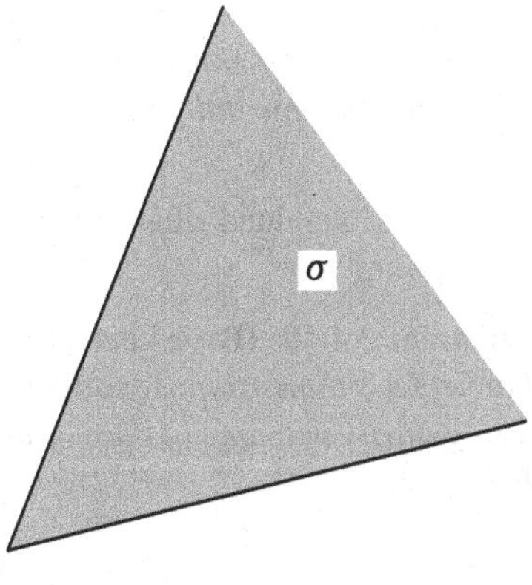

Fig. 7.1 Original cone

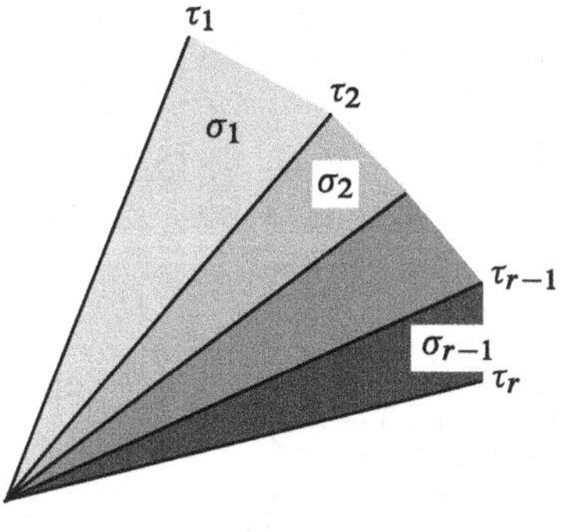

Fig. 7.2 Subdivision

i.e., the dual graph is in the form of a chain. By contracting (-1)-curves we obtain the minimal resolution whose exceptional divisor is a chain. It is also clear that every prime exceptional divisor is $\mathbb{P}^1$.

(iii) $\Rightarrow$ (i). Given a weighted dual graph with the form of a chain

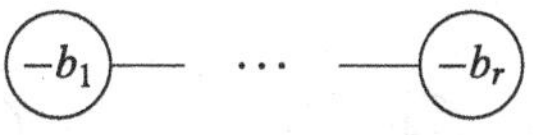

$b_i \geq 2$, let $n/q = b_1 - 1 \lfloor b_2 - \cdots - 1 \lfloor b_r$, $(n, q) = 1$, then by Theorem 7.4.16 the weighted dual graph of the minimal resolution of $X_{n,q}$ coincides with this graph. By [La1], a graph of chain form is *taut* (the singularity whose minimal resolution has the exceptional divisor consisting of $\mathbb{P}^1$ with the form of the dual graph is analytically unique). Therefore, the singularity corresponding to the chain is a cyclic quotient singularity (actually, the equivalence between (i) and (iii) were proved by Brieskorn [Br] before Laufer's result [La1]).

Then, how about characterizations of 2-dimensional quotient singularities? The following is Brieskorn's characterization:

Theorem 7.4.18 (Brieskorn [Br, Satz 2.8]). *For a 2-dimensional normal singularity (X, x), the following are equivalent:*

(i) (X, x) *is a quotient singularity.*
(ii) *There exist a non-singular* (Y, y) *and a finite morphism* $(Y, y) \to (X, x)$.
(iii) *The fundamental group* $\pi_{X,x} := \varprojlim_U \pi_1(U - \{x\})$ *is finite.*

For the weighted dual graph of a quotient singularity, we have the following description:

Theorem 7.4.19 (Brieskorn [Br, Satz 2.10]). *The dual graph of the minimal resolution of a 2-dimensional quotient singularity is one of the following. On the contrary, a singularity with one of these graph is unique and it is a quotient singularity. Here,* $b, b_i \geq 2$*:*

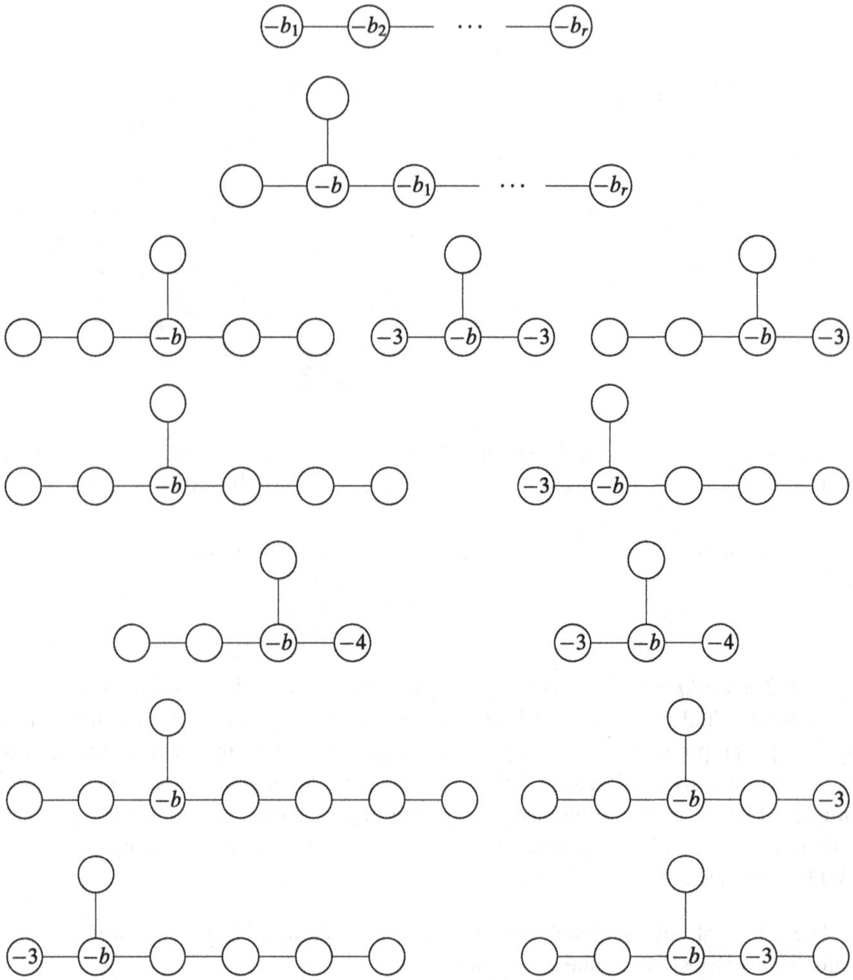

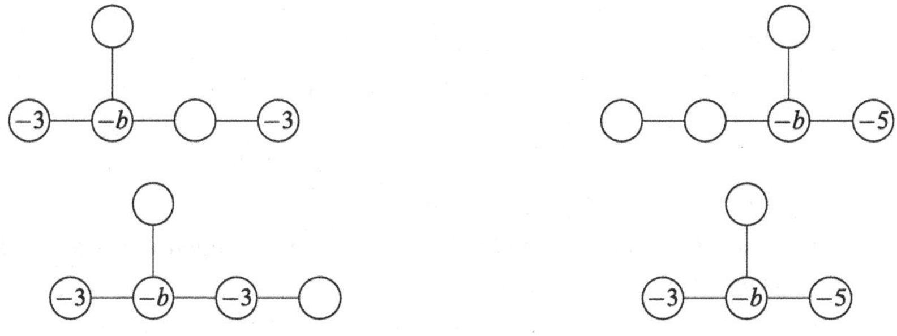

7.5 Rational Double Point

A singular point (X, x) with $\mathrm{mult}(X, x) = 2$ (cf. Corollary 7.3.5) is called a *double point*. In particular, a 2-dimensional rational singular point is called a *Du Val singularity*, Klein singularity, or simple singularity. Varieties of the names may let you imagine varieties of characterization of these singularities. In this section, we study characterizations of rational double points. Some parts of the following can be seen in [Du]. The paper also contains some characterizations which are not contained in the following:

Theorem 7.5.1. *For a 2-dimensional normal singularity (X, x), the following are equivalent:*

(i) (X, x) *is rational double.*
(ii) (X, x) *is a rational hypersurface singularity.*
(iii) (X, x) *is a rational Gorenstein singularity.*
(iv) (X, x) *is a canonical singularity.*
(v) $\kappa_\gamma(X, x) = -\infty$.
(vi) Let $f : Y \to X$ *be the minimal resolution, then* $K_Y = f^*K_X$ *holds.*
(vii) Let $f : Y \to X$ *be the minimal resolution, then every exceptional prime divisor* E_i *satisfies* $E_i^2 = -2$ *and* $E_i \cong \mathbb{P}^1$.
(viii) The exceptional divisor of the minimal resolution of (X, x) *consists of* $\mathbb{P}^1$ *with normal crossings. The weighted dual graph is one of the following (According to the rule, we do not write the weight if* $E_i^2 = -2$ *(Definition 7.1.19)):*

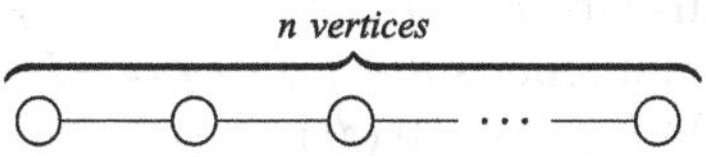

A_n $(n \geq 1)$

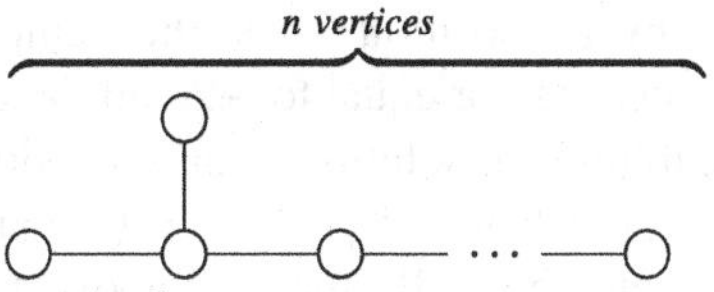

D_n $(n \geq 4)$

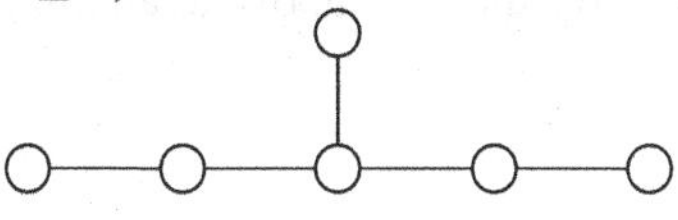

E_6

E_7 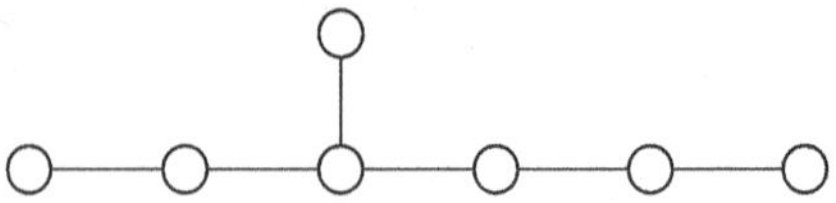

E_8 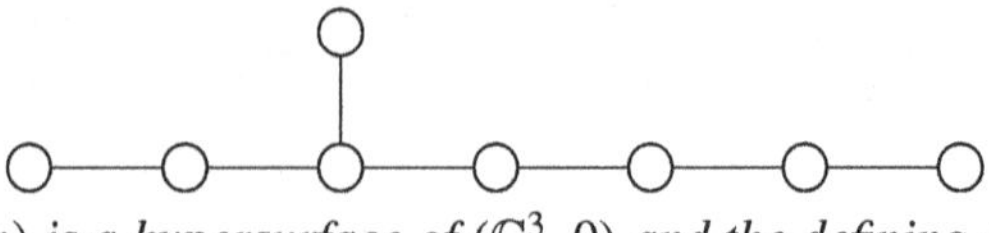

(ix) *(X, x) is a hypersurface of $(\mathbb{C}^3, 0)$ and the defining equation is one of the following:*

$$\begin{array}{ll} A_n\ (n \geq 1) & x^2 + y^2 + z^{n+1} = 0. \\ D_n\ (n \geq 4) & x^2 + y^2 z + z^{n-1} = 0. \\ E_6 & x^2 + y^3 + z^4 = 0. \\ E_7 & x^2 + y^3 + yz^3 = 0. \\ E_8 & x^2 + y^3 + z^5 = 0. \end{array}$$

(x) *(X, x) is a quotient singularity and a Gorenstein singularity.*
(xi) *For some finite subgroup $G \subset SL(2, \mathbb{C})$, $(X, x) \simeq (\mathbb{C}^2/G, 0)$.*

Proof. (i) $\Rightarrow$ (ii). As (X, x) is a rational double point, by Corollary 7.3.5 (i), we have $-Z_f^2 = 2$, therefore by (ii) of the corollary, it follows that $\mathrm{emb}(X, x) = 2 + 1$.

(ii) $\Rightarrow$ (iii). In general, a hypersurface singularity is a Gorenstein singularity. Indeed, for a hypersurface $X \subset \mathbb{C}^{n+1}$, as $\mathbb{C}^{n+1}$ is a Gorenstein variety, by Proposition 5.3.12 it follows that X is also a Gorenstein variety.

(iii) $\Rightarrow$ (iv). It is obvious by Corollary 6.2.15.

(iv) $\Leftrightarrow$ (v). The implication $\Rightarrow$ follows from Proposition 6.3.12. For the implication $\Leftarrow$, we should note that $\gamma_1(X, x) = p_g(X, x) = 0$, which yields that (X, x) is rational. Therefore by Theorem 7.3.2 it turns out to be a $\mathbb{Q}$-Gorenstein singularity and we can apply Proposition 6.3.12.

(iv) $\Leftrightarrow$ (vi). The implication $\Rightarrow$ follows from the definition of a canonical singularity and Proposition 7.1.17. For the implication $\Leftarrow$, it is sufficient to prove that the singularity (X, x) is rational, indeed a rational singularity is a $\mathbb{Q}$-Gorenstein singularity and condition (vi) completes the condition of a canonical singularity. As a 2-dimensional normal singularity is Cohen–Macaulay, by Theorem 6.2.14 it is sufficient to prove the equality $f_*\omega_Y = \omega_X$. Assume $\subsetneq$, then there is a 2-form θ on Y such that $\nu_{E_i}(\theta) < 0$. Then there is a divisor D without exceptional divisor as a component, two divisors $Z_1 \geq 0$, $Z_2 > 0$ without common components and we have the supports on $f^{-1}(x)$ satisfying $\omega_Y(D + Z_1 - Z_2) \cong \mathcal{O}_Y$. Therefore we obtain $K_Y \sim -D - Z_1 + Z_2$. Here, it follows that $K_Y \cdot Z_2 = -D \cdot Z_2 - Z_1 \cdot Z_2 + Z_2^2 < 0$, which is a contradiction to $K_Y \cdot E_i = 0$ $(\forall i)$.

(vi) $\Leftrightarrow$ (vii). Note that by Proposition 7.2.8, the number $E_i^2 + K_Y \cdot E_i$ is in general an even number greater than or equal to -2 and the equality $= -2$ holds if and only if $E_i \cong \mathbb{P}^1$. On the minimal resolution Y, an exceptional divisor E_i satisfies $E_i^2 \leq -2$. Condition (vi) is equivalent to $K_Y \cdot E_i = 0$ $(\forall i)$ and this is equivalent to $E_i^2 + K_Y \cdot E_i = -2$ by the note above. By the note above again, this is equivalent to $E_i^2 = -2$ and $E_i \cong \mathbb{P}^1$.

(vii) $\Leftrightarrow$ (viii). The implication $\Leftarrow$ is obvious. For the proof of $\Rightarrow$ it is sufficient to prove that any graph not in (viii) is not negative definite. We should note that a subgraph of negative definite graph is also negative definite. First, if there are components E_1, E_2 that intersect non-normally, then $(E_1 + E_2)^2 \geq 0$, which is a contradiction to the negative definiteness. Therefore we have only to consider the exceptional divisor with normal crossings. If there is no prime divisor in $f^{-1}(x)_{\mathrm{red}} = \sum_{i=1}^{r} E_i$ that intersects three other components, then the graph becomes either

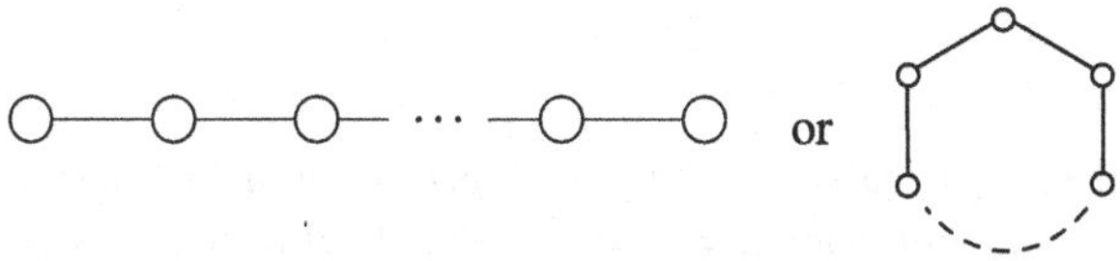

Here, the right one satisfies $\left(\sum_{i=1}^{r} E_i\right)^2 = 0$, which is a contradiction to the negative definiteness. Next, assume that there is a prime divisor intersecting more than four components. Then the graph contains the following subgraph:

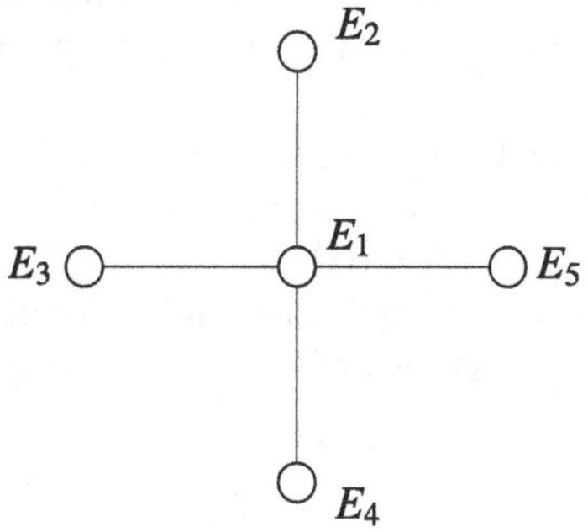

This graph satisfies $\left(2E_1 + \sum_{i=2}^{5} E_i\right)^2 = 0$, which is also a contradiction to the negative definiteness. If there is a prime divisor intersecting exactly three other components, then the number of such prime divisors is only one. Actually, if there are more than two such prime divisors, the graph contains the following subgraph:

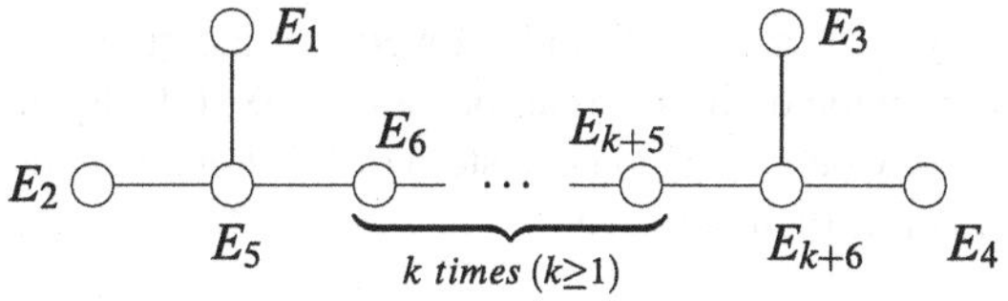

This subgraph satisfies $\left(\sum_{i=1}^{4} E_i + 2\sum_{i=5}^{k+6} E_i\right)^2 = 0$, a contradiction to the negative definiteness. In the same way, we can check that the following graphs are also non-negative definite:

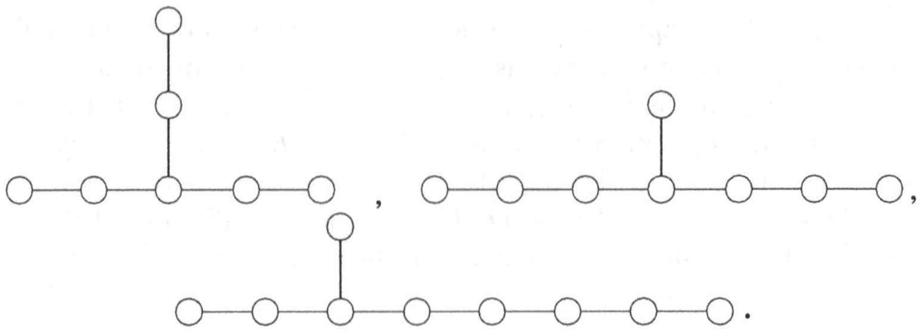

As a consequence, the negative definite graph is only in the list in (viii).

(viii) $\Leftrightarrow$ (ix). The implication $\Leftarrow$ is proved by checking each singularity defined by the equations in (ix). Actually, each singularity in (ix) can be resolved by successive blow-ups at points and the minimal resolution is obtained by this way. The weighted dual graph of the exceptional divisor turns out to be one of the list in (viii). Conversely, by Laufer [La1], the graphs in (viii) are all taut (the singularity corresponding to the graph is analytically unique). Therefore the equation defining the singularity is given in the list in (ix). (Historically, the proof by Brieskorn [Br] was earlier.)

(ix) $\Rightarrow$ (i). By the form of equations in (ix) the singularities are double points and by the equivalence (ix) $\Leftrightarrow$ (iv) the singularities are rational. This completes the proof of the equivalence among (i) $\sim$ (ix).

(x) $\Rightarrow$ (iii). A quotient singularity is rational by Corollary 7.4.10.

(ix) $\Leftrightarrow$ (xi). The finite subgroups of $SL(2, \mathbb{C})$ are classified as follows, up to conjugate:

$C_n \ (n \geq 1)$	:	cyclic group of order n,
$D_n \ (n \geq 4)$	:	binary dihedral group of order $4n$,
T	:	binary tetrahedral group of order 24,
O	:	binary octahedral group of order 48,
I	:	binary icosahedral group of order 120.

Klein [Kl] let these groups act on $\mathbb{C}^2$ and showed that the quotients become hypersurfaces in $\mathbb{C}^3$ and then the defining equations are in (ix) (cf. [Sp]).

(xi) $\Rightarrow$ (x). Let x, y be a coordinate system of $\mathbb{C}^2$. Let an element $g \in SL(2, \mathbb{C})$ act on the form $dx \wedge dy$, then we have

$$g(dx \wedge dy) = dgx \wedge dgy = \det g \, dx \wedge dy = dx \wedge dy,$$

therefore the form $dx \wedge dy$ does not have zeros on $\mathbb{C}^2/G$, it becomes a free generator of $\omega_{\mathbb{C}^2/G}$. By this fact and also by the fact that 2-dimensional normal singularity is Cohen–Macaulay, $\mathbb{C}^2/G$ has Gorenstein singularities.

This completes the proof of the equivalence (i) $\sim$ (xi). (The equivalence (x) $\Leftrightarrow$ (xi) can also be proved directly from [WKi].)

7.6 Elliptic Singularities

A singularity located next to a rational singularity is an elliptic singularity. In this section we will study this singularity.

Definition 7.6.1. A 2-dimensional normal singularity (X, x) is called an *elliptic singularity* if the equality $p_g(X, x) = 1$ holds.

The name "elliptic" singularity makes people imagine an elliptic curve. Does this singularity really correspond to an elliptic curve? Next we define the singularities that really correspond to elliptic curves.

Definition 7.6.2. A 2-dimensional normal singularity (X, x) is called a *simple elliptic singularity* if the exceptional curve E of the minimal resolution $f : Y \to X$ is an irreducible non-singular elliptic curve (i.e., a non-singular curve E is called an *elliptic curve* if $H^1(E, \mathcal{O}_E) = \mathbb{C}$, which is equivalent to $K_E \sim 0$).

Proposition 7.6.3. *A simple elliptic singularity (X, x) is an elliptic singularity. Moreover, it is a Gorenstein singularity and $\kappa_\delta(X, x) = 0$.*

Proof. Let $f : Y \to X$ be the minimal resolution and E the exceptional divisor, then by the exact sequence:

$$\begin{array}{ccccc} R^1 f_* \mathcal{O}_Y & \longrightarrow & H^1(E, \mathcal{O}_E) & \longrightarrow & R^2 f_* \mathcal{O}_Y(-E) = 0 \\ & & \| & & \\ & & \mathbb{C} & & \end{array}$$

we have $p_g(X, x) \geq 1$. Now if $\nu_E(\theta) \geq 0$ holds for every element $\theta \in \omega_X$, then by Theorem 6.2.14 the singularity (X, x) becomes rational, which is a contradiction. Therefore there is an element $\theta \in \omega_X$ such that $\nu_E(\theta) \leq -1$. By this θ we have a linear equivalence $K_Y \sim D + \nu_E(\theta)E$, where D is an effective divisor without E as a component.

By the adjunction formula, we have $D \cdot E + (\nu_E(\theta) + 1)E^2 = 0$. Here, by $D \cdot E \geq 0$ and $(\nu_E(\theta) + 1)E^2 \geq 0$, the both equalities hold and we have $D \cap E = \emptyset$ and $\nu_E(\theta) = -1$. By the former equality we have that (X, x) is a Gorenstein singularity and by the latter equality we have $K_Y = f^* K_X - E$, which implies that (X, x) is log canonical and not log terminal. Hence by Proposition 6.3.12 (iii) it follows that $\kappa_\delta(X, x) = 0$ and, moreover, $\delta_m(X, x) = 1$ for every $m \in \mathbb{N}$. In particular, $\delta_1(X, x) = p_g(X, x) = 1$.

Theorem 7.6.4 (Saito [Sa, 1.9]). *Let (X, x) be a simple elliptic singularity. If $E^2 = -3, -2, -1$, we call the singularity $\widetilde{E}_6$-type $\widetilde{E}_7$-type, $\widetilde{E}_8$-type, respectively. These are all hypersurfaces and the defining equations are, after appropriate coordinate transformations, as follows (here $\lambda \in \mathbb{C}$, $\lambda \neq 0, 1$):*

$$\widetilde{E}_6;\ y(y - x)(y - \lambda x) - xz^2 = 0,$$
$$\widetilde{E}_7;\ yx(y - x)(y - \lambda x) - z^2 = 0,$$
$$\widetilde{E}_8;\ y(y - x^2)(y - \lambda x^2) - z^2 = 0.$$

Conversely, a hypersurface simple elliptic singularity is one of them.

Simple elliptic singularities are typical elliptic singularities, but these are not all elliptic singularities. The following is another group of elliptic singularities. These are minimal elliptic singularities introduced by Laufer.

Definition 7.6.5. Let $f : Y \to X$ be the minimal resolution of a 2-dimensional normal singularity (X, x), $E = \sum_{i=1}^{r} E_i$ the exceptional divisor and Z_f the fundamental cycle. We call (X, x) a *minimally elliptic singularity* if $p(Z_f) = 1$ and every connected proper subdivisor of E is an exceptional divisor of a rational singularity.

Actually, this singularity becomes an elliptic singularity and it has the following characterization.

Theorem 7.6.6 (Laufer [La2]). *For the minimal resolution $f : Y \to X$ of a 2-dimensional normal singularity (X, x) and the fundamental cycle Z_f, the following are equivalent:*

(i) (X, x) is a minimal elliptic singularity.
(ii) The equality $E_i \cdot Z_f = -E_i \cdot K_Y$ holds for every exceptional prime divisor E_i.
(iii) $K_Y \sim -Z_f$.
(iv) (X, x) is a Gorenstein singularity and $p_g(X, x) = 1$ holds.

Example 7.6.7. A simple elliptic singularity is a minimal elliptic singularity.

Let us see another typical elliptic singularity.

Definition 7.6.8. If the exceptional divisor $E = \sum_{i=1}^{r} E_i$ of the minimal resolution $f : Y \to X$ of a 2-dimensional normal singularity (X, x) satisfies the following, then we call (X, x) a *cusp singularity*.

The total exceptional divisor E is an irreducible rational curve with an ordinary node or the equalities $E_i \cong \mathbb{P}^1$ $(\forall i = 1, \ldots, r)$ hold and E is of normal crossings with the dual graph as the following cyclic form (ignoring the weight):

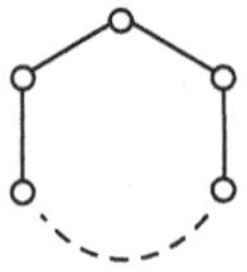

Proposition 7.6.9. *A cusp singularity (X, x) is a minimal elliptic singularity and the equality $\kappa_\delta(X, x) = 0$ holds.*

Proof. As it is clear that the coefficient of Z_f at each irreducible component is 1, the equality $p(Z_f) = 1$ follows. As every proper connected subdivisor is of a chain form, by Theorem 7.4.19 it becomes the exceptional divisor of a quotient singularity.

Therefore (X, x) is minimal elliptic. By Proposition 7.6.6 (iii) it follows that $K_Y = f^*K_X - Z_f$, then the singularity (X, x) is log canonical and not log terminal. Hence by Proposition 6.3.12 (iii) we obtain $\kappa_\delta(X, x) = 0$.

Actually, a Gorenstein singularity with $\kappa_\delta(X, x) = 0$ is either simple elliptic or cusp, which will be proved in a forthcoming section (7.8).

7.7 2-Dimensional Du Bois Singularities

In this section we introduce a result in [I3] about 2-dimensional Du Bois singularities.

A Du Bois singularity is introduced by Steenbrink and the definition will be given in the next chapter. A 2-dimensional normal singularity (X, x) is a Du Bois singularity if and only if for a good resolution of the singularity $f : Y \to X$ with the reduced total exceptional divisor $E = f^{-1}(x)_{\mathrm{red}}$ (red means to take the reduced structure), the canonical map $R^1 f_*\mathcal{O}_Y \to H^1(E, \mathcal{O}_E)$ is isomorphic (Proposition 8.1.13). In this section we regard this as the definition of a Du Bois singularity. First, by the surjectivity of $R^1 f_*\mathcal{O}_Y \twoheadrightarrow H^1(E, \mathcal{O}_E)$ for a general 2-dimensional normal singularity, we obtain the following:

Proposition 7.7.1. *A rational singularity is a Du Bois singularity.*

Definition 7.7.2. For a resolution $f : Y \to X$ of a 2-dimensional normal singularity (X, x) a 2-form $\theta \in \Gamma(X, \omega_X) = \Gamma(X \setminus \{x\}, \omega_{X\setminus\{x\}})$ is called of *general type* if for each exceptional prime divisor E_i with respect to f, the equality $\nu_{E_i}(\theta) = \min\{\nu_{E_i}(\eta) \mid \eta \in \Gamma(X, \omega_X)\}$ holds.

If $\nu_{E_i}(\theta) \neq \nu_{E_i}(\theta')$, then $\nu_{E_i}(\theta + \theta') = \min\{\nu_{E_i}(\theta), \nu_{E_i}(\theta')\}$ holds, on the other hand, if $\nu_{E_i}(\theta) = \nu_{E_i}(\theta')$, then for general ε we have $\nu_{E_i}(\theta + \varepsilon\theta') = \nu_{E_i}(\theta)$. By this observation we can see that the set of general 2-forms is dense in the vector space $\Gamma(X, \omega_X)$.

The following is another characterization of a 2-dimensional normal Du Bois singularity.

Theorem 7.7.3. *Let (X, x) be a 2-dimensional normal singularity, $f : Y \to X$ a good resolution of the singularity and E the reduced total exceptional divisor $f^{-1}(x)_{\mathrm{red}}$. Then (X, x) is a Du Bois singularity if and only if a general 2-form has poles on E at most of order 1. This is also equivalent to $\omega_X = f_*\omega_Y(E)$.*

Proof. The condition being a Du Bois singularity is equivalent to $R^1 f_*\mathcal{O}_Y \simeq H^1(E, \mathcal{O}_E)$. By Corollary 3.5.15 we have that $R^1 f_*\mathcal{O}_Y$ is dual to $H^1_E(Y, \omega_Y)$. On the other hand, by the exact sequence

$$
\begin{array}{l}
0 \longrightarrow \Gamma_E(Y, \omega_Y) \longrightarrow \Gamma(Y, \omega_Y) \longrightarrow \Gamma(Y \setminus E, \omega_Y) \longrightarrow H^1_E(Y, \omega_Y) \\
\qquad\qquad\quad \| \\
\qquad\qquad\quad 0 \\
\longrightarrow H^1(Y, \omega_Y) = 0
\end{array}
$$

it is dual to $\Gamma(Y \setminus E, \omega_Y)/\Gamma(Y, \omega_Y)$. Here, the vanishing $H^1(Y, \omega_Y) = 0$ follows from Theorem 6.1.12.

On the other hand, $H^1(E, \mathcal{O}_E)$ is dual to $\Gamma(E, \omega_E)$, but by the exact sequence:

$$
0 \longrightarrow \Gamma(Y, \omega_Y) \longrightarrow \Gamma\big(Y, \omega_Y(E)\big) \longrightarrow \Gamma(E, \omega_E) \longrightarrow H^1(Y, \omega_Y) = 0
$$

it is equal to $\Gamma(Y, \omega_Y(E))/\Gamma(Y, \omega_Y)$. Therefore the condition being a Du Bois singularity is equivalent to $\Gamma(Y, \omega_Y(E)) = \Gamma(Y \setminus E, \omega_Y)$.

The next Proposition shows the relation between being a Du Bois singularity and plurigenera.

Proposition 7.7.4. *If a 2-dimensional normal singularity (X, x) satisfies $\delta_m(X, x) \leq 1$ ($\forall m \in \mathbb{N}$), then (X, x) is a Du Bois singularity. In particular, if (X, x) is a Gorenstein singularity, the converse also holds.*

Proof. Assume a singularity (X, x) is not Du Bois. Then for a good resolution $f : Y \to X$ of (X, x), there is a 2-form $\theta \in \omega_X$ with the pole of order ≥ 2 on an exceptional prime divisor E_i. Take an element $h \in \mathfrak{m}_x$ and let $\nu_{E_i}(h) = n > 0$, then $\nu_{E_i}(h\theta^{m-1}) \leq n - 2(m-1)$. Then if m is sufficiently big, we have $n - 2(m-1) < -(m-1)$, therefore we obtain

$$0 \neq h\theta^{m-1} \in \omega_X^{[m]}/f_*\omega_Y^{\otimes m}((m-1)E).$$

On the other hand, θ^{m-1} is not zero in this quotient module. Now we have two elements θ^{m-1} and $h\theta^{m-1}$ whose orders of poles on E_i are mutually different, which means that the two elements are linearly independent over $\mathbb{C}$. This completes the proof that $\delta_m(X, x) \geq 2$. The last statement follows from Theorem 8.1.17.

Corollary 7.7.5. *A log canonical singularity is a Du Bois singularity. In particular, a simple elliptic singularity and a cusp singularity are Du Bois singularities.*

Let us study a resolution of a Du Bois singularity. First we prepare the following lemma:

Lemma 7.7.6. *Let $f : Y \to X$ be a good resolution of a 2-dimensional normal singularity (X, x) and E the reduced total exceptional divisor $f^{-1}(x)_{\text{red}}$. Let (X', x') be the normal singularity obtained by contracting a connected subdivisor E' of E and let $g : Y \to X'$ be the contraction morphism. Let $E^* := E - E'$ and denote the reduced divisor $g(E^*)_{\text{red}}$ by $\overline{E}^*$. Let $p := \#\{P \in E^* \mid g(P) = x'\}$, then we have*

$$h^1(E, \mathcal{O}_E) \leq h^1(E', \mathcal{O}_{E'}) + h^1(\overline{E}^*, \mathcal{O}_{\overline{E}^*}). \tag{7.4}$$

Here, the equality holds if and only if $h^0(\overline{E}^*, g_*\mathcal{O}_{E^*}/\mathcal{O}_{\overline{E}^*}) = p - 1$.

Proof. As E is of normal crossings, there is the following exact sequence:

$$0 \longrightarrow \mathcal{O}_E \longrightarrow \mathcal{O}_{E'} \oplus \mathcal{O}_{E^*} \longrightarrow \mathcal{O}_{E' \cap E^*} \longrightarrow 0.$$

By this we obtain the following exact sequence:

$$\begin{aligned} 0 \longrightarrow \Gamma(E, \mathcal{O}_E) \longrightarrow \Gamma(E', \mathcal{O}_{E'}) \oplus \Gamma(E^*, \mathcal{O}_{E^*}) &\longrightarrow \Gamma(E' \cap E^*, \mathcal{O}_{E' \cap E^*}) \\ \longrightarrow H^1(E, \mathcal{O}_E) \longrightarrow H^1(E', \mathcal{O}_{E'}) \oplus H^1(E^*, \mathcal{O}_{E^*}) &\longrightarrow 0. \end{aligned}$$

Let q be the number of connected components of E^*, then we obtain

$$h^1(E, \mathcal{O}_E) = h^1(E', \mathcal{O}_{E'}) + h^1(E^*, \mathcal{O}_{E^*}) + p - q. \tag{7.5}$$

Next consider the restriction morphism $g' := g|_{E^*} : E^* \to \overline{E}^*$, then by the exact sequence

$$0 \longrightarrow \mathcal{O}_{\overline{E}^*} \longrightarrow g'_*\mathcal{O}_{E^*} \longrightarrow g'_*\mathcal{O}_{E^*}/\mathcal{O}_{\overline{E}^*} \longrightarrow 0$$

we obtain the following exact sequence:

$$\begin{aligned} 0 \longrightarrow \Gamma(\overline{E}^*, \mathcal{O}_{\overline{E}^*}) &\longrightarrow \Gamma(E^*, \mathcal{O}_{E^*}) \longrightarrow \Gamma(\overline{E}^*, g'_*\mathcal{O}_{E^*}/\mathcal{O}_{\overline{E}^*}) \\ &\longrightarrow H^1(\overline{E}^*, \mathcal{O}_{\overline{E}^*}) \longrightarrow H^1(E^*, \mathcal{O}_{E^*}) \longrightarrow 0. \end{aligned}$$

Here, as

$$h^0(\overline{E}^*, g'_*\mathcal{O}_{E^*}/\mathcal{O}_{\overline{E}^*}) \geq p - 1, \tag{7.6}$$

we obtain

$$h^1(E^*, \mathcal{O}_{E^*}) + p - q \leq h^1(\overline{E}^*, \mathcal{O}_{\overline{E}^*}). \tag{7.7}$$

By (7.5) and (7.7), it follows that

$$h^1(E, \mathcal{O}_E) \leq h^1(E', \mathcal{O}_{E'}) + h^1(\overline{E}^*, \mathcal{O}_{\overline{E}^*}),$$

where the equality holds if and only if the equality in (7.6) holds.

Theorem 7.7.7. *Let $f : Y \to X$ be a resolution of a normal 2-dimensional Du Bois singularity (X, x), E' a connected subdivisor of $E = f^{-1}(x)_{\mathrm{red}}$ and (X', x') a normal singularity obtained by contraction of E'. Then (X', x') is also a Du Bois singularity and the equality in (1) of Lemma 7.7.6 holds.*

Proof. Denote the canonical morphism $(X', x') \to (X, x)$ by h. By Leray's spectral sequence

$$E_2^{p,q} = R^p h_*(R^q g_*\mathcal{O}_Y) \Rightarrow E^{p+q} = R^{p+q} f_*\mathcal{O}_Y$$

we obtain the following exact sequence (Proposition 3.6.2):

$$0 \longrightarrow R^1 h_*\mathcal{O}_{X'} \longrightarrow R^1 f_*\mathcal{O}_Y \longrightarrow h_*(R^1 g_*\mathcal{O}_Y) \longrightarrow R^2 h_*\mathcal{O}_{X'} = 0.$$

Therefore it follows that $\dim R^1 f_*\mathcal{O}_Y = \dim R^1 h_*\mathcal{O}_{X'} + \dim h_* R^1 g_*\mathcal{O}_Y$. As (X, x) is a Du Bois singularity, the left-hand side $= h^1(E, \mathcal{O}_E)$. On the other hand, as X', Y are 2-dimensional, we have

$$\dim R^1 h_* \mathcal{O}_{X'} \geq h^1(\overline{E}^*, \mathcal{O}_{\overline{E}^*}), \tag{7.8}$$

$$\dim h_* R^1 g_* \mathcal{O}_Y \geq h^1(E', \mathcal{O}_{E'}), \tag{7.9}$$

then by Lemma 7.7.6 the equalities in (7.4) and (7.5) hold. In particular, the equality in (7.5) yields $R^1 g_* \mathcal{O}_Y \cong H^1(E', \mathcal{O}_{E'})$.

Theorem 7.7.8. *A resolution of a Du Bois singularity is a good resolution.*

Proof. Let $h : X' \to X$ be a resolution of a Du Bois singularity (X, x). Then, X' is obtained by contracting some exceptional curves of a good resolution $f : Y \to X$ of the singularity (X, x). Let $g : Y \to X'$ be the contraction morphism and $x'_1, \cdots, x'_r \in X'$ the contracted points. By Theorem 7.7.7, we have

$$h^0(\overline{E}^*, g_* \mathcal{O}_{E^*}/\mathcal{O}_{\overline{E}^*}) = \sum_{i=1}^{r} (p_i - 1),$$

where p_i is the number of points on E^* corresponding to x'_i.

This holds only if the divisor $\overline{E}^*$ has normal crossings at each x'_i. Indeed, denote the multiplicity of $\overline{E}^*$ at each x'_i by m_i, then as g is the composite of some blow-ups at points and by Lemma 7.7.9, we obtain

$$\dim(g_* \mathcal{O}_{E^*}/\mathcal{O}_{\overline{E}^*})_{x'_i} \geq \frac{1}{2} m_i(m_i - 1) \geq \frac{1}{2} p_i(p_i - 1) \geq p_i - 1,$$

where the first equality holds if and only if by the blow-up at x'_i the singularity of $\overline{E}^*$ is resolved.

The second equality holds if and only if x'_i is an ordinary singularity at $\overline{E}^*$ (i.e., the analytic neighborhood of $\overline{E}^*$ consists of m_i non-singular branches).

The third equality holds if and only if $p_i = 2$ or 1. Hence all the equalities hold if and only if $\overline{E}^*$ is an ordinary node at x'_i, i.e., normal crossing double.

Lemma 7.7.9. *Let m be the multiplicity of a 1-dimensional singularity $(C, x) \subset (\mathbb{C}^2, 0)$ and $\pi : \overline{C} \to C$ the blow-up at x. Then*

$$\dim_{\mathbb{C}} \pi_* \mathcal{O}_{\overline{C}}/\mathcal{O}_C = \frac{1}{2} m(m - 1).$$

Proof. By adding a sufficiently high-degree term to the equation of C and then homogenizing it, we may regard C as an irreducible curve in $\mathbb{P}^2$ with only one singular point at 0. Let $\pi : S \to \mathbb{P}^2$ be the blow-up of $\mathbb{P}^2$ at 0 and $\overline{C}$ the strict transform of C on S. Then $\pi^* C = \overline{C} + mE$, where E is the exceptional curve with respect to π.

$$\begin{aligned} h^1(\overline{C}, \mathcal{O}_{\overline{C}}) &= 1 + \frac{\overline{C}^2 + K_S \cdot \overline{C}}{2} \\ &= 1 + \frac{(\pi^*C - mE)^2 + (\pi^*K_{\mathbb{P}^2} + E) \cdot (\pi^*C - mE)}{2} \\ &= 1 + \frac{1}{2}(C^2 + K_{\mathbb{P}^2} \cdot C) - \frac{1}{2}m(m-1) \\ h^1(C, \mathcal{O}_C) &= 1 + \frac{1}{2}(C^2 + K_{\mathbb{P}^2} \cdot C). \end{aligned}$$

By the following exact sequence:

$$\begin{array}{ccccccc} 0 \longrightarrow & \Gamma(C, \mathcal{O}_C) & \longrightarrow & \Gamma(\overline{C}, \mathcal{O}_{\overline{C}}) & \longrightarrow & \pi_*\mathcal{O}_{\overline{C}}/\mathcal{O}_C & \longrightarrow H^1(C, \mathcal{O}_C) \\ & \| & & \| & & & \\ & \mathbb{C} & & \mathbb{C} & & & \\ \longrightarrow & H^1(\overline{C}, \mathcal{O}_{\overline{C}}) & \longrightarrow & 0, & & & \end{array}$$

we obtain $\dim \pi_*\mathcal{O}_{\overline{C}}/\mathcal{O}_C = \frac{1}{2}m(m-1)$.

By the weight of the exceptional divisor we obtain a sufficient condition for a singularity to be a Du Bois singularity.

Theorem 7.7.10. *Let $f : Y \to X$ be a good resolution of a 2-dimensional normal singularity (X, x). Let E be the reduced total exceptional divisor $f^{-1}(x)_{\text{red}}$ and decompose it as $E = \sum_{i=1}^r E_i$. Assume for each i*

$$E_i^2 < -2\Big(E_i \sum_{j \neq i} E_j + \max\{g(E_i) - 1, 0\}\Big),$$

then (X, x) is a Du Bois singularity.

Proof. A general 2-form θ induces an isomorphism $\mathcal{O}_Y \xrightarrow{\sim} \omega_Y(-D - \sum \nu_{E_i}(\theta)E_i)$, therefore $K_Y \sim D + \sum \nu_{E_i}(\theta)E_i$. Here, D is an effective divisor without a component E_i $(i = 1 \cdots r)$. Let $a_i = \nu_{E_i}(\theta)$ and assume $a_1 = \min\{a_i\}$. Consider the equality

$$K_Y \sim a_1 E_1 + a_1 \sum_{j \neq 1} E_j + \sum_{j \neq 1} (a_j - a_1) E_j + D$$

then apply the adjunction formula on it and we obtain

$$\begin{aligned} 2g(E_1) - 2 &= K_Y \cdot E_1 + E_1^2 \\ &= (a_1 + 1)E_1^2 + a_1 \sum_{j \neq 1} E_j \cdot E_1 + \sum_{j \neq 1} (a_j - a_1) E_j \cdot E_1 + D \cdot E_1. \end{aligned}$$

Here, by $\sum_{j \neq 1} (a_j - a_1) E_j \cdot E_1 \geq 0$, $D \cdot E_1 \geq 0$ we obtain the following:

$$2g(E_1) - 2 \geq (a_1 + 1)E_1^2 + a_1 \sum_{j \neq 1} E_j \cdot E_1.$$

Now assume $a_1 \leq -2$, then we obtain

$$E_1^2 \geq 2\{g(E_1) - 1\}/(a_1 + 1) - \{a_1/(a_1 + 1)\} \sum_{j \neq 1} E_j \cdot E_1$$
$$\geq -2\Big(\sum_{j \neq 1} E_j \cdot E_1 + \max\{g(E_1) - 1, 0\}\Big).$$

But this contradicts the assumption of the theorem. Hence $a_1 \geq -1$, and by Theorem 7.7.3 the singularity (X, x) must be a Du Bois singularity.

This theorem shows a sufficient condition for a singularity to be a Du Bois singularity in terms of numerical conditions. How about the converse of this theorem? The following shows that the converse does not hold.

Example 7.7.11. ([13, 4.2]). Take a 2-dimensional normal singularity (X, x) with the following good resolution: The reduced exceptional divisor $E = f^{-1}(x)_{\text{red}}$ is written as $E = E_1 + E_2$, E_1 with a non-singular elliptic curve, $E_2 \cong \mathbb{P}^1$ and $E_1 \cdot E_2 = 1$, $E_1^2 = -1$. In this case the singularity (X, x) is a Du Bois singularity if and only if $\mathcal{O}_{E_1}(E_1) \not\cong \mathcal{O}_{E_1}(-E_2)$. It implies that according to the intersection point of E_1 and E_2, the singularity is a Du Bois singularity or not.

7.8 Classification of 2-Dimensional Singularities by κ_δ

In this section we classify 2-dimensional normal singularities according to the invariant κ_δ introduced in Sect. 6.3. The main goal is the following theorem:

Theorem 7.8.1. *2-dimensional normal singularities are classified by κ_δ as follows:*

κ_δ	*Gorenstein singularities*	*General singularities*
$-\infty$	*Rational double singularities*	*Quotient singularities*
0	*Simple elliptic singularities* *Cusp singularities*	*Quotients by simple elliptic singularities or cusp singularities by finite cyclic group G, where G acts freely outside the singularities*
2	*Gorenstein singularities other than above*	*Singularities other than above*

Proof. (Proof of "rational double $\Leftrightarrow \kappa_\delta = -\infty$") The implication $\Rightarrow$ follows from Theorem 7.5.1, (i) $\Rightarrow$ (iii), (i) $\Rightarrow$ (v) and the inequalities $\kappa_\gamma \geq \kappa_\delta$. The opposite implication $\Leftarrow$ follows from Theorem 7.5.1, (iii) $\Rightarrow$ (i) and the equality $\delta_1 = p_g = 0$.

(Proof of "quotient singularity $\Leftrightarrow \kappa_\delta = -\infty$") By Theorem 7.4.11, a quotient singularity is equivalent to a log terminal singularity. By Proposition 6.3.12, a log terminal singularity satisfies $\kappa_\delta = -\infty$. Conversely, if a singularity satisfies $\kappa_\delta = -\infty$, then it satisfies $p_g = \delta_1 = 0$, which means the singularity is rational. Therefore by Theorem 7.3.2, it is a $\mathbb{Q}$-Gorenstein singularity. Therefore, by Proposition 6.3.12, it is a log terminal singularity.

(Proof of "simple elliptic or cusp singularity $\Leftrightarrow \kappa_\delta = 0$, Gorenstein singularity") The implication $\Rightarrow$ follows from Proposition 7.6.3 and Proposition 7.6.9. Let us show the implication $\Leftarrow$. If a Gorenstein singularity (X, x) satisfies $\kappa_\delta(X, x) = 0$, then by Proposition 6.3.12 (X, x) is a log canonical singularity and satisfies $\delta_m(X, x) = 1$ $(\forall m \in \mathbb{N})$. Therefore by Proposition 7.7.4, (X, x) is a Du Bois singularity. Let $f : Y \to X$ be the minimal resolution of the singularity, then by Theorem 7.7.8, f is a good resolution. Let $E = \sum E_i$ be the reduced total exceptional divisor, then by $K_Y = f^*K_X - E$, we have $K_E \sim 0$. Such a normal crossing divisor E is a non-singular elliptic curve or an irreducible rational curve with a node or a cycle of $E_i \cong \mathbb{P}^1$'s $(\forall i = 1, \cdots, r)$.

(Proof of "quotient of a simple elliptic singularity or a cusp singularity by G $\Leftrightarrow \kappa_\delta = 0$") The implication $\Rightarrow$ follows from Theorem 6.2.9 (ii). The implication $\Leftarrow$ is proved as follows: By the following lemma, a singularity (X, x) with the property $\kappa_\delta(X, x) = 0$ is a $\mathbb{Q}$-Gorenstein singularity. Therefore, let $\pi : X' \to X$ be the canonical cover of X, then X' has Gorenstein log canonical singularities. By the above discussions the singularities on X' are simple elliptic or cusp singularities. Hence X is a quotient of these singularities.

Lemma 7.8.2. **([I6, 3.1]).** *If a 2-dimensional normal singularity (X, x) satisfies $\kappa_\delta(X, x) = 0$, then it is a $\mathbb{Q}$-Gorenstein singularity. Moreover, if r is the minimum positive integer such that $\delta_m(X, x) \neq 0$, then the index of (X, x) is r.*

Chapter 8
Higher-Dimensional Singularities

In any collection of data, the figure most obviously correct, beyond all need of checking, is the mistake [Mur].

A singularity of dimension higher than 2 is called a higher-dimensional singularity. In this section we mostly discuss higher-dimensional singularities. Unless otherwise stated, singularities are always of dimension $n \geq 2$. Varieties are all integral algebraic varieties over $\mathbb{C}$ and the singularities considered are on such varieties.

8.1 Mixed Hodge Structure and Du Bois Singularities

In this section, as stated above, varieties are integral algebraic varieties over $\mathbb{C}$, but we consider the analytic structure on them, i.e., for an algebraic variety X we think of the topology as analytic spaces instead of Zariski topology on X. We also think that the structure sheaf $\mathscr{O}_X$ is the analytic structure sheaf $\mathscr{O}_X^{\text{hol}}$. By GAGA [Se3], if X is complete, we should note that

$$H^i(X, \mathscr{O}_X^{\text{hol}}) = H^i(X, \mathscr{O}_X^{\text{alg}}).$$

In this section we omit "hol" in $\mathscr{O}^{\text{hol}}$.

Now we introduce a Hodge structure on a cohomology group.

Definition 8.1.1. If a finite-dimensional real vector space H and the decomposition $H_{\mathbb{C}} = \bigoplus_{p+q=n} H^{p,q}$ of $H_{\mathbb{C}} := H \otimes_{\mathbb{R}} \mathbb{C}$ satisfies $H^{p,q} = \overline{H^{q,p}}$, then we call the decomposition a *Hodge structure of weight* n.

S. Ishii, *Introduction to Singularities*,
https://doi.org/10.1007/978-4-431-56837-7_8

Definition 8.1.2. If a finite-dimensional real vector space H, an increasing sequence $W = \{W_k\}_{k\in\mathbb{Z}}$ of subspaces of H and a decreasing sequence $F = \{F^k\}_{k\in\mathbb{Z}}$ of subspaces of $H_{\mathbb{C}}$ satisfy the following, we call (H, W, F) a *mixed Hodge structure*:

For each k, $Gr_k^W(H) := W_k(H)/W_{k-1}(H)$ has a Hodge structure of weight k such that $F^p(Gr_k^W(H)_{\mathbb{C}}) = \bigoplus_{p'\geq p} H^{p',q}$. In this case F is called the *Hodge filtration* and W the *weight filtration*. If in particular for k the equalities $Gr_i^W(H) = 0$ $(i \neq k)$ hold, we say that the *mixed Hodge structure is of pure weight* k, or the *pure Hodge structure is of weight k*.

Definition 8.1.3. Let $(H, W, F), (H', W', F')$ be mixed Hodge structures. A linear map $f : H \to H'$ is called a *morphism of a mixed Hodge structure* if f has compatibility with both filtrations, i.e., $f(W_i) \subset W_i'$ and $f(F^i) \subset F'^i$). In particular, by the linear map $Gr_n^W(f) : Gr_n^W(H) \to Gr_n^W(H')$ induced from f, it follows that $Gr_n^W(f)(H^{p,q}) \subset H'^{p,q}$.

Proposition 8.1.4. *Let $(H, W, F), (H', W', F')$ and (H'', W'', F'') be mixed Hodge structures. Let $H' \xrightarrow{f} H \xrightarrow{g} H''$ be a sequence of mixed Hodge structures. If this sequence is exact, then the following are also exact:*

$$\begin{array}{ccccc} Gr_i^{W'} H' & \longrightarrow & Gr_i^W H & \longrightarrow & Gr_i^{W''} H'', \\ Gr_{F'}^i H' & \longrightarrow & Gr_F^i H & \longrightarrow & Gr_{F''}^i H'', \\ H'^{p,q} & \longrightarrow & H^{p,q} & \longrightarrow & H''^{p,q}. \end{array}$$

Example 8.1.5. Let X be a complete non-singular algebraic variety, then for every non-negative integer k the cohomology group $H^k(X, \mathbb{R})$ has the pure Hodge structure of weight k (Example 3.6.8). In this case, the Hodge filtration on $H^{p+q}(X, \mathbb{C})$ is the filtration induced from the Hodge spectral sequence:

$$E_1^{p,q} = H^q(X, \Omega^p) \Longrightarrow E^{p+q} = H^{p+q}(X, \mathbb{C}).$$

Theorem 8.1.6 (Deligne [De]). *Let X be an algebraic variety and k a non-negative integer, then $H^k(X) = H^k(X, \mathbb{C})$ has a mixed Hodge structure and satisfies the following conditions:*

(i) If X is a complete non-singular variety, then the mixed Hodge structure coincides with the structure as in Example 8.1.5.
(ii) The linear map $f^* : H^k(Y) \to H^k(X)$ induced from a morphism $f : X \to Y$ of varieties is a morphism of mixed Hodge structures.
(iii) Let $Y \subset X$ be a close (or open) subvariety, then the relative cohomology $H^k(X, Y) := H^k(X, Y; \mathbb{R})$ also has a mixed Hodge structure and the following sequence becomes an exact sequence of mixed Hodge structures:

$$\cdots \longrightarrow H^k(X) \longrightarrow H^k(Y) \longrightarrow H^{k+1}(X, Y) \longrightarrow H^{k+1}(X) \longrightarrow \cdots$$

(iv) Let $f : (X, Y) \to (X', Y')$ be a morphism of pairs, then the induced map $f^* : H^n(X', Y') \to H^n(X, Y)$ is a morphism of mixed Hodge structures.

(v) *If X is an n-dimensional non-singular algebraic variety, then the pair (p, q) of numbers such that $H_k^{p,q} \neq 0$ which appear in $\{Gr_i^W(H^k(X))_{\mathbb{C}}\}_{i\in\mathbb{Z}}$ lie within the hatched areas of the following figure:*

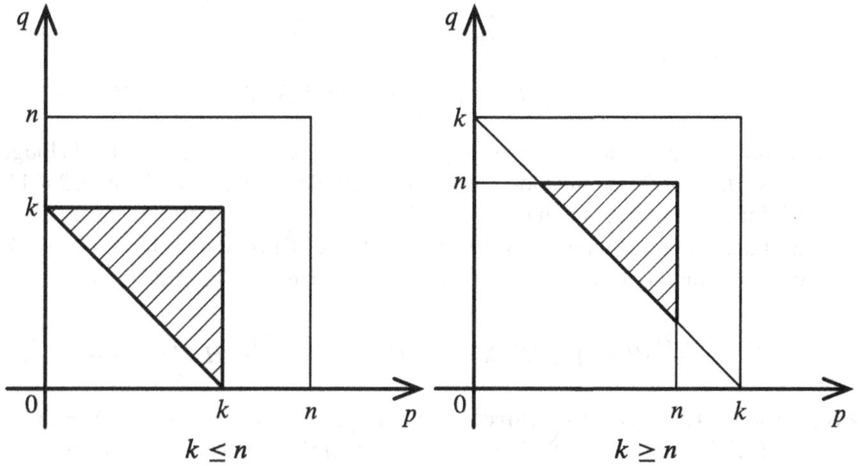

(vi) *If X is an n-dimensional complete algebraic variety, then the pair (p, q) of numbers such that $H_k^{p,q} \neq 0$ which appear in $\{Gr_i^W(H^k(X))_{\mathbb{C}}\}_{i\in\mathbb{Z}}$ lie within the hatched areas of the following figure:*

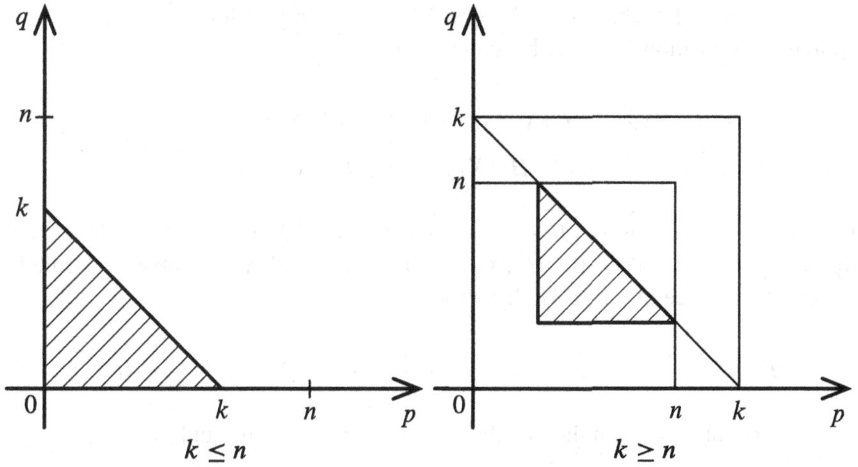

Proposition 8.1.7. *Let $f : \widetilde{X} \to X$ be a proper morphism of algebraic varieties not necessarily irreducible such that its restriction $f^{-1}(U) \to U$ on a dense open subset U is isomorphic. Assume that $\widetilde{X}$ is embedded into a non-singular variety. Let $Y = X \setminus U$ and $\widetilde{Y} = f^{-1}(Y)$, then there exists an exact sequence of mixed Hodge structures:*

$$\cdots \longrightarrow H^k(X) \longrightarrow H^k(Y) \oplus H^k(\widetilde{X}) \longrightarrow H^k(\widetilde{Y}) \longrightarrow H^{k+1}(X) \longrightarrow \cdots .$$

Proof. Consider the following commutative diagram of mixed Hodge structures:

$$\begin{array}{ccccccccc}
\cdots \longrightarrow & H^k(\widetilde{X}) & \xrightarrow{\widetilde{\alpha}} & H^k(\widetilde{Y}) & \xrightarrow{\widetilde{\beta}} & H^{k+1}(\widetilde{X},\widetilde{Y}) & \xrightarrow{\widetilde{\gamma}} & H^{k+1}(\widetilde{X}) \\
& {\scriptstyle\lambda}\uparrow & & \uparrow{\scriptstyle\mu} & & \uparrow{\scriptstyle\nu} & & \uparrow{\scriptstyle\lambda} \\
\cdots \longrightarrow & H^k(X) & \xrightarrow{\alpha} & H^k(Y) & \xrightarrow{\beta} & H^{k+1}(X,Y) & \xrightarrow{\gamma} & H^{k+1}(X).
\end{array}$$

By Theorem 8.1.6 (iii) the horizontal sequences are exact sequences of mixed Hodge structures. Here, as $\widetilde{X}$ is embedded into a non-singular variety, by Theorem 2.4.11, $\widetilde{Y}$ is a deformation retract to an open neighborhood of $\widetilde{X}$. Therefore by [Spa, Ch.4, §8, Th.9] the canonical map $\nu : H^k(X,Y) \to H^k(\widetilde{X},\widetilde{Y})$ is isomorphic for every k. Hence, we obtain the following exact sequence of mixed Hodge structures:

$$\cdots \to H^k(X) \stackrel{(\alpha,\lambda)}{\longrightarrow} H^k(Y) \oplus H^k(\widetilde{X}) \stackrel{\mu-\widetilde{\alpha}}{\longrightarrow} H^k(\widetilde{Y}) \xrightarrow{\gamma\circ\nu^{-1}\circ\widetilde{\beta}} H^{k+1}(X) \longrightarrow \cdots .$$

Proposition 8.1.8 (Mayer–Vietoris). *For a quasi-projective variety $X = X_1 \cup X_2$ and closed subsets $X_i \subsetneq X$ $(i = 1, 2)$, we obtain the following exact sequence of mixed Hodge structures:*

$$\longrightarrow H^k(X) \longrightarrow H^k(X_1) \oplus H^k(X_2) \longrightarrow H^k(X_1 \cap X_2) \longrightarrow H^{k+1}(X) \longrightarrow \cdots$$

Proof. Let $X \subset \mathbb{P}^N$, then $X_1 \amalg X_2 \subset \mathbb{P}^N \amalg \mathbb{P}^N$. Therefore for $f : X_1 \amalg X_2 \to X$, applying Proposition 8.1.7, we have the following:

$$\begin{aligned}\cdots \longrightarrow H^k(X) &\stackrel{(\alpha,\lambda)}{\longrightarrow} H^k(X_1 \cap X_2) \oplus H^k(X_1 \amalg X_2) \\ &\stackrel{\mu\circ p_1-\widetilde{\alpha}\circ p_2}{\longrightarrow} H^k(X_1 \cap X_2) \oplus H^k(X_1 \cap X_2) \xrightarrow{\gamma\circ\nu^{-1}\circ\widetilde{\beta}} \cdots\end{aligned}$$

Here, by the projection $p_2 : H^k(X_1 \cap X_2) \oplus H^k(X_1 \amalg X_2) \twoheadrightarrow H^k(X_1 \amalg X_2)$ and by $p_1 - p_2 : H^k(X_1 \cap X_2) \oplus H^k(X_1 \cap X_2) \to H^k(X_1 \cap X_2)$ we obtain the following exact sequence of mixed Hodge structures:

$$\longrightarrow H^k(X) \xrightarrow{\lambda=p_2\circ(\alpha,\lambda)} H^k(X_1 \amalg X_2) \xrightarrow{(p_1-p_2)\circ\widetilde{\alpha}} H^k(X_1 \cap X_2) \stackrel{\psi}{\longrightarrow} H^{k+1}(X),$$

where ψ is obtained from the fact that $\gamma \circ \nu^{-1} \circ \widetilde{\beta}$ factors through $p_1 - p_2$:

$$\gamma \circ \nu^{-1} \circ \widetilde{\beta} = \psi \circ (p_1 - p_2).$$

As is seen in Example 8.1.5, the Hodge filtration of a complete non-singular algebraic variety is introduced from the spectral sequence obtained on the De Rham complex $\Omega^\bullet_X$ which gives the resolution $0 \to \mathbb{C} \to \Omega^\bullet_X$ of the constant sheaf $\mathbb{C}$ and the stupid filtration F on the complex.

For a not necessarily complete non-singular variety, the construction of the Hodge filtration by Deligne is similar. (Precisely, the Hodge filtration is obtained from the spectral sequence of a suitable filtered complex on a compactification $\widetilde{X}$ of X.)

Now how about taking off the condition "non-singular"? According to the construction of the mixed Hodge structure by Deligne [De], the definition of the Hodge filtration is not like this. Therefore a natural question arises: Is there a filtered complex which is a resolution of $\mathbb{C}$ and whose spectral sequence gives the Hodge filtration for singular varieties? Du Bois gave an answer as follows:

Theorem 8.1.9 (Du Bois [DB]). *For an algebraic variety X there exists a filtered complex $(\underline{\Omega}^\bullet_X, F)$ satisfying the following:*

(i) $\underline{\Omega}^\bullet_X$ *is a resolution of the constant sheaf* $\mathbb{C}$ *(i.e., the sequence* $0 \to \mathbb{C} \to \underline{\Omega}^\bullet_X$ *is exact).*

(ii) Let $\Omega^\bullet_X$ *be the De Rham complex on* X *and* σ *its stupid filtration, then there is a morphism of filtered complexes:*

$$(\Omega^\bullet_X, \sigma) \xrightarrow{\lambda} (\underline{\Omega}^\bullet_X, F).$$

In particular, if X *is non-singular, then this is a filtered quasi-isomorphism.*

(iii) In particular, if X *is complete, the spectral sequence:*

$$E_1^{p,q} = \mathbb{R}^{p+q}\Gamma(Gr_F^p \underline{\Omega}^\bullet_X) \Longrightarrow \mathbb{R}^{p+q}\Gamma(\underline{\Omega}^\bullet_X) = H^{p+q}(X, \mathbb{C})$$

defined by $(\underline{\Omega}^\bullet_X, F)$ *degenerates at* E_1 *and the induced filtration on* $H^{p+q}(X, \mathbb{C})$ *is the same as the Hodge filtration by Deligne.*

(iv) Let $f : \widetilde{X} \to X$ *be a proper morphism of algebraic varieties,* $Y \subset X$ *a closed subvariety. Let* $\widetilde{Y} = f(Y)$, $f' = f \mid_{\widetilde{Y}}$*. If* $f \mid_{\widetilde{X}\setminus\widetilde{Y}} : \widetilde{X} \setminus \widetilde{Y} \xrightarrow{\sim} X \setminus Y$ *is isomorphic, then there is the following triangle [Ha1] in the filtered derived category:*

$$0 \longrightarrow \underline{\Omega}^\bullet_X \longrightarrow \underline{\Omega}^\bullet_Y \oplus \mathbb{R}f_*\underline{\Omega}^\bullet_{\widetilde{X}} \longrightarrow \mathbb{R}f'_*\underline{\Omega}^\bullet_{\widetilde{Y}} \longrightarrow 0,$$

which means for each p*:*

$$\begin{aligned} \cdots \longrightarrow \mathscr{H}^i(Gr_F^p \underline{\Omega}^\bullet_X) &\longrightarrow \mathscr{H}^i\big(Gr_F^p(\underline{\Omega}^\bullet_Y \oplus \mathbb{R}f_*\underline{\Omega}^\bullet_{\widetilde{X}})\big) \\ &\longrightarrow \mathscr{H}^i\big(Gr_F^p(\mathbb{R}f'_*\underline{\Omega}^\bullet_{\widetilde{Y}})\big) \longrightarrow \mathscr{H}^{i+1}\big(Gr_F^p \underline{\Omega}^\bullet_X\big) \longrightarrow \cdots \end{aligned}$$

is an exact sequence.

Now we focus a singularity which has a good property with respect to the Du Bois complex.

Definition 8.1.10. A singularity (X, x) is called a *Du Bois singularity* if the morphism

$$Gr_\sigma^0 \Omega^\bullet_{X,x} \xrightarrow{Gr^0(\lambda)} Gr_F^0 \underline{\Omega}^\bullet_{X,x}$$

of complexes induced from λ in Theorem 8.1.9 (ii) becomes a quasi-isomorphism. Noting that $Gr^0_\sigma \Omega^\bullet_X = \mathcal{O}_X$, the above quasi-isomorphism is equivalent to saying that the sequence $0 \to \mathcal{O}_{X,x} \to Gr^0_F \underline{\Omega}^\bullet_{X,x}$ is exact.

If all points on X are Du Bois singularities, then X is called a *Du Bois variety*. Henceforth, according to Du Bois's paper, we denote $Gr^0_F \underline{\Omega}^\bullet_X$ by $\underline{\Omega}^0_X$.

The following are basic properties of Du Bois singularities.

Proposition 8.1.11. *(i) A non-singular point is a Du Bois singularity.*

(ii) Let X be a normal crossing divisor on a non-singular variety, then X is a Du Bois variety.

(iii) Let X be a complete Du Bois variety and F the Hodge filtration of $H^i(X, \mathbb{C})$, then there exists an isomorphism:

$$H^i(X, \mathcal{O}_X) \simeq Gr^0_F H^i(X, \mathbb{C}).$$

Proof. (i) follows immediately from Theorem 8.1.9 (ii). Statement (ii) is proved by (i) and Theorem 8.1.9 (iv). Let us see the proof for a simple case. Let $X = X_1 \cup X_2$ with X_i non-singular. Let $f : X_1 \amalg X_2 \to X_1 \cup X_2$ be the normalization. Let $Y := X_1 \cap X_2$. Then by Theorem 8.1.9 (iv), we have the following exact sequence:

$$\begin{aligned} 0 &\longrightarrow \mathcal{H}^0(\underline{\Omega}^0_X) \longrightarrow \mathcal{H}^0(\underline{\Omega}^0_Y) \oplus \mathcal{H}^0(\mathbb{R} f_* \underline{\Omega}^0_{X_1 \amalg X_2}) \longrightarrow \mathcal{H}^0(\mathbb{R} f_* \underline{\Omega}^0_{Y \amalg Y}) \\ &\longrightarrow \mathcal{H}^1(\underline{\Omega}^0_X) \longrightarrow \mathcal{H}^1(\underline{\Omega}^0_Y) \oplus \mathcal{H}^1(\mathbb{R} f_* \underline{\Omega}^0_{X_1 \amalg X_2}) \longrightarrow \mathcal{H}^1(\mathbb{R} f_* \underline{\Omega}^0_{Y \amalg Y}) \\ &\longrightarrow \cdots\cdots \end{aligned}$$

By statement (i) this sequence becomes as follows:

$$\begin{aligned} 0 &\longrightarrow \mathcal{H}^0(\underline{\Omega}^0_X) \longrightarrow \mathcal{O}_Y \oplus f_*(\mathcal{O}_{X_1} \oplus \mathcal{O}_{X_2}) \longrightarrow f_*(\mathcal{O}_Y \oplus \mathcal{O}_Y) \\ &\longrightarrow \mathcal{H}^1(\underline{\Omega}^0_X) \longrightarrow 0, \\ &\qquad \mathcal{H}^i(\underline{\Omega}^0_X) = 0 \quad (i \geq 2). \end{aligned}$$

On the other hand, by the following exact sequence:

$$0 \longrightarrow \mathcal{O}_X \longrightarrow \mathcal{O}_Y \oplus f_*\mathcal{O}_{X_1} \oplus f_*\mathcal{O}_{X_2} \longrightarrow f_*\mathcal{O}_Y \oplus f_*\mathcal{O}_Y \longrightarrow 0$$

we obtain $\mathcal{H}^0(\underline{\Omega}^0_X) \cong \mathcal{O}_X$, $\mathcal{H}^i(\underline{\Omega}^0_X) = 0$ $(i \geq 1)$.

Statement (iii) is proved as follows: By Theorem 8.1.9 (iii) we have $Gr^0_F H^i(X, \mathbb{C}) = E_1^{0,i} = \mathbb{R}^i\Gamma(\underline{\Omega}^0_X)$. Here the right-hand side is isomorphic to $H^i(X, \mathcal{O}_X)$ since X is a Du Bois variety.

Proposition 8.1.12. *Let $f : \widetilde{X} \to X$ be a partial resolution of a normal isolated singularity (X, x). If the fiber $E = f^{-1}(x)_{\rm red}$ is a Du Bois variety, for every $i > 0$ the natural homomorphism:*

$$(R^i f_* \mathcal{O}_{\widetilde{X}})_x \longrightarrow H^i(E, \mathcal{O}_E)$$

is surjective.

Proof. Let X be a sufficiently small Stein neighborhood of x. The commutative diagram

$$\begin{array}{ccccc} 0 & \longrightarrow & \mathbb{C}_{\widetilde{X}} & \longrightarrow & \mathcal{O}_{\widetilde{X}} \\ & & \downarrow & & \downarrow \\ 0 & \longrightarrow & \mathbb{C}_E & \longrightarrow & \mathcal{O}_E \end{array}$$

induces the commutative diagram

$$\begin{array}{ccc} H^i(\widetilde{X}, \mathbb{C}) & \longrightarrow & H^i(\widetilde{X}, \mathcal{O}_{\widetilde{X}}) \\ \alpha \downarrow & & \downarrow \beta \\ H^i(E, \mathbb{C}) & \longrightarrow & H^i(E, \mathcal{O}_E) \\ & \searrow^{\gamma} & \downarrow \delta \\ & & Gr_F^0 H^i(E, \mathbb{C}). \end{array}$$

Here, F is the Hodge filtration of $H^i(E, \mathbb{C})$. By Theorem 2.4.11 the fiber E is a deformation retract of $\widetilde{X}$, therefore α is isomorphic. As $F^0 H^i(E, \mathbb{C}) = H^i(E, \mathbb{C})$, the map γ is surjective and as E is a complete Du Bois variety, δ is isomorphic. By this the surjectivity of β follows.

How does the Du Bois property appear on a resolution of a singularity? Let us see for a normal isolated singularity.

Proposition 8.1.13. *Let $f : \widetilde{X} \to X$ be a partial resolution of a normal isolated singularity (X, x). Assume both $\widetilde{X}$ and $E = f^{-1}(x)_{\mathrm{red}}$ are Du Bois varieties. Then the following are equivalent:*

(i) (X, x) is a Du Bois singularity;
(ii) the natural homomorphism

$$(R^i f_* \mathcal{O}_{\widetilde{X}})_x \longrightarrow H^i(E, \mathcal{O}_E)$$

is an isomorphism for every $i > 0$.

Proof. Apply Theorem 8.1.9 (iv) for f. As $\widetilde{X}$ and E are Du Bois varieties, we have the following exact sequence:

$$\begin{array}{lllll} 0 \longrightarrow \mathcal{H}^0(\underline{\Omega}_X^0) & \longrightarrow & f_*\mathcal{O}_{\widetilde{X}} \oplus \mathbb{C}_x & \longrightarrow & f_*\mathcal{O}_E \\ \longrightarrow \mathcal{H}^1(\underline{\Omega}_X^0) & \longrightarrow & R^1 f_*\mathcal{O}_{\widetilde{X}} & \longrightarrow & R^1 f_*\mathcal{O}_E \\ & & \cdots\cdots\cdots & & \\ \longrightarrow \mathcal{H}^i(\underline{\Omega}_X^0) & \longrightarrow & R^i f_*\mathcal{O}_{\widetilde{X}} & \longrightarrow & R^i f_*\mathcal{O}_E \\ \longrightarrow \cdots & & & & \end{array}$$

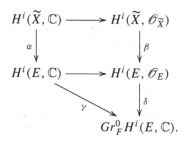

Here, (X, x) is a Du Bois variety if and only if

$$\mathcal{H}^0(\underline{\Omega}^0_X)_x \simeq \mathcal{O}_{X,x}, \quad \mathcal{H}^i(\underline{\Omega}^0_X)_x = 0 \quad (\forall i > 0).$$

This is equivalent to the isomorphism in (ii) of the proposition.

As an easy application of this proposition we obtain the following:

Corollary 8.1.14. *A rational isolated singularity is a Du Bois singularity.*

Proof. For a good resolution $f : \widetilde{X} \to X$ of the singularity the equality $R^i f_* \mathcal{O}_{\widetilde{X}} = 0$ $(i > 0)$ holds, then by Proposition 8.1.12, the homomorphism $R^i f_* \mathcal{O}_{\widetilde{X}} \to H^i(E, \mathcal{O}_E)$ $(i > 0)$ is isomorphic. Hence, by Proposition 8.1.13, the singularity is Du Bois.

For non-isolated rational singularities, the same statement also holds.

Theorem 8.1.15 (Kovács [Kov]). *Let X be a variety with at worst rational singularities, then X is a Du Bois variety.*

The following shows the inheritance of the Du Bois property under the quotient of a group action.

Proposition 8.1.16. *If a normal isolated singularity (Y, y) is a Du Bois singularity, then the quotient (X, x) of (Y, y) by a finite group action is also a Du Bois singularity if it is isolated.*

Proof. We may assume that the point y is fixed by the action of G. Take X to be a sufficiently small neighborhood of x and let $f : \widetilde{X} \to X$ be a good resolution of the singularity (X, x). As G acts on $Y \times_X \widetilde{X}$ canonically, by Theorem 4.3.4 we can take a good resolution $g : \widetilde{Y} \to Y \times_X \widetilde{X}$ of $Y \times_X \widetilde{X}$ such that G acts on $\widetilde{Y}$ canonically. As G acts on $\widetilde{X}$ trivially, we obtain the following commutative diagram:

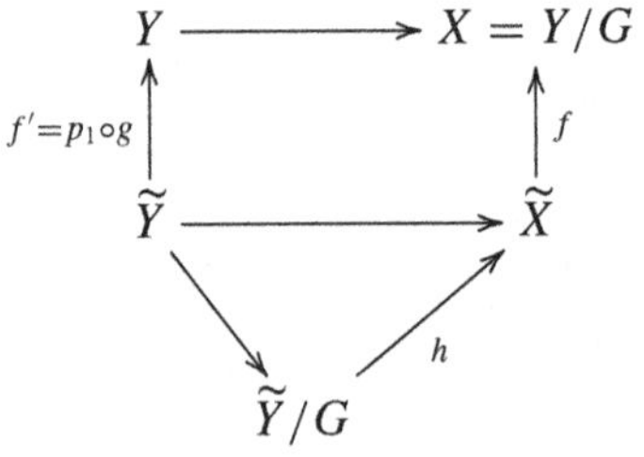

Here, h is a proper morphism isomorphic on open subsets, $\widetilde{X}$ is non-singular and $\widetilde{Y}/G$ has at worst quotient singularities. Therefore we obtain $R^i h_* \mathcal{O}_{\widetilde{Y}/G} = 0\,(i > 0)$. Then in Leray's spectral sequence:

$$E_2^{p,q} = H^p(\widetilde{X}, R^q h_* \mathcal{O}_{\widetilde{Y}/G}) \Longrightarrow H^{p+q}(\widetilde{Y}/G, \mathcal{O}_{\widetilde{Y}/G})$$

we have $E_2^{p,q} = 0\ (q > 0)$, therefore we obtain $H^i(\widetilde{X}, \mathcal{O}_{\widetilde{X}}) \cong H^i(\widetilde{Y}/G, \mathcal{O}_{\widetilde{Y}/G})$. On the other hand, we know $H^i(\widetilde{Y}/G, \mathcal{O}_{\widetilde{Y}/G}) = H^i(\widetilde{Y}, \mathcal{O}_{\widetilde{Y}})^G$. Now let $E = f^{-1}(x)_{\mathrm{red}}$

and $F = f'^{-1}(y)_{\rm red}$, then the surjective morphism $F \twoheadrightarrow E$, and there is a homomorphism $H^i(E, \mathcal{O}_E) \to H^i(F, \mathcal{O}_F)$. Hence we obtain the following commutative diagram:

$$\begin{array}{ccc} R^i f'_* \mathcal{O}_{\widetilde{Y}}^G & \xrightarrow{\alpha} & H^i(F, \mathcal{O}_F)^G & \\ \wr\downarrow & & \uparrow & \\ R^i f_* \mathcal{O}_{\widetilde{X}} & \xrightarrow[\beta]{} & H^i(E, \mathcal{O}_E) & \longrightarrow 0. \end{array}$$

Here, as (Y, y) is a Du Bois singularity, we have $R^i f'_* \mathcal{O}_{\widetilde{Y}} \xrightarrow{\sim} H^i(F, \mathcal{O}_F)$ $(i > 0)$ which yields the injectivity of α, therefore we have that β is injective.

The following shows the relation of the Du Bois property and the invariant δ_m.

Theorem 8.1.17 (Ishii [I2], Watanabe [W2]). *Let (X, x) be a normal isolated 1-Gorenstein singularity. Then, the following are equivalent:*

(i) *(X, x) is a Du Bois singularity.*
(ii) $\kappa_\delta(X, x) \leq 0$.
(iii) $\delta_m(X, x) \leq 1$ $(\forall m \in \mathbb{N})$.
(iv) *(X, x) is a log canonical singularity.*

Proof. As (X, x) is a 1-Gorenstein, by Proposition 6.3.12, the equivalence among (ii), (iii), (iv) is obvious.

(i) $\Rightarrow$ (iv): Let $f : \widetilde{X} \to X$ be a good resolution of the singularity (X, x). Let $E = f^{-1}(x)_{\rm red}$, then, as (X, x) is a Du Bois singularity, in particular we have the following isomorphism:

$$R^{n-1} f_* \mathcal{O}_{\widetilde{X}} \simeq H^{n-1}(E, \mathcal{O}_E). \tag{8.1}$$

Here, in the same way as in the proof of Theorem 7.7.3, we can see that $R^{n-1} f_* \mathcal{O}_{\widetilde{X}}$ is dual to $H^1_E(\widetilde{X}, \omega_{\widetilde{X}})$. Then, by the exact sequence:

$$\begin{array}{l} 0 \longrightarrow \underset{\substack{\| \\ 0}}{\Gamma_E(\widetilde{X}, \omega_{\widetilde{X}})} \longrightarrow \Gamma(\widetilde{X}, \omega_{\widetilde{X}}) \longrightarrow \Gamma(\widetilde{X} \setminus E, \omega_{\widetilde{X}}) \longrightarrow H^1_E(\widetilde{X}, \omega_{\widetilde{X}}) \\ \longrightarrow H^1(\widetilde{X}, \omega_{\widetilde{X}}) = 0, \end{array}$$

it is also dual to $\Gamma(\widetilde{X} \setminus E, \omega_{\widetilde{X}}) / \Gamma(\widetilde{X}, \omega_{\widetilde{X}})$. On the other hand, $H^{n-1}(E, \mathcal{O}_E)$ is dual to $\Gamma(E, \omega_E)$, which is isomorphic to $\Gamma(\widetilde{X}, \omega_{\widetilde{X}}(E)) / \Gamma(\widetilde{X}, \omega_{\widetilde{X}})$ by the exact sequence:

$$0 \longrightarrow \Gamma(\widetilde{X}, \omega_{\widetilde{X}}) \longrightarrow \Gamma(\widetilde{X}, \omega_{\widetilde{X}}(E)) \longrightarrow \Gamma(E, \omega_E) \longrightarrow H^1(\widetilde{X}, \omega_{\widetilde{X}}) = 0.$$

Therefore (1) is equivalent to

$$\Gamma(\widetilde{X}, \omega_{\widetilde{X}}(E)) = \Gamma(\widetilde{X} \setminus E, \omega_{\widetilde{X}}).$$

If we define $K_{\widetilde{X}} = f^*K_X + \sum_{i=1}^{r} m_i E_i$, the equality is equivalent to $m_i \geq -1$ ($\forall i = 1, \cdots, r$).

(iv) $\Rightarrow$ (i): If we assume (iv), then by the argument above it follows that $R^{n-1} f_* \mathcal{O}_{\widetilde{X}} \xrightarrow{\sim} H^{n-1}(E, \mathcal{O}_E)$. Take i such that $0 < i < n-1$. Let ω be a free generator of ω_X. Then, as $\omega \in \Gamma(\widetilde{X}, \omega_{\widetilde{X}}(E))$, we have a homomorphism $\mathcal{O}_{\widetilde{X}} \xrightarrow{\cdot\omega} \omega_{\widetilde{X}}(E)$ which is isomorphic outside an E. Consider the following diagram:

$$\begin{array}{ccccc} H^i_E(\widetilde{X}, \mathcal{O}_{\widetilde{X}}) & \longrightarrow & H^i(\widetilde{X}, \mathcal{O}_{\widetilde{X}}) & \xrightarrow{\alpha} & H^i(\widetilde{X} \setminus E, \mathcal{O}_{\widetilde{X}}) \\ \downarrow & & \downarrow{\scriptstyle \beta} & & {\scriptstyle \gamma}\downarrow \\ H^i_E(\widetilde{X}, \omega_{\widetilde{X}}(E)) & \longrightarrow & H^i(\widetilde{X}, \omega_{\widetilde{X}}(E)) & \xrightarrow{\gamma} & H^i(\widetilde{X} \setminus E, \omega_{\widetilde{X}}(E)). \end{array}$$

Here, by duality and the Grauert–Riemenschneider vanishing theorem, we have $H^i_E(\widetilde{X}, \mathcal{O}_{\widetilde{X}}) = 0$ $(0 < i < n-1)$. Then, by the diagram, α is injective. On the other hand, δ is isomorphic, therefore β is injective. Next, consider the following diagram:

$$\begin{array}{ccc} H^i(\widetilde{X}, \mathcal{O}_{\widetilde{X}}) & \xrightarrow{\beta} & H^i(\widetilde{X}, \omega_{\widetilde{X}}(E)) \\ \downarrow{\scriptstyle \varphi} & & \downarrow{\scriptstyle \psi} \\ H^i(E, \mathcal{O}_E) & \xrightarrow{\omega} & H^i(E, \omega_E). \end{array}$$

By the Grauert–Riemenschneider vanishing theorem $H^i(\widetilde{X}, \omega_{\widetilde{X}}) = 0$ $(i > 0)$, and ψ turns out to be isomorphic. Therefore we obtain that φ is injective.

The following is an application of the theorem:

Corollary 8.1.18. *An isolated log canonical singularity is a Du Bois singularity.*

Proof. Let $\pi : Y \to X$ be the canonical cover of an isolated log canonical singularity (X, x), then X is regarded as the quotient of Y. As Y is a 1-Gorenstein isolated log canonical singularity, by Theorem 8.1.17, Y has a Du Bois singularity. Applying Proposition 8.1.16, we obtain the statement.

This result is generalized as follows:

Theorem 8.1.19 (Kollár, Kovács [KK]). *A log canonical singularity is a Du Bois singularity.*

Example 8.1.20. The converse of Corollary 8.1.18 does not hold. For example, 2-dimensional rational singularities are all Du Bois. But there exists a rational singularity which is not log canonical. For example, consider the singularity whose weighted graph of the exceptional curve on the minimal resolution is as follows:

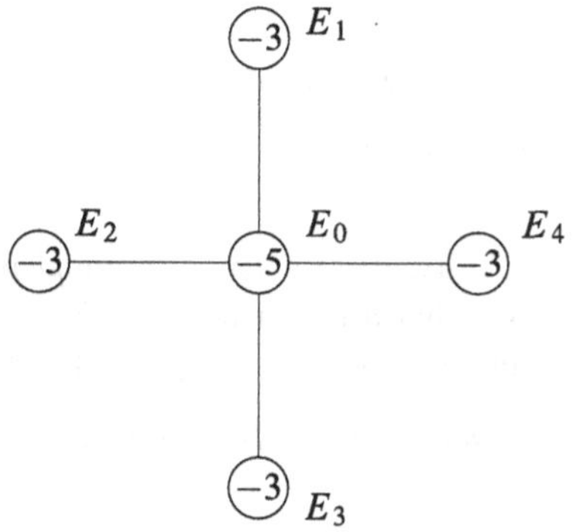

Here each irreducible component is isomorphic to $\mathbb{P}^1$. Then the fundamental cycle Z_f is reduced and $p(Z_f) = 0$, which shows that the singularity is rational. On the other hand, let $f : \widetilde{X} \to X$ be the minimal resolution, then the equality $K_{\widetilde{X}} = f^*K_X - \frac{13}{11}E_0 - \frac{8}{11}\sum_{i=1}^4 E_i$ holds, which yields that the singularity is not log canonical.

8.2 Minimal Model Problem

For a 2-dimensional singularity the minimal resolution always exists. But for a 3- or higher-dimensional singularity, there is no minimal resolution in general. The following shows this fact.

Example 8.2.1. Let X be a hypersurface defined by the equation $xy - zw = 0$ in $\mathbb{A}^4_{\mathbb{C}}$. Then X has an isolated singularity. Let $D_1, D_2 \subset X$ be divisors on X defined by $x = z = 0$ and $y = z = 0$, respectively. Let $b_1 : Y_1 \to X$ and $b_2 : Y_2 \to X$ be the blow-ups by the ideals (x, z) and (y, z), respectively. Then both Y_i are non-singular and there exist $\ell_i \subset Y_i$, $\ell_i \cong \mathbb{P}^1$ such that $b_i \mid_{Y_i \setminus \ell_i} : Y_i \setminus \ell_i \xrightarrow{\sim} X \setminus \{0\}$, $i = 1, 2$. Therefore b_i are both resolutions of the singularity of X. Let $E_i := b_i^{-1}(D_i)$, $i = 1, 2$, then $-E_i$ are Cartier divisors and b_i-ample ([EGA II, 8.1.7). Hence, $-E_i \cdot \ell_i > 0$, in particular $\ell_i \subset E_i$, $i = 1, 2$. On the other hand, denoting the strict transform of D_i in Y_j $(i \neq j)$ by $[D_i]$, we obtain that $\ell_j \not\subset [D_i]$ and $\ell_j \cap [D_i] \neq \emptyset$. Clearly Y_1 and Y_2 are not isomorphic over X.

Here, assume that there exists the minimal resolution $f : Y \to X$ of the singularity $(X, 0)$. Let $\sigma_i : Y_i \to Y$ be the morphism over X obtained by the minimality of f. As σ_i are proper morphisms, these are surjective and the images $\ell := \sigma_1(\ell_1) = \sigma_2(\ell_2)$ must be an irreducible curve. Therefore the morphisms σ_i are finite morphisms and then by the Zariski Main Theorem [Ha2, V, 5.2] these are isomorphic, which yields an isomorphism between Y_1 and Y_2, a contradiction.

Now, instead of minimal resolution, we will consider a minimal model with "mild singularities".

Definition 8.2.2. A partial resolution $f : Y \to X$ of a normal singularity (X, x) is called a *minimal model* if the following hold:

(1) Y has at worst terminal singularities.
(2) K_Y is f-nef.

For 2-dimensional singularity, there exists a minimal model by the following proposition. We can also see that a minimal model is a generalization of the 2-dimensional minimal resolution to the case of higher-dimensional singularities.

Proposition 8.2.3. *For a 2-dimensional normal singularity (X, x), the minimal resolution is a minimal model.*

Proof. The morphism $f : Y \to X$ is the minimal resolution, if and only if the following hold:

(1′) Y is non-singular.
(2′) The exceptional set on Y does not contain (-1)-curves.

Condition (1′) is equivalent to (1) by Corollary 7.1.18. On the other hand, an irreducible exceptional divisor C is a (-1)-curve if and only if $K_Y \cdot C < 0$, therefore we obtain that (2′) is equivalent to (2).

Next we will consider a slightly different model.

Definition 8.2.4. A partial resolution $f : Y \to X$ of a normal singularity (X, x) is called a *canonical model* if the following hold:

(1) Y has at worst canonical singularities.
(2) K_Y is f-ample.

Here, a $\mathbb{Q}$-Cartier divisor D is called *f-ample* if there exist a neighborhood $U \subset X$ of x and a closed immersion $\iota : f^{-1}(U) \hookrightarrow \mathbb{P}^N_{\mathbb{C}} \times U$, $m > 0$ and a hypersurface $H \subset \mathbb{P}^N_{\mathbb{C}}$ such that $mD \sim \iota^*(H \times U)$.

Proposition 8.2.5. *If a canonical model $f : Y \to X$ exists, then it is unique up to isomorphisms over X.*

Proof. Let $f_i : Y_i \to X$, $(i = 1, 2)$ be canonical models. As K_{Y_i} are f_i-ample, Y_i is isomorphic to $\mathrm{Proj}\big(\bigoplus_{m\geq 0} f_{i_*}\mathcal{O}(mK_{Y_i})\big)$ over X $(i = 1, 2)$ ([EGA III, 2.3.4(ii)]). Now take a resolution $f : Y \to X$ of the singularities of X such that it factors through f_i $(i = 1, 2)$. Define h_i so that $f = f_i \circ h_i$, $(i = 1, 2)$, then Y_i has at worst canonical singularities $h_{i_*}\mathcal{O}_Y(mK_Y) = \mathcal{O}_{Y_i}(mK_{Y_i})$. Therefore both Y_1 and Y_2 are isomorphic to $\mathrm{Proj} \bigoplus_{m\geq 0} f_*\mathcal{O}(mK_Y)$ over X.

The canonical model and a minimal model are isomorphic with X_{reg} over X_{reg} by the definitions. The following proposition shows the existence of the canonical model for 2-dimensional case.

Proposition 8.2.6. *Let $g : \widetilde{X} \to X$ be the minimal resolution of a normal 2-dimensional singularity (X, x). Let $h : \widetilde{X} \to Y$ be the contraction of all (-2)-curves in $g^{-1}(x)$. Then $f : Y \to X$ is the canonical model. Here a (-2)-curve C means $C^2 = -2$ and $C \cong \mathbb{P}^1$.*

Proof. First note that a subdivisor of the exceptional divisor has the negative definite intersection matrix, therefore we can contract the subdivisor. When we contract the subdivisor consisting of (−2)-curves, by Theorem 7.5.1 we obtain rational double points. Therefore Y has at worst canonical singularities. As $K_{\widetilde{X}} = h^* K_Y$, for every exceptional prime divisor E_i for f, the equality holds:

$$K_Y \cdot E_i = K_{\widetilde{X}} \cdot E'_i \quad (E'_i \text{ is the strict transform of } E_i).$$

As E'_i is not a (-2)-curve, we have $K_{\widetilde{X}} \cdot E'_i > 0$. Therefore K_Y is ample on $E = f^{-1}(x)_{\text{red}}$. Hence for $m \gg 0$, $\mathcal{O}_Y(mK_Y) \otimes_{\mathcal{O}_Y} \mathcal{O}_E$ is generated by $\Gamma(E, \mathcal{O}(mK_Y) \otimes \mathcal{O}_E)$. On the other hand, for $m \gg 0$, $(m-1)K_Y - E$ is f-nef, by Kawamata–Viehweg's vanishing theorem (Theorem 6.2.16), it follows that $R^1 f_* \mathcal{O}_Y(mK_Y - E) = 0$. By the exact sequence

$$f_* \mathcal{O}_Y(mK_Y) \longrightarrow \Gamma(E, \mathcal{O}_Y(mK_Y) \otimes \mathcal{O}_E) \longrightarrow 0$$

we have the following exact sequence:

$$f^* f_* \mathcal{O}_Y(mK_Y) \longrightarrow \Gamma(E, \mathcal{O}_Y(mK_Y) \otimes \mathcal{O}_E) \longrightarrow 0. \tag{8.2}$$

Then the cokernel of the canonical homomorphism $f^* f_* \mathcal{O}_Y(mK_Y) \to \mathcal{O}_Y(mK_Y)$ has the support without intersection with E. Therefore by replacing X by a sufficiently small neighborhood x we obtain the surjection

$$f^* f_* \mathcal{O}_Y(mK_Y) \twoheadrightarrow \mathcal{O}_Y(mK_Y).$$

This gives a proper morphism $\Phi : Y \to \mathbb{P}^N \times X$ over X such that $mK_Y = \Phi^*(H \times X)$, where H is a hyperplane of $\mathbb{P}^N$. The morphism Φ is a closed immersion over $X \setminus \{x\}$. On the other hand, by the fact that $mK_Y \mid_E$ is ample and the exactness of (8.2), $\Phi \mid_E$ is a closed immersion. Therefore Φ is a finite morphism and mK, the pull-back of a p_2-ample divisor $H \times X$ by Φ, is $f = p_2 \circ \Phi$-ample. This complete the proof of that $f : Y \to X$ is the canonical model.

Remark 8.2.7. In the proofs of Propositions 8.2.5 and 8.2.6, we used Proj and some properties of ample divisors, which we did not introduce in this book. The reader can refer to [Ha2] or [EGA II, III, IV].

How about the existence of a minimal model and the canonical model for cases of three or more dimensions? This problem is called the *relative Minimal Model Problem* and is considered as the relative version of the Minimal Model Problem. The Minimal Model Problem is a little bit away from the main topics of this book, but here we introduce it, because it seems useful for the reader to know it. The following definition is the most classical definition. After this definition was given around 1980, the notion of a minimal model was generalized to a pair consisting of projective variety and boundary with certain singularities. For further information see, for example, [KM].

Definition 8.2.8. A complete variety Y is called a *minimal model* (resp. *canonical model*) if the following are satisfied:

(1) Y has at worst terminal (resp. canonical) singularities.
(2) K_Y is nef (resp. ample).

Minimal Model Problem For every complete variety X such that $\kappa(X) \geq 0$, is there a minimal model (resp. canonical model) birationally equivalent to X?

For the 3-dimensional case, the relative Minimal Model Problem and Minimal Model Problem are affirmatively solved by contributions by Reid, Shokurov, Kawamata, and Kollár, and conclusively by Mori.

Theorem 8.2.9 (Mori [Mo1]).

(i) For every 3-dimensional normal (not necessarily isolated) singularity (X, x), there exists a minimal model and the canonical model. Furthermore, we can take a $\mathbb{Q}$-factorial minimal model.
(ii) For every 3-dimensional complete variety X such that $\kappa(X) \geq 0$, there is a minimal model and the canonical model birational equivalent to X. Furthermore, we can take a $\mathbb{Q}$-factorial minimal model.

The relative Minimal Model Problem and the Minimal Model Problem for dimensions higher than 3 has made much progress recently, as follows:

Theorem 8.2.10 (Birkar, Cascini, Hacon, McKernan [BCHM]).

(i) Every normal singularity (X, x) on an affine variety X has a minimal model and the canonical model.
(ii) For every projective non-singular variety X such that $\kappa(X) = \dim X$ there exists a minimal model birationally equivalent to X.

Now for the canonical model (or a minimal model) $f : Y \to X$, we compare K_Y and f^*K_X. Here we note that f^*K_X is defined if X is $\mathbb{Q}$-Gorenstein. In this case, we have the same as the 2-dimensional result (Proposition 7.1.17). To see this we prepare a divisor supported on the exceptional divisor.

Lemma 8.2.11. *Let $f : Y \to X$ be a projective partial resolution of a normal singularity (X, x). If an f-nef $\mathbb{Q}$-divisor Δ has the support on the exceptional divisor of f, then either $\Delta = 0$ or $-\Delta$ is effective with the support whole on $f^{-1}(S)$ for some closed subset $S \subset X$.*

In particular, if Δ is f-ample, the support of Δ is whole on the exceptional divisor of f.

Proof. Let $E_1, \cdots, E_r$ be the exceptional prime divisors of f, then, as f is projective, by Proposition 4.3.7, f turns out to be the blow-up by an ideal $I \subset \mathcal{O}_X$. Therefore, there exists an effective Cartier divisor L on Y such that $I\mathcal{O}_Y = \mathcal{O}_Y(-L)$, which is an f-ample invertible sheaf [EGA, II, 8.1.7].Reference [EGA, II, 8.1.7] is cited in text but not provided in the reference list. Please provide references in the list or delete these citations. As Δ is f-nef, for every $a \geq 0$ the divisor $\Delta - aL$ is f-nef. Let $\Delta = \sum_{i=1}^{r} a_i E_i$ and $L = \sum_{i=1}^{t} m_i E_i$ $(t \geq r,\ m_i > 0)$, where $E_{r+1}, \cdots, E_t$ are the components which are not exceptional. If there is an i such that $a_i \geq 0$, then we have $a := \max_{1 \leq i \leq r}\{a_i/m_i\} \geq 0$ and we can represent:

$$\Delta - aL = -\sum_{i=r+1}^{t} am_i E_i + \sum_{i=1}^{r}(a_i - am_i)E_i,$$

where $-am_i \leq 0$ $(i = r+1, \cdots, t)$, $a_i - am_i \leq 0$ $(i = 1, \cdots, r)$. Now assume $a = a_i/m_i$ holds for every $i = 1, \cdots, r$. If $-\sum_{i=r+1}^{t} am_i E_i < 0$, then there is an irreducible curve C on $\bigcup_{i=1}^{r} E_i$ such that $f(C) =$ one point and $(\Delta - aL) \cdot C = -\sum am_i E_i \cdot C < 0$, which is a contradiction. Then the equality $\sum_{i=r+1}^{t} am_i E_i = 0$ must hold. If $a \neq 0$, by $\sum_{i=r+1}^{t} m_i E_i = 0$ it follows that $\Delta = aL$, which is a contradiction to the fact that Δ is f-nef. Hence $a = 0$, then by $a = a_i/m_i$ $(i = 1, \cdots, r)$ we obtain $a_i = 0$.

Assume there is j $(1 \leq j \leq r)$ such that $a > a_j/m_j$. Let $S = \bigcup_{a > a_j/m_j} f(E_j)$. If there is an E_i in $f^{-1}(S)$ such that $a = a_i/m_i$, take an E_i among them such that $E_i \cap E_j \neq \emptyset$ $(a > a_j/m_j)$. Next take a curve $C \subset E_i$ such that $C \cap E_j \neq \emptyset, C \not\subset E_j$ and $f(C) =$ one point. Then $(\Delta - aL) \cdot C < 0$, which is a contradiction to f-nef. Therefore every prime divisor E_j on $f^{-1}(S)$ satisfies $a > a_j/m_j$.

Let $T = f$(total exceptional divisor), then by the definition of a it follows that $S \neq T$ and for a neighborhood U of a general point of $T \setminus S$, for every exceptional divisor E_i in $f^{-1}(U)$ the equality $a = a_i/m_i$ holds. Therefore by the previous discussion, $a_i = 0$ holds for such an E_i. As a consequence, for every E_j in $f^{-1}(S)$, the inequality $a_j < 0$ holds. It is easy to see that $f^{-1}(S)$ has no irreducible component with codimension ≥ 2.

The following theorem is a generalization of Proposition 7.1.17.

Theorem 8.2.12. *Let (X, x) be a normal $\mathbb{Q}$-Gorenstein singularity.*

(i) Let $f : Y \to X$ be the canonical model and $E_1, \cdots, E_r$ the exceptional prime divisors of f, then the following holds:

$$K_Y = f^* K_X + \sum_{i=1}^{r} m_i E_i, \quad m_i < 0\ (\forall i)$$

(ii) If $f : Y \to X$ is a projective minimal model and $E_1, \cdots, E_r$ the exceptional prime divisors of f, then the following holds:

$$K_Y = f^* K_X + \sum_{i=1}^{r} m_i E_i,$$

where either $m_i < 0$ $(\forall i)$ or $m_i = 0$ $(\forall i)$.

Proof. It is sufficient to apply Lemma 8.2.11 to $\Delta = \sum_{i=1}^{r} m_i E_i$.

8.3 Higher-Dimensional Canonical Singularities, Terminal Singularities

In this section we do not assume the isolatedness of a singularity and see the properties of canonical or terminal singularities: properties on the singular locus and hyperplane sections. In particular we study the 3-dimensional case closely.

Proposition 8.3.1. *Let X be a quasi-projective variety of dimension ≥ 2. Let H be a general hyperplane section of X, then the following hold:*

(i) If X is a normal and Cohen–Macaulay variety, H is also a normal and Cohen–Macaulay variety.
(ii) If X has at worst canonical (resp. terminal) singularities, then H has also at worst canonical (resp. terminal) singularities.

Proof. (i) If X is a Cohen–Macaulay variety, by Proposition 5.3.12, H is also a Cohen–Macaulay variety. Since X is normal, $\operatorname{codim}_X X_{\text{sing}} \geq 2$. On the other hand, H is general, we may assume that H does not contain an irreducible component of X_{sing}. Therefore $\operatorname{codim}_H H \cap X_{\text{sing}} \geq 2$. By Bertini's theorem [S, p. 24], we may assume that H is non-singular outside $H \cap X_{\text{sing}}$. Then, by Serre's criteria (Theorem 3.5.16), X is normal.

(ii) First H is normal by (i). Let $f : \widetilde{X} \to X$ be a good resolution of the singularities of X, E_i $(i = 1, \cdots, r)$ the prime exceptional divisors of f. As H is general, we may assume that $f(E_i) \not\subset H$ for every i. Then the strict transform $\widetilde{H}$ of H coincides with $f^*(H)$ and it is non-singular by Bertini's theorem. If we write $K_{\widetilde{X}} = f^*K_X + \sum_{i=1}^r m_i E_i$, then we obtain

$$K_{\widetilde{H}} = K_{\widetilde{X}} + \widetilde{H} \mid_{\widetilde{H}} = \left(f^*(K_X + H) + \sum m_i E_i\right) \mid_{\widetilde{H}}$$
$$= f'^*(K_H) + \sum_{i=1}^r m_i (E_i \mid_{\widetilde{H}}),$$

where $f' = f \mid_{\widetilde{H}}$. Here, if X has canonical (resp. terminal) singularities, then $m_i \geq 0$ (resp. $m_i > 0$), which yields that H has also canonical (resp. terminal) singularities.

Corollary 8.3.2. *For a quasi-projective variety X, the following hold:*

(i) If X has at worst canonical singularities, then outside a closed subset of codimension 3, X has at worst Gorenstein singularities.
(ii) If X has at worst terminal singularities, then $\operatorname{codim} X_{\text{sing}} \geq 3$. *In particular, 3-dimensional terminal singularities are isolated.*

Proof. (i) Let $Z \subset X$ be the subset consisting of non-Gorenstein singularities, then it is known that Z is a closed subset. Let H_1 be a general hyperplane section, let H_2 be a general hyperplane sections, and take H_i like this to obtain the sequence:

$X \supset H_1 \supset H_2 \supset \cdots \supset H_{n-2}$ ($n = \dim X$). By Proposition 8.3.1, H_{n-2} is a surface with at worst canonical singularities, therefore it is a Gorenstein variety. By $H_{n-2} \cap Z = \emptyset$, we have $\operatorname{codim} Z \geq 3$.

(ii) Take the singular locus X_{sing} as Z, and follow the same discussion as in (i).

Remark 8.3.3. Let (X, x) be a canonical (resp. terminal) singularity and $\rho : \overline{X} \to X$ the canonical covering. By Proposition 6.2.3 and Theorem 6.2.9 (i), $\overline{X}$ has at worst 1-Gorenstein canonical (resp. terminal) singularities. By this, some discussions are reduced to the case of 1-Gorenstein singularities. Therefore studying 1-Gorenstein canonical (resp. terminal) singularities may be useful for studying general canonical (resp. terminal) singularities. Here, we should note that 1-Gorenstein canonical is equivalent to Gorenstein rational by Corollary 6.2.15.

Proposition 8.3.4. *(i) Let (X, x) be a Gorenstein rational singularity. A general hyperplane section (H, x) of X passing through the point x is a Gorenstein singularity which is either rational or satisfying $\mathfrak{m}_{H,x}\omega_H = f_*\omega_{\widetilde{H}}$. Here, $f : \widetilde{H} \to H$ is a resolution of the singularities of H.*
In particular, if $\dim(X, x) = 3$, (H, x) is a Du Val singularity or a minimal elliptic singularity.

(ii) For a normal singularity (X, x), if there is a hyperplane section (H, x) which is Gorenstein and rational, then (X, x) is also a Gorenstein rational singularity.

Proof. In general, (X, x) is a Gorenstein singularity if and only if (H, x) is a Gorenstein singularity (Proposition 5.3.12).

(i) Let $\sigma : Y_1 \to X$ be the blow-up by the maximal ideal $\mathfrak{m}_{X,x}$, $\varphi : Y \to Y_1$ a resolution of the singularities of Y_1, and $f := \sigma \circ \varphi$. As f factors through σ, the sheaf $\mathfrak{m}_{X,x}\mathcal{O}_Y$ is invertible. So it can be written as $\mathcal{O}_Y\left(-\sum_{i=1}^r s_i E_i\right)$ ($s_i > 0$, E_i is an exceptional prime divisor such that $f(E_i) = \{x\}$). Since H is general, the strict transform $\widetilde{H}$ of H on Y satisfies the following equality:

$$f^*H = \widetilde{H} + \sum_{i=1}^r s_i E_i.$$

Again as f factors through σ, if we consider the linear system of hyperplanes H passing through x, the corresponding linear system consisting of $\widetilde{H}$ for general H has no base point on $f^{-1}(x)$. Then, by Bertini's theorem, a general member $\widetilde{H}$ has no singular point on $f^{-1}(x)$. By this $f' := f \mid_{\widetilde{H}} : \widetilde{H} \to H$ is considered as a resolution of the singularities of (H, x). If (X, x) is a Gorenstein canonical singularity, we have

$$K_Y = f^*K_X + \sum_{i=1}^r m_i E_i + \sum_{j=1}^t n_j F_j, \quad (m_i \geq 0,\ n_j \geq 0)$$

$$K_Y + \widetilde{H} = f^*(K_X + H) + \sum_{i=1}^r (m_i - s_i) E_i + \sum n_j F_j,$$

therefore the following holds:

$$K_{\tilde{H}} = f'^{*}(K_H) + \sum_{i=1}^{r}(m_i - s_i)E'_i + \sum n_j F'_j,$$

where $\{F_j\}$ is an exceptional prime divisor except for $\{E_i\}$ and let $E'_i = E_i \mid_{\tilde{H}}$, $F'_j = F_j \mid_{\tilde{H}}$. Thus for every $\theta \in \omega_H$, we have $\nu_{E'_i}(\theta) \geq m_i - s_i$, $\nu_{F'_j}(\theta) \geq n_j \geq 0$.

As every element $h \in \mathfrak{m}_{H,x}$ comes from $\mathfrak{m}_{X,x}$, by $\nu_{E'_i}(h) \geq s_i$ we have

$$\nu_{E'_i}(h\theta) \geq m_i \geq 0, \ \nu_{F'_j}(h\theta) \geq n_j \geq 0,$$

which implies $h\theta \in f'_*\omega_{\tilde{H}}$. Hence we obtain $\mathfrak{m}_{H,x}\omega_H \subset f'_*\omega_{\tilde{H}}$. On the other hand, the inclusion $f'_*\omega_{\tilde{H}} \subset \omega_H$ is obvious. By $\dim_{\mathbb{C}} \omega_H/\mathfrak{m}_{H,x}\omega_H = 1$, either $f'_*\omega_{\tilde{H}} = \mathfrak{m}_{H,x}\omega_H$ or $f'_*\omega_{\tilde{H}} = \omega_H$ holds. The latter equality implies that (H, x) is a rational singularity (Theorem 6.2.14). In particular, when $\dim(X, x) = 3$, then (H, x) is a 2-dimensional normal singularity, therefore it is isolated. As is seen in the proof of Theorem 8.1.17, the sheaf $\omega_H/f'_*\omega_{\tilde{H}}$ and $R^1 f'_*\mathcal{O}_{\tilde{H}}$ are dual to each other, which implies $\dim R^1 f'_*\mathcal{O}_{\tilde{H}} = \dim \omega_H/f'_*\omega_{\tilde{H}} = 0$ or 1. Therefore (H, x) is a Du Val singularity or a minimal elliptic singularity.

(ii) The following lemma gives a flat morphism $\pi : \mathscr{X} \to \mathbb{A}^1_{\mathbb{C}}$ such that $X_0 := \pi^{-1}(0) \simeq H \times \mathbb{A}^1_{\mathbb{C}}$, $X_t := \pi^{-1}(t) \simeq X \quad (t \neq 0)$. By Elkik's theorem (Theorem 9.1.12) if X_0 has at worst rational singularities, then X_t also has at worst rational singularities.

Lemma 8.3.5. *Let X be an affine variety and H a hyperplane on X. Then there exists a flat morphism $\pi : \mathscr{X} \to \mathbb{A}^1_{\mathbb{C}}$ such that $\mathscr{X}_0 := \pi^{-1}(0) \cong H \times \mathbb{A}^1$, $\mathscr{X}_t := \pi^{-1}(t) \cong X$ $(t \neq 0)$.*

Proof. Let $I \subset \mathbb{C}[X_1, \cdots, X_N]$ be the defining ideal of $X \subset \mathbb{A}^N_{\mathbb{C}}$ and H is defined by the equation $X_N = 0$. Define a ring homomorphism $\varphi : \mathbb{C}[X_1, \cdots, X_N] \to \mathbb{C}[Y_1, \cdots, Y_{N+1}]$ by $X_i \mapsto Y_i$ $(i \leq N-1)$ $X_N \mapsto Y_N Y_{N+1}$. Let $J \subset \mathbb{C}[Y_1, \cdots, Y_{N+1}]$ be the ideal generated by $\varphi(I)$ and let $\mathscr{X} \subset \mathbb{A}^{N+1}$ be the variety defined by J, then the morphism $\pi : \mathscr{X} \to \mathbb{A}^1_{\mathbb{C}}$, $(y_1, \cdots, y_{N+1}) \to y_{N+1}$ has the required property. Indeed $\mathscr{X}_0 \subset \mathbb{A}^N_{\mathbb{C}}$ is defined by the image of I by the homomorphism $\varphi_0 : \mathbb{C}[X_1, \cdots, X_N] \to \mathbb{C}[Y_1, \cdots, Y_N]$, $X_i \mapsto Y_i$ $(i \leq N-1)$, $X_N \mapsto 0$, therefore $\mathscr{X}_0$ is isomorphic to $H \times \mathbb{A}^1_{\mathbb{C}}$. On the other hand, $\mathscr{X}_t \subset \mathbb{A}^N_{\mathbb{C}}$ $(t \neq 0)$ is defined by the image of I by the homomorphism, $\varphi_t : \mathbb{C}[X_1, \cdots, X_N] \to \mathbb{C}[Y_1, \cdots, Y_N]$, $X_i \mapsto Y_i$ $(i \leq N-1)$ $X_N \mapsto tY_N$, and $\mathscr{X}_t$ is isomorphic to X.

Next we define a singularity which plays an important role.

Definition 8.3.6. A 3-dimensional normal singularity (X, x) is called a *compound Du Val singularity* of there is a hypersurface H such that (H, x) is a Du Val singularity (*cDV singularity* for short).

Theorem 8.3.7. *For a 3-dimensional 1-Gorenstein singularity (X, x) the following are equivalent:*

(i) (X, x) is a terminal singularity.
(ii) (X, x) is an isolated cDV singularity.

Proof. (i)⇒(ii) By Corollary 8.3.2, (X, x) is an isolated singularity. Let (H, x) be a general hyperplane section, and $h = 0$ is the defining equation of H. Let $f : Y \to X$ be a resolution of the singularities of X factoring through the blow-up by the maximal ideal $\mathfrak{m}_{X,x}$. Let $\widetilde{H}$ be the strict transform of H on Y, then $f' := f \mid_{\widetilde{H}} : \widetilde{H} \to H$ is a resolution of the singularities of H. If we represent $f^{-1}(x)_{\text{red}} = \sum_{i=1}^{r} E_i$, we have

$$K_Y = f^* K_X + \sum_{i=1}^{r} m_i E_i, \quad m_i > 0\ (\forall i).$$

On the other hand, by $\widetilde{H} = f^*H - \sum_{i=1}^{r} \nu_{E_i}(h) E_i$ and the adjunction formula:

$$K_{\widetilde{H}} = f'^* K_H + \sum \bigl(m_i - \nu_{E_i}(h)\bigr) E_i \mid_{\widetilde{H}}. \tag{8.3}$$

Here, if (H, x) is a minimal elliptic singularity, then by [Re1, 2.9] there exists E'_i among $\{E'_i := E_i \mid_{\widetilde{H}}\}$ such that E'_i appears on the canonical model and for a general $\overline{h} \in \mathfrak{m}_{H,x}$ $\nu_{E'_i}(\overline{h})$ is the coefficient of E'_i in the fundamental cycle $Z_{f'}$. By Theorem 7.6.6 this is also the coefficient of E'_i in $-K_{\widetilde{H}}$. Then by (8.3) the equality $m_i - \nu_{E_i}(h) = -\nu_{E'_i}(\overline{h})$ holds. As $h \in \mathfrak{m}_{X,x}$, $\overline{h} \in m_{H,x}$ are general and E_i and $\widetilde{H}$ intersect normally, we have $\nu_{E_i}(h) = \nu_{E'_i}(\overline{h})$. Therefore $m_i = 0$, which contradicts the fact that (X, x) is a terminal singularity.

(ii) ⇒ (i) Take $f : Y \to X$, $\widetilde{H}$ and $\{E_i\}$ as above. Consider the following exact sequence:

$$0 \longrightarrow \omega_Y \longrightarrow \omega_Y(\widetilde{H}) \longrightarrow \omega_{\widetilde{H}} \longrightarrow 0.$$

By the Grauert–Riemenschneider vanishing theorem, we obtain the following diagram:

$$\begin{array}{ccccccccc}
0 & \longrightarrow & f_*\omega_Y & \longrightarrow & f_*\omega_Y(\widetilde{H}) & \longrightarrow & f_*\omega_{\widetilde{H}} & \longrightarrow & R^1 f_*\omega_Y = 0 \\
 & & \downarrow \alpha & & \downarrow \beta & & \downarrow \gamma & & \\
0 & \longrightarrow & \omega_X & \longrightarrow & \omega_X(H) & \longrightarrow & \omega_H & \longrightarrow & 0.
\end{array}$$

By the assumption, the singularity (H, x) is rational. By Proposition 8.3.4, the singularity (X, x) is also a rational singularity, therefore the inclusions α, γ are identities. Therefore β is also an identity. This implies that for every $\theta \in \omega_X$ the element $\frac{\theta}{h} \in \omega_X(H)$ belongs to $f_*\omega_Y(\widetilde{H})$, so the inequality $\nu_{E_i}\left(\frac{\theta}{h}\right) \geq 0$ holds for every E_i, which shows that (X, x) is a terminal singularity.

As is seen in Remark 8.3.3 and Theorem 8.3.7, a 3-dimensional terminal singularity is the quotient of a cDV singularity which is a hypersurface singularity by a cyclic group. Then we will study a slightly wider class of singularities: the quotients of a hypersurface singularities by a cyclic group. As a non-singular point is also regarded as a hypersurface singularity, we will also study a cyclic quotient singularity.

Lemma 8.3.8. *Let $(Y, 0) \subset (\mathbb{C}^{n+1}, 0)$ be an n-dimensional hypersurface singularity, G a finite group acting on Y and (X, x) the quotient of $(Y, 0)$ by G. Then there exists a small finite subgroup G' of $GL(n+1, \mathbb{C})$ acting on Y such that $(Y, 0)/G' \simeq (X, x)$.*

Proof. If the stabilizer $G_0 := \{g \in G \mid g0 = 0\}$ of $0 \in Y$ does not coincide with G, then $X = V/G_0$ for an appropriate neighborhood V of 0. Therefore we may assume that $G_0 = G$. As 0 is fixed by the action of G, the maximal ideal $\mathfrak{m}$ of the point 0 is fixed by the action of G. Therefore every element $g \in G$ induces an isomorphism $g_1 : \mathfrak{m}/\mathfrak{m}^2 \xrightarrow{\sim} \mathfrak{m}/\mathfrak{m}^2$ of the vector spaces. Let $x_0, \cdots, x_n$ be a coordinate system of $\mathbb{C}^{n+1}$, then g_1 gives an automorphism of the vector space generated by $x_0, \cdots, x_n$. By the isomorphism $\widehat{\mathcal{O}}_{\mathbb{C}^{n+1},0} \cong \mathbb{C}[[x_0, \cdots, x_n]]$, the isomorphism $g^* : \widehat{\mathcal{O}}_{Y,0} \to \widehat{\mathcal{O}}_{Y,0}$ can be extended to a ring homomorphism $g_\infty : \widehat{\mathcal{O}}_{\mathbb{C}^{n+1},0} \to \widehat{\mathcal{O}}_{\mathbb{C}^{n+1},0}$. Here, as g_1 is isomorphic, g_∞ is isomorphic. By this fact and by Theorem 7.4.5 the action on $(Y, 0)$ can be lifted to the linear action on $(\mathbb{C}^{n+1}, 0)$. By Theorem 7.4.8 we may assume that G is a small linear subgroup.

Lemma 8.3.9. *Let $G \subset GL(n+1, \mathbb{C})$ be a small cyclic group of order r and let*

$$g = \begin{pmatrix} \varepsilon^{a_0} & & 0 \\ & \ddots & \\ 0 & & \varepsilon^{a_n} \end{pmatrix},$$

where ε is an rth root of unity and $0 \le a_i \le r-1$. Let $\overline{M} := \mathbb{Z}^{n+1}$, $\overline{N} := \mathrm{Hom}_{\mathbb{Z}}(\overline{M}, \mathbb{Z})$ and $N := \overline{N} + \frac{1}{r}(a_0, \cdots, a_n)\mathbb{Z}$. Let σ be the positive octant $\mathbb{R}^{n+1}_{\ge 0} \subset N_{\mathbb{R}} = \overline{N}_{\mathbb{R}}$ and Δ be the fan corresponding to σ. In this case

$$\mathbb{C}^{n+1}/G \cong T_N(\Delta).$$

Proof. Let $M := \mathrm{Hom}_{\mathbb{Z}}(N, \mathbb{Z})$, then it has the natural inclusion $M \subset \overline{M}$. The affine coordinate ring of $\mathbb{C}^{n+1}/G$ is $\mathbb{C}[x_0, \cdots, x_n]^G = \mathbb{C}[\overline{M} \cap \sigma^\vee]^G$. Here, the monomial $x^{\boldsymbol{m}} = x_1^{m_1} \cdot \cdots \cdot x_n^{m_n}$ is G-invariant if and only if $(\boldsymbol{a}, \boldsymbol{m}) = \sum a_i m_i \equiv 0 \pmod{r}$, because $g^*(x^{\boldsymbol{m}}) = \varepsilon^{\Sigma a_i m_i} x^{\boldsymbol{m}}$. On the other hand, an element $\boldsymbol{m} \in \overline{M}$ belongs to M if and only if it maps all elements of N into $\mathbb{Z}$, i.e., $(\frac{1}{r}(a_0, \cdots, a_n), \boldsymbol{m}) \in \mathbb{Z}$ and this is equivalent to $(\boldsymbol{a}, \boldsymbol{m}) \equiv 0 \pmod{r}$. Therefore $\mathbb{C}[\overline{M} \cap \sigma^\vee]^G = \mathbb{C}[M \cap \sigma^\vee]$ and this coincides with the affine coordinate ring of $T_N(\Delta)$.

Now we define the weight of a function.

Definition 8.3.10. Let $M = \mathbb{Z}^{n+1}$, $N = \mathrm{Hom}_{\mathbb{Z}}(M, \mathbb{Z})$ and σ a rational strongly convex polyhedral cone in $N_{\mathbb{R}}$. For $a \in \sigma$, $f \in \mathbb{C}[M \cap \sigma^{\vee}]$ define

$$\boldsymbol{a}(f) := \min\{(\boldsymbol{a}, \boldsymbol{m}) \mid x^{\boldsymbol{m}} \in f\},$$

where the notation $x^{\boldsymbol{m}} \in f$ means that the monomial $x^{\boldsymbol{m}}$ appears in f with a non-zero coefficient.

An element $\boldsymbol{a} \in N$ is called *primitive* if on the half line $\mathbb{R}_{\geq 0}\boldsymbol{a}$ there is no point of N between 0 and $\boldsymbol{a}$.

Now we will study properties of hypersurface singularities by using a weighted blow-up (Example 4.4.20).

Proposition 8.3.11. *Let $(Y, 0) \subset (\mathbb{C}^{n+1}, 0)$ be a normal hypersurface singularity defined by an equation $f = 0$. Let $g = \begin{pmatrix} \varepsilon^{a_0} & & 0 \\ & \ddots & \\ 0 & & \varepsilon^{a_n} \end{pmatrix}$ (ε is a primitive rth root of unity and $0 \leq a_i \leq r-1$) generate a small cyclic group G acting on $(Y, 0)$.*

Let $(X, x) := (Y, 0)/G$ and we use the notation in Lemma 8.3.9. For every primitive element $\boldsymbol{a} \in N \cap \sigma$ let $\Delta(\boldsymbol{a})$ be the star-shaped subdivision of Δ by $\boldsymbol{a}$. Let X_a be the strict transform of $X \subset \mathbb{C}^{n+1}/G$ by the morphism

$$T_N(\Delta(\boldsymbol{a})) \xrightarrow{\varphi_a} T_N(\Delta) = \mathbb{C}^{n+1}/G$$

and let Γ be the exceptional divisor of φ_a. Then the following hold:

(i) $X_a = \varphi_a^* X - \boldsymbol{a}(f)\Gamma$, $K_{T_N(\Delta(a))} = \varphi_a^* K_{T_N(\Delta)} + (\boldsymbol{a}(x_0 \cdots x_n) - 1)\Gamma$.

(ii) Let $a := \boldsymbol{a}(x_0 \cdots x_n) - \boldsymbol{a}(f) - 1 \leq 0$ (resp. < 0), then there exists a resolution $\psi : \widetilde{X} \to X_a$ of the singularities of $X_{\boldsymbol{a}}$ such that under the representation

$$K_{\widetilde{X}} = (\varphi_a \circ \psi)^* K_X + \sum m_i E_i$$

the components E_i corresponding to the components of $\Gamma \cap X_{\boldsymbol{a}}$ have the coefficients $m_i \leq 0$ (resp. < 0). Therefore, in particular (X, x) is not a terminal singularity (resp. not a canonical singularity).

Proof. (i) As the polynomial $f^r \in \mathbb{C}[M \cap \sigma^{\vee}]$ defines a Cartier divisor rX,

$$\varphi_{\boldsymbol{a}}^*(rX) = rX_{\boldsymbol{a}} + \nu_\Gamma(f^r)\Gamma.$$

Therefore if one proves $\nu_\Gamma(g) = \boldsymbol{a}(g)$ for every $g \in \mathbb{C}[M \cap \sigma^{\vee}]$, then the first equality of (i) follows. Let $\tau = \mathbb{R}_{\geq 0}\boldsymbol{a}$, then $\Gamma \cap U_\tau = \mathrm{orb}\,\tau$ and $\Gamma \cap U_\tau \subset U_\tau$ are isomorphic to $(\mathbb{C}^*)^n \times \{0\} \subset (\mathbb{C}^*)^n \times \mathbb{C}$, which implies that $\Gamma \cap U_\tau$ is defined by one equation $\zeta = 0$ on U_τ. If we decompose $g = \zeta^m h$ $h \notin (\zeta)$ in the affine coordinate ring of U_τ, it is sufficient to prove that $\boldsymbol{a}(g) = m$. It is clear that $\boldsymbol{a}(g) = m\boldsymbol{a}(\zeta) + \boldsymbol{a}(h)$. By [Od1, 5.3, Remark], the ideal (ζ) is equal to the ideal generated by $\{\eta \in \mathbb{C}[\tau^{\vee} \cap M] \mid \boldsymbol{a}(\eta) > 0\}$. Therefore $\boldsymbol{a}(h) = 0$. On the other hand, ζ is a generator of the ideal and a is a primitive element, therefore there exists an element $\boldsymbol{b} \in M$ such that $(\boldsymbol{a}, \boldsymbol{b}) = 1$. (Indeed, as $\Delta(\boldsymbol{a})$ has a unimodular subdivision, there is a unimodular matrix with a as the first column. Let ${}^t\boldsymbol{b}$ be the first row of the inverse matrix.) Hence $\boldsymbol{a}(\zeta) = 1$. By this, it follows that $\boldsymbol{a}(g) = m$.

For the second equality in (i), note that on a toric variety $T_N(\Sigma)$ the equality

$$-K_{T_N(\Sigma)} = \sum_{\eta \in \Sigma(1)} \overline{\text{orb}\, \eta}$$

holds, where $\Sigma(1)$ is the set of 1-dimensional cones in Σ [Od1, 6.6]. In particular for $T_{\overline{N}}(\Delta) = \mathbb{C}^{n+1}$,

$$-K_{T_{\overline{N}}(\Delta)} = \sum_{i=0}^{n} \overline{\text{orb}\, \mathbb{R}_{\geq 0} \boldsymbol{e}_i} = \{x_0 \cdot \cdots \cdot x_n = 0\},$$

where $\boldsymbol{e}_0, \cdots, \boldsymbol{e}_n$ are the n-dimensional unit vectors. On the other hand, $\pi : T_{\overline{N}}(\Delta) \to T_N(\Delta)$ has no ramification divisor, because the group G is small, therefore we have $rK_{T_{\overline{N}}(\Delta)} = \pi^*(rK_{T_N(\Delta)})$. Therefore $-rK_{T_N(\Delta)}$ is a Cartier divisor defined by $x_0^r \cdot \cdots \cdot x_n^r$. Then, apply the previous discussion for $f = x_0 \cdot \cdots \cdot x_n$ and we obtain $\varphi_{\boldsymbol{a}}^*(-K_{T_N(\Delta)}) = \sum_{i=0}^{n} \overline{\text{orb}\, \mathbb{R}_{\geq 0} \boldsymbol{e}_i} + \boldsymbol{a}(x_0 \cdot \cdots \cdot x_n)\Gamma$. On the other hand, as $-K_{T_N(\Delta(\boldsymbol{a}))} = \sum_{i=0}^{n} \overline{\text{orb}\, \mathbb{R}_{\geq 0} \boldsymbol{e}_i} + \Gamma$, it follows that

$$K_{T_N(\Delta(\boldsymbol{a}))} = \varphi_{\boldsymbol{a}}^*(K_{T_N(\Delta)}) + (\boldsymbol{a}(x_0 \cdot \cdots \cdot x_n) - 1)\Gamma.$$

(ii) By the two equalities in (i) we obtain

$$K_{T_N(\Delta(\boldsymbol{a}))} + X_{\boldsymbol{a}} = \varphi_{\boldsymbol{a}}^*(K_{T_N(\Delta)} + X) + a\Gamma.$$

First, consider the case that there is a point in $\Gamma \cap X_a$ such that $X_{\boldsymbol{a}}$ is normal at the point. Then, $X_{\boldsymbol{a}}$ is non-singular along an open subset in $\Gamma \cap X_{\boldsymbol{a}}$, therefore, a resolution $\psi : Y \to X_{\boldsymbol{a}}$ of the singularities of $X_{\boldsymbol{a}}$ is isomorphic at a general point of $\Gamma \cap X_a$. Hence on the neighborhood of the point we have:

$$K_Y = (\varphi_{\boldsymbol{a}} \circ \psi)^*(K_X) + a\Gamma_X, \quad a \leq 0 \text{ (or } a < 0).$$

Next, consider the case that X_a is not normal along $\Gamma \cap X_{\boldsymbol{a}}$. The variety $T_N(\Delta(a))$ is isomorphic to $S \times \mathbb{C}^{n-1}$ (S is a surface with a cyclic quotient singularity) on a neighborhood of each point outside a closed subset of codimension 3. Therefore, at a neighborhood of a general point of $\Gamma \cap X_a$ we can take a resolution $\alpha : Z \to T_N(\Delta(\boldsymbol{a}))$ of the singularities of $T_N(\Delta(\boldsymbol{a}))$ such that $K_Z = \alpha^* K_{T_N(\Delta(\boldsymbol{a}))} - D$, $D \geq 0$ (Proposition 7.1.17). Let $X' \subset Z$ be the strict transform of $X_{\boldsymbol{a}}$, then $X' = \alpha^* X_a - D'$, $D' \geq 0$. Here, as X' is a divisor on a non-singular variety Z, it is a Gorenstein variety, and by the adjunction formula, it follows that

$$\begin{aligned} K_{X'} &= K_Z + X' \mid_{X'} = \alpha^*(K_{T_N(\Delta(\boldsymbol{a}))} + X_a) - (D + D') \mid_{X'} \\ &= \alpha^* \cdot \varphi_{\boldsymbol{a}}^* K_X + a\alpha^*\Gamma - (D + D') \mid_{X'}. \end{aligned}$$

Now let $\pi : \overline{X} \to X'$ be the normalization of X', then $K_{\overline{X}} < \pi^* K_{X'}$. Here, $h = \varphi_{\boldsymbol{a}} \circ \alpha \circ \pi$ is a resolution of the singularities of X outside the closed subset of $\overline{X}$ of codimension 2, therefore at the points if we write

$$K_{\overline{X}} = h^* K_X + \sum m_i E_i,$$

then for the divisor E_i corresponding to a component of $\alpha^*\Gamma \cap X'$, $m_i \leq 0$, if $a \leq 0$ and $m_i < 0$ if $a < 0$.
By this we obtain necessary conditions for a singularity to be terminal or canonical.

Theorem 8.3.12. *Let (X, x), $(Y, 0)$, f, G be as in Proposition 8.3.11. If (X, x) is a terminal singularity (resp. canonical singularity), then for a primitive element $\boldsymbol{a} \in N \cap \sigma$ the following holds:*

$$\boldsymbol{a}(x_0 \cdot \cdots \cdot x_n) > \boldsymbol{a}(f) + 1$$
$$(\textit{resp. } \boldsymbol{a}(x_0 \cdot \cdots \cdot x_n) \geq \boldsymbol{a}(f) + 1).$$

Proof. It is clear from Proposition 8.3.11 (ii).

Definition 8.3.13. Let $G \subset GL(n, \mathbb{C})$ be a small cyclic group generated by $g = \begin{pmatrix} \varepsilon^{a_1} & & 0 \\ & \ddots & \\ 0 & & \varepsilon^{a_n} \end{pmatrix}$, where ε is a primitive rth root of unity and $0 \leq a_i \leq r - 1$ $(i = 1, \cdots, n)$.
In this case the singularity $(X, x) = (\mathbb{C}^n, 0)/G$ is said to be of *type* $\frac{1}{r}(a_1, \cdots, a_n)$.

The necessary and sufficient condition for a cyclic quotient singularity to be terminal (resp. canonical) is given below:

Theorem 8.3.14. *A quotient singularity (X, x) of type $\frac{1}{r}(a_1, \cdots, a_n)$ is terminal (resp. canonical) if and only if*

$$\frac{1}{r}\sum_{i=1}^{n} \overline{ka_i} > 1 \quad (\textit{or} \geq 1) \quad (k = 1, 2, \cdots, r - 1).$$

Here, $\overline{\alpha}$ is an integer such that $0 \leq \overline{\alpha} < r$ and $\alpha \equiv \overline{\alpha}$ (mod. r).

Proof. Use the notation in Lemma 8.3.9 and Proposition 8.3.11. By the second equality in Proposition 8.3.11 (i), if (X, x) is terminal (resp. canonical), then for a primitive element $\boldsymbol{a} \in \sigma^\vee \cap M$ the inequality $\boldsymbol{a}(x_1 \cdot \cdots \cdot x_n) > 1$ (resp. ≥ 1) holds. In particular, for $\boldsymbol{a} = \frac{1}{r}(\overline{ka_1}, \cdots, \overline{ka_n})$, we obtain the required inequality.

Let us prove the converse. Let (N, Σ) be a unimodular subdivision of (N, Δ) and $\varphi : T_N(\Sigma) \to T_N(\Delta) = X$ the corresponding morphism. In the same way as in the proof of Proposition 8.3.11, we obtain

$$K_{T_N(\Sigma)} = \varphi^* K_{T_N(\Delta)} + \sum_{\boldsymbol{p} \in \Sigma[1] \setminus \Delta[1]} (\boldsymbol{p}(x_1 \cdot \cdots \cdot x_n) - 1)\overline{\mathrm{orb}\, \mathbb{R}_{\geq 0} \boldsymbol{p}},$$

where $\Sigma[1], \Delta[1]$ are the sets of primitive elements $p \in N$ such that $\mathbb{R}_{\geq 0}\boldsymbol{p} \in \Sigma(1)$ or $\Delta(1)$, respectively. Therefore if $\boldsymbol{p}(x_1 \cdot \cdots \cdot x_n) > 1$ (resp. ≥ 1), $\forall \boldsymbol{p} \in \Sigma[1] \setminus \Delta[1]$, then (X, x) is terminal (resp. canonical).

Now define $\boldsymbol{p} = (p_1, \cdots, p_n)$. If there is i such that $p_i \geq 1$, then by $\boldsymbol{p} \in \sigma \cap N$ and $\boldsymbol{p} \neq \boldsymbol{e}_1, \cdots, \boldsymbol{e}_n$ (unit vectors) the inequality $\boldsymbol{p}(x_1 \cdot \cdots \cdot x_n) > 1$ obviously holds.

Conversely, if for every i $p_i < 1$, then, as $\boldsymbol{p}$ is an element of N, it is expressed as $\frac{1}{r}(\overline{ka_1}, \overline{ka_2}, \cdots, \overline{ka_n})$. By the assumption, for such a $\boldsymbol{p}$ the inequality $\boldsymbol{p}(x_1 \cdot \cdots \cdot x_n) > 1$ (resp. ≥ 1) holds. Therefore $X = T_N(\Delta)$ has terminal (resp. canonical) singularities.

Now we will determine 3-dimensional terminal singularities.

As is seen in Remark 8.3.3 and Theorem 8.3.7, a 3-dimensional terminal singularity is either non-singular or the quotient of an isolated cDV singularity by a cyclic group. First we will determine terminal singularities which are quotients of non-singular point.

Lemma 8.3.15. *Let (X, x) be a cyclic quotient singularity of type $\frac{1}{r}(a, b, c)$ and let $d = a + b + c$. Then (X, x) is a terminal singularity if and only if*

$$\overline{ak} + \overline{bk} + \overline{ck} = \overline{dk} + r, \quad \overline{dk} > 0,$$

holds for $k = 1, \cdots, r - 1$.

Proof. First, assume that these equalities and inequalities hold. Then we have $\frac{1}{r}(\overline{ak} + \overline{bk} + \overline{ck}) > 1$, $k = 1, \cdots, r - 1$, therefore by Theorem 8.3.14, (X, x) is a terminal singularity.

Conversely, assume that (X, x) is a terminal singularity. As $r < \overline{ak} + \overline{bk} + \overline{ck} \leq 3r - 3$, either

$$\overline{ak} + \overline{bk} + \overline{ck} = \overline{dk} + r, \quad \text{or}$$
$$\overline{ak} + \overline{bk} + \overline{ck} = \overline{dk} + 2r$$

holds. But actually the latter equality does not hold. Indeed if it holds, let $k' := k(r - 1)$, then we have $\overline{ak'} + \overline{bk'} + \overline{ck'} = \overline{ak(r - 1)} + \overline{bk(r - 1)} + \overline{ck(r - 1)} =$

$\overline{(-ak)} + \overline{(-bk)} + \overline{(-ck)} = (r - \overline{ak}) + (r - \overline{bk}) + (r - \overline{ck}) = 3r - (\overline{ak} + \overline{bk} + \overline{ck}) \leq r$, a contradiction. By $r < \overline{ak} + \overline{bk} + \overline{ck}$, the inequality $\overline{dk} > 0$ is also clear.

The following lemma is also used for the theorem determining 3-dimensional cyclic quotient terminal singularities. But we skip the proof because it is technical, see for details [Re2, Appendix to §5].

Lemma 8.3.16 (Terminal lemma). *Let n, m be positive integers such that $n \equiv m$ (mod. 2).*
Let $\frac{1}{r}(a_1, a_2, \cdots, a_n; b_1, \cdots, b_m)$ be a combination of integers r, a_i, b_j $(1 \leq i \leq n,\ 1 \leq j \leq m)$. Assume that a_i, b_j are prime to r. Then the following are equivalent:

(i) $\displaystyle\sum_{i=1}^{n} \overline{a_i k} = \sum_{j=1}^{m} \overline{b_j k} + \frac{n-m}{2} \cdot r, \quad k = 1, \cdots, r-1.$

(ii) *The set $\{a_i, -b_j\}$ consisting of $n+m$ integers is divided into $(n+m)/2$ pairs $(a_i, a_{i'})$, $(b_j, b_{j'})$, $(a_i, -b_j)$ such that the sum of two numbers of each pair is 0 modulo r.*

Theorem 8.3.17. *A cyclic quotient singularity $(X, x) = (\mathbb{C}^3, 0)/G$, $(\#G = r)$ is a terminal singularity if and only if it is of type $\frac{1}{r}(a, -a, 1)$, where a and r are mutually prime.*

Proof. By Lemma 8.3.15, the singularity of type $\frac{1}{r}(a, b, c)$ is terminal if and only if the following hold:

$$\overline{ak} + \overline{bk} + \overline{ck} = \overline{dk} + r, \quad \overline{dk} > 0, \quad k = 1, \cdots, r-1. \tag{8.4}$$

As the type $\frac{1}{r}(a, -a, 1)$ satisfies this condition, the singularity of this type is terminal.

Conversely, assume that the quotient singularity (X, x) of type $\frac{1}{r}(a, b, c)$ is terminal. Then as (X, x) is an isolated singularity, the action of G is free outside 0. Therefore a, b, c are prime to r. By (8.4), the equality in (i) of the terminal lemma with $n = 3, m = 1$ holds. Therefore (ii) of the terminal lemma follows, which implies $a + b \equiv 0$ (mod. r), $c - d \equiv 0$ (mod. r). Hence, $\frac{1}{r}(a, b, c) = \frac{1}{r}(a, -a, c)$. Here, as c is prime to r, we may let c be 1.

Next we show the classification theorem of 3-dimensional terminal singularities which are cyclic quotient singularities of singular hypersurfaces $(Y, 0) \subset (\mathbb{C}^4, 0)$. This was proved by Mori by using Theorem 8.3.12, which led to the solution of the 3-dimensional Minimal Model Problem.

Theorem 8.3.18 ([Mo2, Theorem 12, 23, 25]).

(i) *Let $(Y, 0) \subset (\mathbb{C}^4, 0)$ be a cDV singularity (not non-singular) defined by $f = 0$ and G a cyclic group of order r acting on $(Y, 0) \subset (\mathbb{C}^4, 0)$. If $(X, x) = (Y, 0)/G$ is a terminal singularity, then the type $\frac{1}{r}(a_1, a_2, a_3, a_4)$ of $\mathbb{C}^4/G$ and the standard form of f are as follows:*

	r	Type		f	Condition
(1)	Arbitrary	$\frac{1}{r}(a, -a, 1, 0)$		$xy + g(z^r, t)$	$g \in \mathfrak{m}^2,\ (a, r) = 1$
(2)	4	$\frac{1}{4}(1, 1, 3, 2)$		$xy + z^2 + g(t)$	$g \in \mathfrak{m}^3$
			or	$x^2 + z^2 + g(y, t)$	$g \in \mathfrak{m}^3$
(3)	2	$\frac{1}{2}(0, 1, 1, 1)$		$x^2 + y^2 + g(z, t)$	$g \in \mathfrak{m}^4$
(4)	3	$\frac{1}{3}(0, 2, 1, 1)$		$x^2 + y^3 + z^3 + t^3$	
			or	$x^2 + y^3 + z^2 t + yg(z, t) + h(z, t)$	$g \in \mathfrak{m}^4,\ h \in \mathfrak{m}^6$
			or	$x^2 + y^3 + z^3 + yg(z, t) + h(z, t)$	$g \in \mathfrak{m}^4,\ h \in \mathfrak{m}^6$
(5)	2	$\frac{1}{2}(1, 0, 1, 1)$		$x^2 + y^3 + yzt + g(z, t)$	$g \in \mathfrak{m}^4$
			or	$x^2 + yzt + y^n + g(z, t)$	$g \in \mathfrak{m}^4,\ n \geq 4$
			or	$x^2 + yz^2 + y^n + g(z, t)$	$g \in \mathfrak{m}^4,\ n \geq 3$
(6)	2	$\frac{1}{2}(1, 0, 1, 1)$		$x^2 + y^3 + yg(z, t) + h(z, t)$	$g, h \in \mathfrak{m}^4,\ h_4 \neq 0.$

(iii) *(Kollár–Shepherd–Barron [KSB, 6.5]) If the defining equation f satisfying one of (1)–(6) gives isolated singularity and if the action of $\mathbb{Z}_r$ is free outside the origin, then the quotient singularity gives a terminal singularity (X, x).*

8.4 Higher-Dimensional Isolated 1-Gorenstein Singularities

In this section, singularities are all assumed to be isolated. An n-dimensional isolated 1-Gorenstein singularity (X, x) is one of the following (Proposition 6.3.12):

(i) $\kappa_\delta(X, x) = -\infty \iff (X, x)$is a rational Gorenstein singularity.
(ii) $\kappa_\delta(X, x) = 0 \iff \delta_m(X, x) = 1, \quad \forall m \in \mathbb{N}.$
$\iff (X, x)$ is a log canonical singularity and not a log terminal singularity.
(iii) $\kappa_\delta(X, x) = n \iff (X, x)$ is not a log canonical singularity.

One of examples of 1-Gorenstein singularity is a hypersurface singularity. In this case the three types above are characterized by the Newton polytope.

Theorem 8.4.1 ([WH, Re2, I10, Cor.1.7]). *Let $(X, x) \subset (\mathbb{C}^{n+1}, 0)$ be a hypersurface isolated singularity defined by an equation $f = 0$. Let $\Gamma_+(f)$ be the Newton polytope, $\partial\Gamma_+(f)$ its boundary and $\Gamma_+(f)^0$ the set of the interior points.*
(I) (i) $\kappa_\delta(X, x) = -\infty,$ *then* $\mathbf{1} = (1, \cdots, 1) \in \Gamma_+(f)^0;$
(ii) $\kappa_\delta(X, x) \leq 0,$ *then* $\mathbf{1} \in \Gamma_+(f).$
(II) *I n particular, if f is non-degenerate, then the following hold:*
(iii) $\kappa_\delta(X, x) = -\infty \iff \mathbf{1} \in \Gamma_+(f)^0;$
(iv) $\kappa_\delta(X, x) = 0 \iff \mathbf{1} \in \partial\Gamma_+(f);$
(v) $\kappa_\delta(X, x) = n \iff \mathbf{1} \notin \Gamma_+(f).$

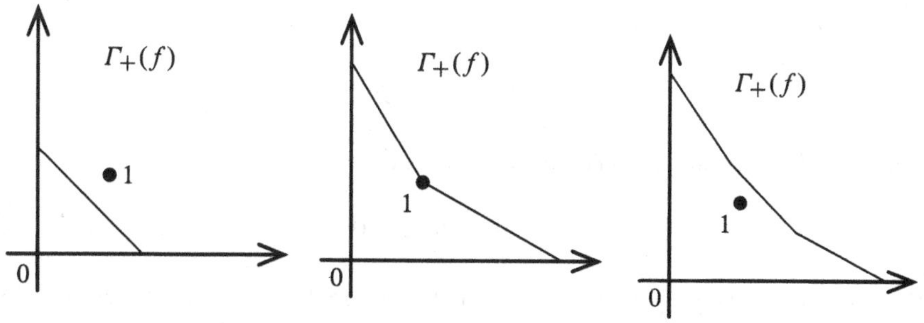

$\kappa_\delta = -\infty$ $\kappa_\delta = 0$ $\kappa_\delta = n$

Example 8.4.2. (i) Let $f = x_0^{a_0} + x_1^{a_1} + \cdots + x_n^{a_n}$, then this is non-degenerate. Let $(X, x) \subset (\mathbb{C}^{n+1}, 0)$ be the hypersurface singularity defined by f. Then the following hold:

$$\kappa_\delta(X, x) = -\infty \Longleftrightarrow \sum_{i=0}^{n} \frac{1}{a_i} > 1;$$

$$\kappa_\delta(X, x) = 0 \Longleftrightarrow \sum_{i=0}^{n} \frac{1}{a_i} = 1;$$

$$\kappa_\delta(X, x) = n \Longleftrightarrow \sum_{n=0}^{n} \frac{1}{a_i} < 1.$$

(ii) Let $f = x_0 x_1 \cdots x_n + x_0^{a_0} + \cdots + x_n^{a_n}$, $\sum_{i=0}^{n} \frac{1}{a_i} < 1$, then f is non-degenerate and the hypersurface singularity $(X, x) \subset (\mathbb{C}^{n+1}, 0)$ has $\kappa_\delta(X, x) = 0$. If the dimension is 2, then these are cusp singularities (Definition 7.6.8).

Definition 8.4.3. For a good resolution $f : Y \to X$ of a normal 1-Gorenstein singularity (X, x) assume

$$K_Y = f^* K_X + \sum_{i \in I} m_i E_i - \sum_{j \in J} m_j E_j,$$
$$I \cap J = \emptyset, \quad m_i \geq 0 \quad (i \in I), \quad m_j > 0 \quad (j \in J).$$

Then the divisor $E_J := \sum_{j \in J} m_j E_j$ is called an *essential part* of f.

Indeed, E_J is essential in the following sense:

Proposition 8.4.4. *Let $f : Y \to X$ be a good resolution of an n-dimensional normal isolated 1-Gorenstein singularity (X, x), and E_J the essential part of f. For an effective divisor D with the support on $E = f^{-1}(x)_{\mathrm{red}}$ the following hold:*

(i) For every D such that $D \geq E_J$

$$h^{n-1}(D, \mathcal{O}_D) = p_g(X, x).$$

(ii) For every D such that $D \not\geq E_J$

$$h^{n-1}(D, \mathcal{O}_D) < p_g(X, x).$$

(iii) If D does not contain a component of E_J, then

$$h^{n-1}(D, \mathcal{O}_D) = 0.$$

Proof. First, as is seen in the proof of Theorem 8.1.17, by duality we obtain $\dim R^{n-1} f_* \mathcal{O}_Y = \dim \omega_X / f_* \omega_Y$.

Now for every D by the exact sequence:

$$0 \longrightarrow f_*\omega_Y \longrightarrow f_*\omega_Y(D) \longrightarrow f_*\omega_D \longrightarrow R^1 f_*\omega_Y = 0$$

it follows that

$$\Gamma(D, \omega_D) = f_*\omega_D = f_*\omega_Y(D)/f_*\omega_Y.$$

Write K_Y as $K_Y = f^*K_X + \sum_{i \in I} m_i E_i - \sum_{j \in J} m_j E_j$, $m_i \geq 0$ $(i \in I)$ $m_j > 0$ $(j \in J)$, then for a generator ω of ω_X we have $\nu_{E_i}(\omega) = m_i$ $(\forall i \in I)$, $\nu_{E_j}(\omega) = -m_j$ $(\forall j \in J)$. Then $\omega_X = f_*\omega_Y(E_J)$.

(i) If $D \geq E_J$, then $\omega_X = f_*\omega_Y(D)$. Therefore, we have

$$\begin{aligned} p_g(X, x) &= \dim R^{n-1} f_*\mathcal{O}_Y = \dim \omega_X / f_*\omega_Y = \dim f_*\omega_Y(D)/f_*\omega_Y \\ &= \dim \Gamma(D, \omega_D), \end{aligned}$$

where the last term equals $h^{n-1}(D, \mathcal{O}_D)$ by Serre's duality theorem [Ha2, III, 7.6].

(ii) If $D \not\geq E_J$, then $\omega \notin f_*\omega_Y(D)$. Therefore we have

$$f_*\omega_Y(D)/f_*\omega_Y \subsetneqq \omega_X / f_*\omega_Y.$$

Hence, $h^{n-1}(D, \mathcal{O}_D) < p_g(X, x)$.

(iii) If D does not contain a component of E_J, then, by $\nu_{E_i}(\omega) = m_i \geq 0$ $(i \in I)$ every element of ω_X does not have a pole on the support of D. Therefore $f_*\omega_Y(D) = f_*\omega_Y$. Hence, we obtain $h^{n-1}(D, \mathcal{O}_D) = \dim f_*\omega_Y(D)/f_*\omega_Y = 0$.

We now study an isolated normal 1-Gorenstein singularity with $\kappa_\delta = 0$. As (X, x) satisfies $\kappa_\delta(X, x) = 0$ if and only if (X, x) is log canonical and not log terminal, it is also equivalent to saying that the essential part E_J is reduced.

In this case the equality $p_g(X, x) = \delta_1(X, x) = 1$ holds, then by Proposition 8.4.4, for every reduced divisor D such that $D \geq E_J$, we have $H^{n-1}(D, \mathcal{O}_D) = \mathbb{C}$. On the other hand, D is a Du Bois variety as it is a normal crossing divisor. Therefore by Proposition 8.1.11 (iii) and Theorem 8.1.6 (vi) we obtain

$$\mathbb{C} = H^{n-1}(D, \mathcal{O}_D) \simeq Gr_F^0 H^{n-1}(D, \mathbb{C}) \cong \bigoplus_{i=0}^{n-1} H_{n-1}^{0,i}(D).$$

Hence, there exists a unique i such that $H_{n-1}^{0,i}(D) \neq 0$, $0 \leq i \leq n-1$.

Definition 8.4.5. Let (X, x) be an isolated normal 1-Gorenstein singularity with $\kappa_\delta(X, x) = 0$. Under the notation above, if $H_{n-1}^{0,i}(D) \neq 0$, we call the singularity (X, x) of *type* $(0, i)$. By the following proposition, type is independent of the choice of a good resolution f.

Proposition 8.4.6. *The type of an isolated normal 1-Gorenstein singularity* (X, x) *with* $\kappa_\delta(X, x) = 0$ *is independent of the choice of a good resolution.*

Proof. It is sufficient to prove for the case that $D = f^{-1}(x)_{\text{red}}$. Let $f_i : Y_i \to X$ ($i = 1, 2$) be two good resolutions. We may assume that there is a morphism $\sigma : Y_2 \to Y_1$ such that $f_2 = f_1 \circ \sigma : Y_2 \xrightarrow{\sigma} Y_1 \xrightarrow{f_1} X$. By embedding X into a projective variety, we may assume that Y_1 and Y_2 are non-singular complete varieties. Let $E^{(i)} := f_i^{-1}(x)_{\text{red}}$, then we have the following exact sequence of mixed Hodge structures:

$$\begin{aligned} \cdots \to H^{n-1}(Y_1, \mathbb{C}) &\longrightarrow H^{n-1}(Y_2, \mathbb{C}) \oplus H^{n-1}(E^{(1)}, \mathbb{C}) \\ &\longrightarrow H^{n-1}(E^{(2)}, \mathbb{C}) \longrightarrow H^n(Y_1, \mathbb{C}). \end{aligned}$$

By taking Gr_F^0 we obtain the following exact sequence:

$$\begin{aligned} Gr_F^0 H^{n-1}(Y_1, \mathbb{C}) &\longrightarrow Gr_F^0 H^{n-1}(Y_2, \mathbb{C}) \oplus Gr_F^0 H^{n-1}(E^{(1)}, \mathbb{C}) \\ &\xrightarrow{\alpha} Gr_F^0 H^{n-1}(E^{(2)}, \mathbb{C}) \xrightarrow{\beta} Gr_F^0 H^n(Y_1, \mathbb{C}). \end{aligned}$$

Here, as Y_j's are complete and non-singular, by Theorem 8.1.6, (v), (vi), it follows that $Gr_F^0 H^i(Y_j, \mathbb{C}) = H^{0,i}(Y_j)$. Then, by applying Theorem 8.1.6, (vi) for $E^{(2)}$ we obtain that β is the zero map, which implies that α is surjective. Now if $Gr_F^0 H^{n-1}(E^{(2)}) = H_{n-1}^{0,i}(E^{(2)})$ ($i < n-1$), then this component comes from $Gr_F^0 H^{n-1}(E^{(1)}, \mathbb{C})$. If $Gr_F^0 H^{n-1}(E^{(2)}, \mathbb{C}) = H_{n-1}^{0,n-1}(E^{(2)})$, then $Gr_F^0 H^{n-1}(E^{(1)}, \mathbb{C})$ must be also of $(0, n-1)$-type. Indeed if it is of $(0, i)$-type ($i < n-1$), then $Gr_F^0 H^{n-1}(Y_1, \mathbb{C})$ has only the $(0, n-1)$-component, which implies that the $(0, i)$-component must be injectively embedded into $Gr_F^0 H^{n-1}(E^{(2)}, \mathbb{C})$, which is a contradiction.

In order to study the shape of the exceptional divisor on a good resolution of a singularity, we associate it to a complex.

Definition 8.4.7. For a simple normal crossing divisor $E = \sum_{i=1}^r E_i$ on a non-singular variety Y, we associate a complex Γ_E called the *dual complex* as follows:

(1) Associate each irreducible component E_i to a vertex $\circ$.
(2) Associate the intersection of two irreducible components E_i, E_j to the segment connecting the vertices corresponding to E_i, E_j:

(3) Associate the intersection of three irreducible components E_i, E_j, E_k to the triangle (2-dimensional simplex) formed by the vertices corresponding to E_i, E_j, E_k:

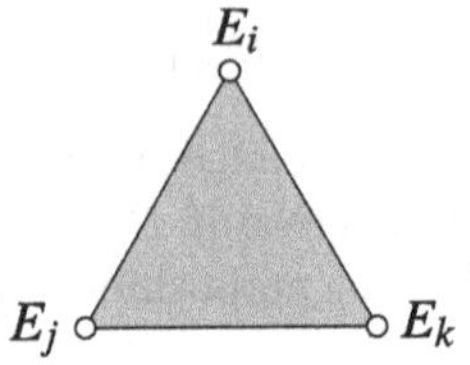

......

i. Associate the intersection of i irreducible components $E_1, \cdots, E_i$ to an $(i-1)$-dimensional simplex formed by the vertices corresponding to $E_1, \cdots, E_i$.

......

In this way we associate an $(n-1)$-dimensional simple normal crossing divisor E to a simplicial complex Γ_E of at most $(n-1)$-dimension.

The dual complex of the essential part of a three-dimensional singularity with $\kappa_\delta = 0$ is studied in [I4]. Below we will see how the type of the singularity reflects the dual complex of the essential part. The following is shown in [I13]. This is also proved in [Fjo] by using recent results on the Minimal Model Problem.

Theorem 8.4.8. *Let $f : Y \to X$ be a good resolution of an n-dimensional 1-Gorenstein normal isolated singularity (X, x) with $\kappa_\delta(X, x) = 0$. Assume the exceptional divisor $E = f^{-1}(x)_{\mathrm{red}}$ is simple normal crossings. If (X, x) is of $(0, i)$-type, then*

$$\dim \Gamma_{E_J} = n - 1 - i.$$

In particular, the essential part of the singularity of $(0, n-1)$-type is irreducible.

For the proof of the theorem, we first show the following lemma:

Lemma 8.4.9. *Let E be a simple normal crossing divisor on an n-dimensional non-singular variety. If $H^{0,i}_{n-1}(E) \neq 0$, then $\dim \Gamma_E \geq n - i - 1$.*

Proof. After renumbering the suffixes if necessary, we prove that there exist $n-i$ irreducible components $E_1, \ldots, E_{n-i}$ such that $E_1 \cap \cdots \cap E_{n-i} \neq \emptyset$. Let E' be a minimal subdivisor of E such that $H^{0,i}_{n-1}(E') \neq 0$. If E' is irreducible, then it is a non-singular variety of dimension $n-1$, therefore we obtain $i = n-1$ by the basic fact in mixed Hodge theory (see, for example, Theorem 8.1.6). Therefore,

$$\dim \Gamma_E \geq 0 = n - (n-1) - 1,$$

i.e., the required inequality becomes trivial. If E' is not irreducible, take an irreducible component $E_1 < E'$ and decompose E' as $E' = E_1 + E_1^\vee$. Then by the minimality of E', we have $H^{0,i}_{n-1}(E_1) = H^{0,i}_{n-1}(E_1^\vee) = 0$. Consider the exact sequence:

$$H^{n-2}(E_1 \cap E_1^{\vee}, \mathbb{C}) \longrightarrow H^{n-1}(E', \mathbb{C}) \longrightarrow H^{n-1}(E_1, \mathbb{C}) \oplus H^{n-1}(E_1^{\vee}, \mathbb{C}).$$

By the above vanishing, the $(0, i)$-component of the center term comes from the left term, therefore $i \leq n-2$ and $H^{0,i}_{n-2}(E_1 \cap E_1^{\vee}) \neq 0$.

Take $E_1^{\dagger}$, a minimal subdivisor of $E_1^{\vee}$ such that $H^{0,i}_{n-2}(E_1 \cap E_1^{\dagger}) \neq 0$. If $E_1 \cap E_1^{\dagger}$ is irreducible, then it is a non-singular variety of dimension $n-2$, therefore we obtain $i = n-2$ by the basic fact in mixed Hodge theory. Therefore,

$$\dim \Gamma_E \geq \dim \Gamma_{E_1+E_1^{\dagger}} \geq 1 = n - (n-2) - 1,$$

i.e., the required inequality holds. If $E^{\dagger} = E_1 \cap E_1^{\dagger}$ is not irreducible, take an irreducible component E_2 of $E^{\dagger}$ such that the decomposition $E^{\dagger} = E_2 + E_2^{\vee}$ gives a non-trivial decomposition $E_1 \cap E_1^{\dagger} = E_1 \cap E_2 + E_1 \cap E_2^{\vee}$. By the same argument as above, we obtain $i \leq n-3$ and $H^{0,i}_{n-3}(E_1 \cap E_2 \cap E_2^{\vee}) \neq 0$. Continue this procedure successively until we eventually obtain

$$H^{0,i}_{i}(E_1 \cap E_2 \cap \cdots \cap E_{n-i-1} \cap E_{n-i-1}^{\vee}) \neq 0,$$

which yields $E_1 \cap E_2 \cap \cdots \cap E_{n-i-1} \cap E_{n-i-1}^{\vee} \neq \emptyset$.

Proof of Theorem 8.4.8. The inequality $\geq$ is proved in Lemma 8.4.9. Assume the strict inequality. Then there exist components $E_1, \ldots E_s, (s > n-i)$ such that $C := E_1 \cap \cdots \cap E_s \neq \emptyset$. We may assume that $E_j \cap C = \emptyset$ for any E_j $(j > s)$. Let $\varphi : Y' \to Y$ be the blow-up at C, E' the reduced total pull-back of E, E_0 the exceptional divisor for φ and E'_j the proper transform of E_j. Then E' is again the essential part on Y' and E' itself is a minimal subdivisor of E' such that $H^{0,i}_{n-1}(E') \neq 0$ by [I2, Corollary 3.9]. Make the procedure of the proof of the lemma, taking E_0 as E_1 in the lemma. Then we obtain $E'_1, \ldots, E'_{n-i-1}$ (by renumbering the suffixes $1, \ldots, s$) such that $H^{0,i}_{i}(E_0 \cap E'_1 \cap \cdots \cap E'_{n-i-1}) \neq 0$. On the other hand, the i-dimensional variety $E_0 \cap E'_1 \cap \cdots \cap E'_{n-i-1}$ is a $(\mathbb{P}^{s-n+i})$ -bundle over C, because it is the exceptional divisor of the blow-up of an $(i+1)$-dimensional variety $E_1 \cap \cdots \cap E_{n-i-1}$ with the $(n-s)$-dimensional center C. By the assumption on s, we note that $s - n + i > 0$. Hence we have $H^i(E_0 \cap E'_1 \cap \cdots \cap E'_{n-i-1}, \mathcal{O}) = 0$. In particular,

$$H^{0,i}_{i}(E_0 \cap E'_1 \cap \cdots \cap E'_{n-i-1}) = 0,$$

a contradiction. This proves the first statement of the theorem.

For the second statement of the theorem, assume $i = n-1$ and then we will prove that E_J is irreducible. Assume that E_J is decomposed as $E_J = E_1 + E_2$, then by the Mayer–Vietoris exact sequence (Proposition 8.1.8) we have the following exact sequence:

$$H^{n-2}(E_1 \cap E_2, \mathbb{C}) \longrightarrow H^{n-1}(E_J, \mathbb{C}) \longrightarrow H^{n-1}(E_1, \mathbb{C}) \oplus H^{n-1}(E_2, \mathbb{C}).$$

Here, by taking Gr^0_F we obtain the exact sequence:

$$H^{n-2}(E_1 \cap E_2, \mathcal{O}_{E_1 \cap E_2}) \longrightarrow H^{n-1}(E_J, \mathcal{O}_{E_J}) \\ \longrightarrow H^{n-1}(E_1, \mathcal{O}_{E_1}) \oplus H^{n-1}(E_2, \mathcal{O}_{E_2}).$$

Now by $p_g(X, x) = 1$ and Proposition 8.4.4, (ii) it follows that $H^{n-1}(E_i, \mathcal{O}_{E_i}) = 0$, $i = 1, 2$. Therefore the $(0, n-1)$-component of the second term comes from the first term. But in the first term, there is no $(0, n-1)$-component by Theorem 8.1.6(vi), a contradiction. Thus E_J is irreducible.

We see what kind of singularities are 2-dimensional normal 1-Gorenstein singularities with $\kappa_\delta = 0$.

Example 8.4.10. Let (X, x) be a 2-dimensional isolated normal 1-Gorenstein singularity with $\kappa_\delta(X, x) = 0$. Then it is Cohen–Macaulay, and therefore it is a Gorenstein singularity. By Theorem 7.8.1, (X, x) is either a simple elliptic singularity or a cusp singularity. The minimal resolutions of both singularities are essential good resolutions, and the essential part E_J is one elliptic curve if (X, x) is simple elliptic and a cycle of finite number of $\mathbb{P}^1$ if (X, x) is a cusp singularity. Therefore the dual graphs of the essential parts are one point $\circ$ and a cycle 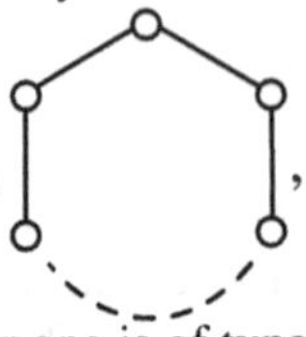, respectively. Then the dimension is 0 and 1, respectively. The former one is of type $(0, 1)$ and the latter is of type $(0, 0)$.

Let us see examples of 3-dimensional isolated normal 1-Gorenstein singularity with $\kappa_\delta = 0$.

Example 8.4.11. (i) Let S be either an abelian surface or a $K3$-surface and (X, x) a cone over S (cf. Example 6.3.13). Then (X, x) is a 3-dimensional isolated normal 1-Gorenstein singularity with $\kappa_\delta(X, x) = 0$ and is of type $(0, 2)$.

(ii) Let $S_1 \subset \mathbb{P}^3$ be a hyperplane and $S_2 \subset \mathbb{P}^3$ a hypersurface of degree 3 such that $S = S_1 + S_2$ is of simple normal crossings. Let $D \subset \mathbb{P}^3$ be a general hypersurface of degree 5 and let $C := S \cap D$. Let $b : Y \to \mathbb{P}^3$ be the blow-up with the center C and E the strict transform of S. Then, there exist a normal variety X and a morphism $f : Y \to X$ such that $f(E) = x$ (one point) and $f \mid_{Y \setminus E} : Y \setminus E \xrightarrow{\sim} X \setminus \{x\}$ isomorphic. In this case (X, x) is a 3-dimensional isolated normal 1-Gorenstein singularity with $\kappa_\delta(X, x) = 0$ by [I2, Lemma 5.1] and it is of type $(0, 1)$. Every good resolution of this singularity has the essential part whose dual graph is a chain.

(iii) Let $S_i \subset \mathbb{P}^3$ be the hyperplane defined by $x_i = 0$ $(i = 0, 1, 2, 3)$. Let $S := S_0 + \cdots + S_3$ and construct Y, (X, x) as in (ii). Then (X, x) is a 3-dimensional isolated normal 1-Gorenstein singularity with $\kappa_\delta(X, x) = 0$ and is of type $(0, 0)$. In this case the essential part is the simplicial decomposition of the sphere S^2.

(iv) Tsuchihashi constructs in [Ts1, Ts2] a higher-dimensional generalization of a cusp singularity (called *Tsuchihashi-cusp*). These singularities are $\kappa_\delta(X, x) = 0$; in particular, for the 3-dimensional case, the dual complex of the essential part is a simplicial subdivision of a compact surface except for a sphere, real projective plane, or Klein bottle. These are examples of type $(0, 0)$.

Definition 8.4.12. Let (X, x) be a 3-dimensional isolated normal 1-Gorenstein singularity with $\kappa_\delta(X, x) = 0$ and of type $(0, 2)$. In this case we define:

If (X, x) is a Gorenstein singularity, we call it a *simple $K3$-singularity*. If (X, x) is not a Gorenstein singularity, we call it a *simple abelian singularity*.

These are 3-dimensional versions of simple elliptic singularities of dimension 2. In the 2-dimensional case, a normal 1-Gorenstein singularity with $\kappa_\delta(X, x) = 0$ and of type $(0, 1)$ is a simple elliptic singularity. In other words, these singularities have a characterization by the exceptional divisors of minimal resolution. The following shows a similar characterization for the 3-dimensional case. Here, we have $\mathbb{Q}$-factorial minimal models instead of the minimal resolution (Theorem 8.2.9, (i)).

Proposition 8.4.13 (Ishii–Watanabe [IW]). *A 3-dimensional normal isolated singularity (X, x) is a simple K3-singularity (resp. simple abelian singularity) if and only if a $\mathbb{Q}$-factorial minimal model $f : Y \to X$ has the irreducible exceptional divisor E which is isomorphic to a normal $K3$-surface (resp. abelian surface). Here a normal $K3$-surface means a normal surface whose minimal resolution is a $K3$-surface.*

What kind of equations define hypersurface singularities of type $(0, n-1)$?

The following proposition answers that question.

Proposition 8.4.14 (Watanabe [W2]). *Let $(X, x) \subset (\mathbb{C}^{n+1}, 0)$ be a hypersurface isolated singularity defined by a non-degenerate polynomial f. The singularity (X, x) satisfies $\kappa_\delta(X, x) = 0$ and is of type $(0, n-1)$, if and only if there is an n-dimensional face of the boundary $\partial\Gamma_+(f)$ of the Newton polytope such that the face contains the point $(1, \cdots, 1)$ in its relative interior.*

In particular, if f is a weighted homogeneous polynomial with weight $\boldsymbol{p} = (p_0, p_1, \cdots, p_n)$ such that $\sum_{i=0}^n p_i = 1$, then the singularity (X, x) satisfies $\kappa_\delta(X, x) = 0$ and is of type $(0, n-1)$. Here, the polynomial $f(x_0, \cdots, x_n)$ is called weighted homogeneous of weight $\boldsymbol{p} = (p_0, \cdots, p_n)$ if for every $\lambda \in \mathbb{C}^*$ the equality $f(\lambda^{p_0}x_0, \lambda^{p_1}x_1, \cdots, \lambda^{p_n}x_n) = \lambda f(x_0, \cdots, x_n)$ holds.

Example 8.4.15. A 2-dimensional simple elliptic singularity on a hypersurface is defined by:

$$\widetilde{E}_6;\ \ y(y-x)(y-\lambda x) - xz^2 = 0.$$
$$\widetilde{E}_7;\ \ yx(y-x)(y-\lambda x) - z^2 = 0.$$
$$\widetilde{E}_8;\ \ y(y-x^2)(y-\lambda x^2) - z^2 = 0.$$

$$\lambda \in \mathbb{C} \quad \lambda \neq 0, 1.$$

These are all weighted homogeneous and the weights are $\left(\frac{1}{3}, \frac{1}{3}, \frac{1}{3}\right)$, $\left(\frac{1}{4}, \frac{1}{4}, \frac{1}{2}\right)$, $\left(\frac{1}{6}, \frac{1}{3}, \frac{1}{2}\right)$, respectively.

Example 8.4.16. For a 3-dimensional hypersurface, simple $K3$-singularities are classified into 95 classes by Yonemura [Y] as follows. Note that these classes coincide with the classes of $K3$ surface classified in [F].

No.	weight $\boldsymbol{p}$	f
1	$\left(\frac{1}{4}, \frac{1}{4}, \frac{1}{4}, \frac{1}{4}\right)$	$x^4 + y^4 + z^4 + w^4$
2	$\left(\frac{1}{3}, \frac{1}{4}, \frac{1}{4}, \frac{1}{6}\right)$	$x^3 + y^4 + z^4 + w^6$
3	$\left(\frac{1}{3}, \frac{1}{3}, \frac{1}{6}, \frac{1}{6}\right)$	$x^3 + y^3 + z^6 + w^6$
4	$\left(\frac{1}{3}, \frac{1}{3}, \frac{1}{4}, \frac{1}{12}\right)$	$x^3 + y^3 + z^4 + w^{12}$
5	$\left(\frac{1}{2}, \frac{1}{6}, \frac{1}{6}, \frac{1}{6}\right)$	$x^2 + y^6 + z^6 + w^6$
6	$\left(\frac{1}{2}, \frac{1}{5}, \frac{1}{5}, \frac{1}{10}\right)$	$x^2 + y^5 + z^5 + w^{10}$
7	$\left(\frac{1}{2}, \frac{1}{4}, \frac{1}{8}, \frac{1}{8}\right)$	$x^2 + y^4 + z^8 + w^8$
8	$\left(\frac{1}{2}, \frac{1}{4}, \frac{1}{6}, \frac{1}{12}\right)$	$x^2 + y^4 + z^6 + w^{12}$
9	$\left(\frac{1}{2}, \frac{1}{4}, \frac{1}{5}, \frac{1}{20}\right)$	$x^2 + y^4 + z^5 + w^{20}$
10	$\left(\frac{1}{2}, \frac{1}{3}, \frac{1}{12}, \frac{1}{12}\right)$	$x^2 + y^3 + z^{12} + w^{12}$
11	$\left(\frac{1}{2}, \frac{1}{3}, \frac{1}{10}, \frac{1}{15}\right)$	$x^2 + y^3 + z^{10} + w^{15}$
12	$\left(\frac{1}{2}, \frac{1}{3}, \frac{1}{9}, \frac{1}{18}\right)$	$x^2 + y^3 + z^9 + w^{18}$
13	$\left(\frac{1}{2}, \frac{1}{3}, \frac{1}{8}, \frac{1}{24}\right)$	$x^2 + y^3 + z^8 + w^{24}$
14	$\left(\frac{1}{2}, \frac{1}{3}, \frac{1}{7}, \frac{1}{42}\right)$	$x^2 + y^3 + z^7 + w^{42}$
15	$\left(\frac{1}{3}, \frac{4}{15}, \frac{1}{5}, \frac{1}{5}\right)$	$x^3 + y^3 z + y^3 w + z^5 - w^5$
16	$\left(\frac{1}{3}, \frac{7}{24}, \frac{1}{4}, \frac{1}{8}\right)$	$x^3 + y^3 w + z^4 + w^8$
17	$\left(\frac{1}{3}, \frac{1}{3}, \frac{1}{5}, \frac{2}{15}\right)$	$x^3 + y^3 + z^5 + xw^5 + yw^5 + zw^6$
18	$\left(\frac{1}{3}, \frac{1}{3}, \frac{2}{9}, \frac{1}{9}\right)$	$x^3 + y^3 + xz^3 + yz^3 + z^4 w + w^9$
19	$\left(\frac{3}{8}, \frac{1}{4}, \frac{1}{4}, \frac{1}{8}\right)$	$x^2 y + x^2 z + x^2 w^2 + y^4 + z^4 + w^6$
20	$\left(\frac{3}{8}, \frac{1}{3}, \frac{1}{4}, \frac{1}{24}\right)$	$x^2 z + x^2 w^6 + y^3 + z^4 + w^{24}$

No.	weight $\boldsymbol{p}$	f
21	$\left(\frac{2}{5}, \frac{1}{5}, \frac{1}{5}, \frac{1}{5}\right)$	$x^2y + x^2z + x^2w + y^5 + z^5 + w^5$
22	$\left(\frac{2}{5}, \frac{1}{3}, \frac{1}{5}, \frac{1}{15}\right)$	$x^2z + x^2w^3 + y^3 + z^5 - w^{15}$
23	$\left(\frac{5}{12}, \frac{1}{4}, \frac{1}{6}, \frac{1}{6}\right)$	$x^2z + x^2w + y^4 + z^6 + w^6$
24	$\left(\frac{5}{12}, \frac{1}{3}, \frac{1}{6}, \frac{1}{12}\right)$	$x^2z + x^2w^2 + y^3 + z^6 + w^{12}$
25	$\left(\frac{4}{9}, \frac{1}{3}, \frac{1}{9}, \frac{1}{9}\right)$	$x^2z +^2 w + y^3 + z^9 - w^9$
26	$\left(\frac{9}{20}, \frac{1}{4}, \frac{1}{5}, \frac{1}{10}\right)$	$x^2w + y^4 + z^5 + w^{10}$
27	$\left(\frac{11}{24}, \frac{1}{3}, \frac{1}{8}, \frac{1}{12}\right)$	$x^2w + y^3 + z^8 + w^{12}$
28	$\left(\frac{10}{21}, \frac{1}{3}, \frac{1}{7}, \frac{1}{21}\right)$	$x^2w + y^3 + z^7 + w^{21}$
29	$\left(\frac{1}{2}, \frac{1}{5}, \frac{1}{6}, \frac{2}{15}\right)$	$x^2 + y^5 + z^6 + yw^6 + z^2w^5$
30	$\left(\frac{1}{2}, \frac{1}{5}, \frac{7}{40}, \frac{1}{8}\right)$	$x^2 + y^5 + z^5w + w^8$
31	$\left(\frac{1}{2}, \frac{5}{24}, \frac{1}{6}, \frac{1}{8}\right)$	$x^2 + y^4z + y^3w^3 + z^6 + w^8$
32	$\left(\frac{1}{2}, \frac{3}{14}, \frac{1}{7}, \frac{1}{7}\right)$	$x^2 + y^4z + y^4w + z^7 - w^7$
33	$\left(\frac{1}{2}, \frac{2}{9}, \frac{1}{6}, \frac{1}{9}\right)$	$x^2 + y^3z^2 + y^4w + z^6 - w^9$
34	$\left(\frac{1}{2}, \frac{7}{30}, \frac{1}{5}, \frac{1}{15}\right)$	$x^2 + y^4w + z^5 + w^{15}$
35	$\left(\frac{1}{2}, \frac{1}{4}, \frac{1}{7}, \frac{3}{28}\right)$	$x^2 + y^4 + z^7 + yw^7 + zw^8$
36	$\left(\frac{1}{2}, \frac{1}{4}, \frac{3}{20}, \frac{1}{10}\right)$	$x^2 + y^4 + yz^5 + z^6w + w^{10}$
37	$\left(\frac{1}{2}, \frac{1}{4}, \frac{3}{16}, \frac{1}{16}\right)$	$x^2 + y^4 + yz^4 + z^5w + w^{36}$
38	$\left(\frac{1}{2}, \frac{4}{15}, \frac{1}{5}, \frac{1}{30}\right)$	$x^2 + y^3z + y^3w^6 + z^5 - w^{30}$
39	$\left(\frac{1}{2}, \frac{5}{18}, \frac{1}{6}, \frac{1}{18}\right)$	$x^2 + y^3z + y^3w^3 + z^6 + w^{18}$
40	$\left(\frac{1}{2}, \frac{2}{7}, \frac{1}{7}, \frac{1}{14}\right)$	$x^2 + y^3z + y^3w^2 + z^7 - w^{34}$

No.	weight $\boldsymbol{p}$	f
41	$\left(\frac{1}{2}, \frac{7}{24}, \frac{1}{8}, \frac{1}{12}\right)$	$x^2 + y^3z + y^2w^5 + z^8 + w^{12}$
42	$\left(\frac{1}{2}, \frac{3}{10}, \frac{1}{10}, \frac{1}{10}\right)$	$x^2 + y^3z + y^3w + z^{10} + w^{10}$
43	$\left(\frac{1}{2}, \frac{11}{36}, \frac{1}{9}, \frac{1}{12}\right)$	$x^2 + y^3w + z^9 + w^{12}$
44	$\left(\frac{1}{2}, \frac{5}{16}, \frac{1}{8}, \frac{1}{16}\right)$	$x^2 + y^2z^3 + y^3w + z^8 + w^{16}$
45	$\left(\frac{1}{2}, \frac{9}{28}, \frac{1}{7}, \frac{1}{28}\right)$	$x^2 + y^3w + z^7 + w^{28}$
46	$\left(\frac{1}{2}, \frac{1}{3}, \frac{1}{11}, \frac{5}{66}\right)$	$x^2 + y^3 + z^{11} + zw^{12}$
47	$\left(\frac{1}{2}, \frac{1}{3}, \frac{2}{21}, \frac{1}{14}\right)$	$x^2 + y^3 + yz^7 + z^9w^2 + w^{14}$
48	$\left(\frac{1}{2}, \frac{1}{3}, \frac{5}{48}, \frac{1}{16}\right)$	$x^2 + y^3 + z^9w + w^{16}$
49	$\left(\frac{1}{2}, \frac{1}{3}, \frac{5}{42}, \frac{1}{21}\right)$	$x^2 + y^3 + z^8w + w^{21}$
50	$\left(\frac{1}{2}, \frac{1}{3}, \frac{2}{15}, \frac{1}{30}\right)$	$x^2 + y^3 + yz^5 + z^7w^2 + w^{30}$
51	$\left(\frac{1}{2}, \frac{1}{3}, \frac{5}{36}, \frac{1}{36}\right)$	$x^2 + y^3 + z^7w + w^{36}$
52	$\left(\frac{1}{3}, \frac{1}{4}, \frac{2}{9}, \frac{7}{36}\right)$	$x^3 + y^4 + xz^3 + zw^4$
53	$\left(\frac{1}{3}, \frac{5}{18}, \frac{2}{9}, \frac{1}{6}\right)$	$x^3 + y^3w + y^2z^2 + xz^3 + z^3w^2 + w^6$
54	$\left(\frac{1}{3}, \frac{2}{7}, \frac{5}{21}, \frac{1}{7}\right)$	$x^3 + y^3w + yz^3 + z^3w^2 - w^7$
55	$\left(\frac{7}{20}, \frac{3}{10}, \frac{1}{4}, \frac{1}{10}\right)$	$x^2y + x^2w^3 + y^3w + z^4 - w^{10}$
56	$\left(\frac{11}{30}, \frac{4}{15}, \frac{1}{5}, \frac{1}{6}\right)$	$x^2y + y^3z + z^5 + w^6$
57	$\left(\frac{3}{8}, \frac{1}{4}, \frac{5}{24}, \frac{1}{6}\right)$	$x^2y + y^4 + xz^3 + z^4w + w^6$
58	$\left(\frac{3}{8}, \frac{5}{16}, \frac{1}{4}, \frac{1}{16}\right)$	$x^2z + x^2w^4 + xy^2 + y^3w + z^4 + w^{16}$
59	$\left(\frac{8}{21}, \frac{1}{3}, \frac{5}{21}, \frac{1}{21}\right)$	$x^2z + x^2w^5 + y^3 + z^4w - w^{21}$
60	$\left(\frac{7}{18}, \frac{1}{3}, \frac{2}{9}, \frac{1}{18}\right)$	$x^2z + x^2w^4 + y^3 + yz^3 + z^4w^2 + w^{18}$

No.	weight $\boldsymbol{p}$	f
61	$\left(\frac{11}{28}, \frac{1}{4}, \frac{3}{14}, \frac{1}{7}\right)$	$x^2z + y^4 + z^4w + w^7$
62	$\left(\frac{2}{5}, \frac{1}{4}, \frac{1}{5}, \frac{3}{20}\right)$	$x^2z + xw^4 + y^4 + yw^5 + z^5 + z^2w^4$
63	$\left(\frac{2}{5}, \frac{3}{10}, \frac{1}{5}, \frac{1}{10}\right)$	$x^2z + x^2w^2 + xy^2 + y^2z^2 + y^3w + z^5 + w^{10}$
64	$\left(\frac{5}{12}, \frac{7}{24}, \frac{1}{6}, \frac{1}{8}\right)$	$x^2z + xy^2 + y^3w + z^6 + w^8$
65	$\left(\frac{14}{33}, \frac{1}{3}, \frac{5}{33}, \frac{1}{11}\right)$	$x^2z + y^3 + z^6w + w^{11}$
66	$\left(\frac{3}{7}, \frac{2}{7}, \frac{1}{7}, \frac{1}{7}\right)$	$x^2z + x^2w + xy^2 + y^3z + y^38w + z^7 + w^7$
67	$\left(\frac{3}{7}, \frac{1}{3}, \frac{1}{7}, \frac{2}{21}\right)$	$x^2z + xw^6 + y^3 + yw^7 + z^7 + zw^9$
68	$\left(\frac{13}{30}, \frac{1}{3}, \frac{2}{15}, \frac{1}{10}\right)$	$x^2z + y^3 + yz^5 + z^6w^2 + w^{10}$
69	$\left(\frac{7}{16}, \frac{1}{4}, \frac{3}{16}, \frac{1}{8}\right)$	$x^2w + xz^3 + y^4 + yz^4 + z^4w^2 + w^8$
70	$\left(\frac{4}{9}, \frac{5}{18}, \frac{1}{6}, \frac{1}{9}\right)$	$x^2w + xy^2 + y^3z + y^2 + z^6 + w^9$
71	$\left(\frac{7}{15}, \frac{4}{15}, \frac{1}{5}, \frac{1}{15}\right)$	$x^2w + xy^2 + y^3z + y^3w^3 + z^5 + w^{15}$
72	$\left(\frac{7}{15}, \frac{1}{3}, \frac{2}{15}, \frac{1}{15}\right)$	$x^2w + xz^4 + y^3 + xy^5 + z^7w + w^{15}$
73	$\left(\frac{1}{2}, \frac{1}{5}, \frac{4}{25}, \frac{7}{50}\right)$	$x^2 + y^5 + yz^5 + zw^6$
74	$\left(\frac{1}{2}, \frac{7}{32}, \frac{5}{32}, \frac{1}{8}\right)$	$x^2 + y^4w + yz^5 + z^4w^3 + w^8$
75	$\left(\frac{1}{2}, \frac{5}{22}, \frac{2}{11}, \frac{1}{11}\right)$	$x^2 + y^4w + y^2z^3 + z^5w + w^{11}$
76	$\left(\frac{1}{2}, \frac{3}{13}, \frac{5}{26}, \frac{1}{13}\right)$	$x^2 + y^4w + yz^4 + z^4w^3 + w^{13}$
77	$\left(\frac{1}{2}, \frac{7}{26}, \frac{5}{26}, \frac{1}{26}\right)$	$x^2 + y^3z + y^3w^5 + z^5w + w^{26}$
78	$\left(\frac{1}{2}, \frac{3}{11}, \frac{2}{11}, \frac{1}{22}\right)$	$x^2 + y^3z + y^3w^4 + yz^4 + z^5w^2 + w^{22}$
79	$\left(\frac{1}{2}, \frac{9}{32}, \frac{5}{32}, \frac{1}{16}\right)$	$x^2 + y^3z + y^2w^7 + z^6w + w^{16}$
80	$\left(\frac{1}{2}, \frac{13}{44}, \frac{5}{44}, \frac{1}{11}\right)$	$x^2 + y^3z + z^8w + w^{11}$

No.	weight $\boldsymbol{p}$	f
81	$\left(\frac{1}{2},\frac{4}{13},\frac{3}{26},\frac{1}{13}\right)$	$x^2+y^3w+yz^6+z^8w+w^{13}$
82	$\left(\frac{1}{2},\frac{7}{22},\frac{3}{22},\frac{1}{22}\right)$	$x^2+y^3w+yz^5+z^7w+w^{22}$
83	$\left(\frac{1}{2},\frac{1}{3},\frac{5}{54},\frac{2}{27}\right)$	$x^2+y^3+yw^9+z^{10}w+z^2w^{11}$
84	$\left(\frac{1}{3},\frac{7}{27},\frac{2}{9},\frac{5}{27}\right)$	$x^3+xz^3+y^3z+yw^4+z^2w^3$
85	$\left(\frac{5}{14},\frac{2}{7},\frac{3}{14},\frac{1}{7}\right)$	$x^2y+x^2w^2+xz^3+y^3w+y^2z^2+z^4w+w^7$
86	$\left(\frac{9}{25},\frac{7}{25},\frac{1}{5},\frac{4}{25}\right)$	$x^2y+xw^4+y^3w+z^5+zw^5$
87	$\left(\frac{5}{13},\frac{4}{13},\frac{3}{13},\frac{1}{13}\right)$	$x^2z+x^2w^3+xy^2+y^3w+yz^3+z^4w+w^{13}$
88	$\left(\frac{11}{27},\frac{1}{3},\frac{5}{27},\frac{2}{27}\right)$	$x^2z+xw^8+y^3+yw^9+z^5w+zw^{11}$
89	$\left(\frac{5}{11},\frac{3}{11},\frac{2}{11},\frac{1}{11}\right)$	$x^2w+xy^2+xz^3+y^3z+y^3w^2+yz^4+z^5w+w^{11}$
90	$\left(\frac{1}{2},\frac{7}{34},\frac{3}{17},\frac{2}{17}\right)$	$x^2+y^4z+y^2w^5+z^5w+zw^7$
91	$\left(\frac{1}{2},\frac{4}{19},\frac{3}{19},\frac{5}{38}\right)$	$x^2+y^4z+yz^5+yw^6+z^3w^4$
92	$\left(\frac{1}{2},\frac{11}{38},\frac{5}{38},\frac{3}{38}\right)$	$x^2+y^3z+yw^9+z^7w+zw^{11}$
93	$\left(\frac{1}{2},\frac{5}{17},\frac{2}{17},\frac{3}{34}\right)$	$x^2+y^3z+yz^6+yw^8+z^7w^2+zw^{10}$
94	$\left(\frac{7}{19},\frac{5}{19},\frac{4}{19},\frac{3}{19}\right)$	$x^2y+xz^3+xw^4+y^3z+y^2w^3+z^4w+zw^5$
95	$\left(\frac{7}{17},\frac{5}{17},\frac{3}{17},\frac{2}{17}\right)$	$x^2z+xy^2+zw^5+y^3w+yz^4+yw^6+z^5w+zw^7$

As is seen above, classes of weights of weighted homogeneous hypersurface singularities with $\kappa_\delta = 0$ are two in the 2-dimensional case and 95 in the 3-dimensional case. Tsuchihashi informed me that the number of classes of weights for the 4-dimensional case is more than 10,000. He also studies simple K3 singularities which are hypersurface sections of toric singularities in [Ts3].

Chapter 9
Deformations of Singularities

Inasmuch as the mathematical theorems are related to reality, they are not sure;
inasmuch as they are sure, they are not related to reality
Einstein [Mur], p. 120

In this chapter we study deformations of singularities by varying parameters. In any deformations, singularities deform under certain orders. We will see this in Sect. 9.1, and in Sect. 9.2 we will see that there exists a big deformation which contains every deformation. We assume that the varieties are defined over $\mathbb{C}$.

9.1 Variation of Properties Under Deformations

Definition 9.1.1. Let (X, x) be a singularity on an integral algebraic variety X. A morphism $\pi : (\mathscr{X}, x) \to (C, 0)$ of germs of algebraic varieties is called a *deformation* of (X, x) if the following hold:

(1) The morphism $\pi : \mathscr{X} \to C$ is a flat morphism; i.e., for every point $p \in \mathscr{X}$ the ring $\mathscr{O}_{\mathscr{X},p}$ is a flat $\mathscr{O}_{C,\pi(p)}$-module.
(2) Let $\mathscr{X}_t := \pi^{-1}(t)$, $t \in C$, then there is an isomorphism $(\mathscr{X}_0, x) \simeq (X, x)$ of germs.

In this case $\mathscr{X}$ is called the *total space* of the deformation and C is called the *base space* of the deformation.

In particular, if C is a non-singular curve, π is called a 1-*parameter deformation.*

Definition 9.1.2. For a deformation $\pi : (\mathscr{X}, x) \to (C, 0)$ of a singularity (X, x), a morphism $f : \mathscr{Y} \to \mathscr{X}$ is called a *simultaneous resolution* of π if taking $\mathscr{X}$, C as

S. Ishii, *Introduction to Singularities*,
https://doi.org/10.1007/978-4-431-56837-7_9

Table 9.1 Behaviors of properties

Property of $(\mathscr{X}_0, x)$	Property of $(\mathscr{X}, x)$	Property of $\mathscr{X}_t$ ($t \in C$, t close to 0)
Non-singular	Non-singular (Theorem 9.1.5)	Non-singular (Theorem 9.1.5)
Cohen–Macaulay	Cohen–Macaulay (Theorem 9.1.6)	Cohen–Macaulay (Theorem 9.1.6)
Gorenstein	Gorenstein (Theorem 9.1.6)	Gorenstein (Theorem 9.1.6)
$\mathbb{Q}$-Gorenstein	Not $\mathbb{Q}$-Gorenstein in general (Example 9.1.7)newline $\mathbb{Q}$-Gorenstein under some conditions (Proposition 9.1.9, Corollary 9.1.11)	Not $\mathbb{Q}$-Gorenstein in general (Example 9.1.8) $\mathbb{Q}$-Gorenstein under some conditions (Proposition 9.1.9, Corollary 9.1.11)
Rational	Rational (Theorem 9.1.12)	Rational (Theorem 9.1.12)
Terminal	Terminal (Theorem 9.1.14)	Terminal (Theorem 9.1.14)
Canonical	Canonical (Theorem 9.1.13)	Canonical (Theorem 9.1.13)
Log terminal	Not log terminal in general (Remark 9.1.15) log terminal under some conditions (Theorem 9.1.16)	Not log terminal in general (Remark 9.1.15)log terminal under some conditions (Theorem 9.1.16, Theorem 9.1.17)
Log canonical	Not log canonical in general (Remark 9.1.15) log canonical under some conditions (Theorem 9.1.17)	Not log canonical in general (Remark 9.1.15)log canonical under some conditions (Theorem 9.1.17)
Quotient singularity	Not quotient singularity in general (Remark 9.1.18) if $\operatorname{codim} X_{\text{sing}} \geq 3$, then $\mathscr{X}$ is $X \times C$ (Theorem 9.1.21)	Quotient singularity if $\dim X = 2$ (Theorem 9.1.19) if $\operatorname{codim} X_{\text{sing}} \geq 3$, then it is isomorphic to X (Theorem 9.1.21)

sufficiently small neighborhoods of x and 0, respectively for every $t \in C$ the restriction morphism $f_t := f \mid_{\mathscr{Y}_t} : \mathscr{Y}_t := (\pi \circ f)^{-1}(t) \to \mathscr{X}_t := \pi^{-1}(t)$ on the fibers of t is a resolution of singularities. If f_t ($\forall t \in C$) is the canonical model, then f is called the *simultaneous canonical model.*

In this section, under a 1-parameter deformation, we consider how a property of the singularity $(X, x) \cong (\mathscr{X}_0, x)$ affects the singularities of $\mathscr{X}_t$ for $t \in C$ sufficiently close to 0. Before the statements of results, the table of the results comes above (Table 9.1).

The deformations of Du Bois singularities are also Du Bois singularities, which is proved recently in [KSS]. The behaviors of invariants of isolated singularities under deformations are listed in Table 9.2 below.

Table 9.2 Invariant's behavior

Invariant	Behavior under deformations
emb	Upper semi-continuous (Theorem 9.1.22)
p_g	Upper semi-continuous (Theorem 9.1.23)
δ_m	Upper semi-continuous (Theorem 9.1.24)
γ_m	Upper semi-continuous (Theorem 9.1.25)
$\kappa_\delta = 0$, 1-Gorenstein type $(0, r)$	Lower semi-continuous under some conditions (Theorem 9.1.26)

Here an invariant P is said to be upper semi-continuous (resp. lower semi-continuous) if for sufficiently small neighborhoods $\mathscr{X}$, C of x, 0, respectively the inequality $P(\mathscr{X}_0, x) \geq P(\mathscr{X}_t, y)$, $y \in \mathscr{X}_t$ holds for $t \in C$ (resp. $P(\mathscr{X}_0, x) \leq P(\mathscr{X}_t, y)$, $y \in \mathscr{X}_t$ holds for $t \in C$).

First we study deformations of a non-singular point.

Theorem 9.1.5. *Let $\pi : (\mathscr{X}, x) \to (C, 0)$ be a 1-parameter deformation of a non-singular point (X, x). Then, $(\mathscr{X}, x)$ is also a non-singular point. If one takes $\mathscr{X}$ and C as sufficiently small neighborhoods of x and 0, respectively, then $\mathscr{X}_t$ $(t \in C)$ are also non-singular.*

Proof. On a non-singular curve C the point 0 is a Cartier divisor, therefore the fiber $\mathscr{X}_0 = \pi^{-1}(0)$ is also a Cartier divisor. Let (h) be the defining ideal of $\mathscr{X}_0$ around x. Then we have $\Omega_{\mathscr{X}_0} = \Omega_{\mathscr{X}}/dh$. By the hypothesis, $\Omega_{\mathscr{X}_0,x}$ is generated by $\dim \mathscr{X}_0$ elements. Then $\Omega_{\mathscr{X},x}$ is generated by $\dim \mathscr{X}_0 + 1 = \dim \mathscr{X}$ elements, which shows that $(\mathscr{X}, x)$ is non-singular. By [EGA, IV, 17.5.1] the morphism π is smooth at x. Then if we take $\mathscr{X}$ and C sufficiently small, then π is smooth on $\mathscr{X}$. Therefore the fiber $\mathscr{X}_t = \pi^{-1}(t)$ of each point $t \in C$ is non-singular.

Next we study the Gorenstein property and Cohen–Macaulay property.

Theorem 9.1.6. *Let $\pi : (\mathscr{X}, x) \to (C, 0)$ be a 1-parameter deformation of a Cohen–Macaulay (resp. Gorenstein) singular point (X, x).*

Then $(\mathscr{X}, x)$ is also a Cohen–Macaulay (resp. Gorenstein) singularity. If one takes $\mathscr{X}$ and C as sufficiently small neighborhoods of x and 0, respectively, then $\mathscr{X}_t$ $(t \in C)$ are also Cohen–Macaulay (resp. Gorenstein) singularity.

Proof. If (X, x) is a Cohen–Macaulay (resp. Gorenstein) singularity, then by Proposition 5.3.12 the singularity $(\mathscr{X}, x)$ is also a Cohen–Macaulay (resp. Gorenstein) singularity. By replacing $\mathscr{X}$, C sufficiently small, we have that $\mathscr{X}$ is also a Cohen–Macaulay (resp. Gorenstein) variety. Therefore, again by Proposition 5.3.12 the fiber $\mathscr{X}_t$ is Cohen–Macaulay (resp. Gorenstein) for every $t \in C$.

In the above discussion we observed that the Gorenstein property is stable under deformations, but the $\mathbb{Q}$-Gorenstein property is not stable. Let us see first that the total space is not necessarily $\mathbb{Q}$-Gorenstein, even if the special fiber is $\mathbb{Q}$-Gorenstein.

Example 9.1.7. Let us construct a 1-parameter deformation $\pi : (\mathscr{X}, x) \to (C, 0)$ of a $\mathbb{Q}$-Gorenstein singularity (X, x) such that $(\mathscr{X}, x)$ is not a $\mathbb{Q}$-Gorenstein singularity.

First we will show that it is sufficient to construct a 1-parameter deformation $\pi : (\mathscr{X}, x) \to (C, 0)$ with the following properties:

(i) The singularity $(\mathscr{X}_0, x)$ is a 2-dimensional quotient singularity which is not canonical.
(ii) The singularity $\mathscr{X}_t$ $(t \neq 0,\ t \in C)$ is non-singular.
(iii) The deformation π has a simultaneous minimal resolution $f : \mathscr{Y} \to \mathscr{X}$.

By (i) the singularity $(\mathscr{X}_0, x)$ is a $\mathbb{Q}$-Gorenstein singularity. If there is an m such that $mK_{\mathscr{X}}$ is a Cartier divisor, then, as the exceptional set of f is of codimension 2, we have

$$mK_{\mathscr{Y}} \sim f^*mK_{\mathscr{X}}.$$

By $m(K_{\mathscr{Y}} + \mathscr{Y}_0) \sim f^*(m(K_{\mathscr{X}} + \mathscr{X}_0))$, it follows that $mK_{\mathscr{Y}_0} \sim f^*(mK_{\mathscr{X}_0})$. But this is a contradiction, because $(\mathscr{X}_0, x)$ is assumed to be non-canonical in (i).

Next we construct a deformation π with the properties (i), (ii), (iii).

Let $U_1, U_2 \cong \mathbb{C}$ and $\mathscr{M} = \mathbb{C} \times U_1 \times \mathbb{P}^1 \cup \mathbb{C} \times U_2 \times \mathbb{P}^1$ with the identification $(t, z_1, \zeta_1) \leftrightarrow (t, z_2, \zeta_2)$, $z_1 = \frac{1}{z_2}$, $\zeta_1 = z_2^4\zeta_2 + tz_2^2$. Define a projection $\widetilde{\pi} : \mathscr{M} \to \mathbb{C}$, $(t, z, \zeta) \mapsto t$, then by [MK, Theorem 4.2]

$$\mathscr{M}_0 \cong \Sigma_4,$$

$$\mathscr{M}_t \cong \Sigma_0 = \mathbb{P}^1 \times \mathbb{P}^1 \quad (t \neq 0).$$

Define the second projection $\varphi : \mathscr{M} \to \mathbb{P}^1 : (t, z, \zeta) \mapsto z$ and let $\varphi^{-1}(P) =: F$ $(P \in \mathbb{P}^1)$.

On the other hand, define a morphism $\psi : \widetilde{\pi}^{-1}(\mathbb{C} \setminus \{0\}) \to \mathbb{P}^1$ by $(t, z, \zeta) \mapsto \frac{z_1^2\zeta_1 - t}{t\zeta_1} = \frac{\zeta_2}{tz_2^2\zeta_2 + t^2}$. Let $H \subset \mathscr{M}$ be the closure of the inverse image $\psi^{-1}(Q)$ $(Q \in \mathbb{P})$. Then the linear system $|H + 2F|$ is base point free and it induces a morphism $\Phi : \mathscr{M} \to \mathbb{P}^5 \times \mathbb{C}$ such that $\widetilde{\pi} = p_2 \circ \Phi$. Let C_0 be the minimal section of the ruled surface $\mathscr{M}_0$ and f the ruling of the ruled surface. Then, as $H + 2F \mid_{\mathscr{M}_0} \sim C_0 + 4f$, $\Phi \mid_{\mathscr{M}_0}$ turns out to be the blow-down of C_0. On the other hand, for $t \neq 0$ $H + 2F \mid_{\mathscr{M}_t}$ is very ample, therefore $\Phi \mid_{\mathscr{M}_t}$ is a closed immersion. Let $\mathscr{X}$ be the image of Φ and let $\pi = p_2 \mid_{\mathscr{X}}$, then $\pi : \mathscr{X} \to \mathbb{C}$ is a deformation of a normal surface. As $C_0^2 = -4$, the surface $\mathscr{X}_0$ has a non-canonical quotient singularity at a point, $\mathscr{X}_t$ $(t \neq 0)$ is non-singular and $\Phi : \mathscr{M} \to \mathscr{X}$ is a simultaneous minimal resolution.

Next we can see an example showing that the $\mathbb{Q}$-Gorenstein property is not preserved under deformations.

Example 9.1.8. By making use of the previous example we can construct a deformation $(\mathscr{Z}, z) \to (C, 0)$ of a $\mathbb{Q}$-Gorenstein singularity $(\mathscr{Z}_0, z)$ such that $\mathscr{Z}_t$ $(t \neq 0)$ is not $\mathbb{Q}$-Gorenstein. In fact let $(\mathscr{X}, x)$ be the total space of the deformation in Example 9.1.7. By Lemma 8.3.5, there is a flat morphism $\pi : \mathscr{Z} \to \mathbb{A}^1$ such that

$\mathscr{Z}_0 \cong \mathscr{X}_0 \times \mathbb{A}^1$, $\mathscr{Z}_t \cong \mathscr{X}$. Here $\mathscr{Z}_0$ is $\mathbb{Q}$-Gorenstein but $\mathscr{Z}_t$ are not $\mathbb{Q}$-Gorenstein for $t \neq 0$.

The following proposition shows that under some conditions the total space $(\mathscr{X}, x)$ of a deformation of a $\mathbb{Q}$-Gorenstein singularity becomes a $\mathbb{Q}$-Gorenstein singularity.

Proposition 9.1.9. *Let $\pi : (\mathscr{X}, x) \to (C, 0)$ be a 1-parameter deformation of a normal Cohen–Macaulay singularity (X, x). Assume that $\mathscr{X}$ and X have at worst Gorenstein singularities outside a closed subset of codimension 3. If $\omega_X^{[m]}$ is invertible, then $\omega_{\mathscr{X}}^{[m]}$ is also invertible.*

Proof. Take an open subset $\mathscr{U} \subset \mathscr{X}$ such that $\mathscr{U}$ and $U = \mathscr{U} \cap \mathscr{X}_0 \subset \mathscr{X}_0$ satisfy that $\operatorname{codim}_{\mathscr{X}}(\mathscr{X} \setminus \mathscr{U}) \geq 3$, $\operatorname{codim}_{\mathscr{X}_0}(\mathscr{X}_0 \setminus U) \geq 3$ and on $\mathscr{U}$ and U, $\mathscr{X}$ and $\mathscr{X}_0$ are Gorenstein, respectively.

Now if $\omega_{\mathscr{X}_0}^{[m]}$ is invertible, then we may assume that $\omega_{\mathscr{X}_0}^{[m]} |_U \cong \mathscr{O}_U$. As $\mathscr{X}_0 \cong X$ is Cohen–Macaulay, for every prime ideal $\mathfrak{p} \subset \mathscr{O}_{\mathscr{X}_0}$ such that $V(\mathfrak{p}) \subset \mathscr{X}_0 \setminus U$ we have depth $\mathscr{O}_{\mathscr{X}_0,\mathfrak{p}} = \dim(\mathscr{O}_{\mathscr{X}_0})_{\mathfrak{p}} \geq 3$. Then by the following lemma, we obtain

$$\omega_{\mathscr{X}}^{[m]} |_{\mathscr{U}} \cong \mathscr{O}_{\mathscr{U}},$$

therefore $\omega_{\mathscr{X}}^{[m]}$ is invertible on $\mathscr{X}$.

Lemma 9.1.10 (Lipman [Lip, Corollary]). *Let $\pi : (\mathscr{X}, x) \to (C, 0)$ be a 1-parameter deformation of (X, x). Let B and A be the affine coordinate rings of $\mathscr{X}$ and $\mathscr{X}_0 = X$, respectively. Let $\mathscr{U}$ be an open subset of $\mathscr{X}$ and $U \subset \mathscr{U}$ an open subset of $\mathscr{X}_0$ such that for every prime ideal $\mathfrak{p} \subset A$ with $V(\mathfrak{p}) \subset \mathscr{X}_0 \setminus U$ the inequality* depth $A_{\mathfrak{p}} \geq 3$ *holds. Let $i : U \hookrightarrow \mathscr{U}$ be the inclusion map. If a locally free sheaf $\mathscr{F}$ on $\mathscr{U}$ satisfies $i^*\mathscr{F} \cong \mathscr{O}_U^{\oplus n}$, then it follows that $\mathscr{F} \cong \mathscr{O}_{\mathscr{U}}^{\oplus n}$. In particular, the canonical map $P_{ic}(\mathscr{U}) \to P_{ic}(U)$ is injective.*

Corollary 9.1.11. *Let $\pi : (\mathscr{X}, x) \to (C, 0)$ be a 1-parameter deformation of a canonical singularity, then $(\mathscr{X}, x)$ is a $\mathbb{Q}$-Gorenstein singularity.*

Proof. Let $\mathscr{Z} \subset \mathscr{X}$ be the set of non-Gorenstein points in $\mathscr{X}$. By Proposition 5.3.12 the intersection $\mathscr{Z} \cap \mathscr{X}_0$ is the set of non-Gorenstein points of $\mathscr{X}_0$. Therefore by Corollary 8.3.2 (i) we have that $\operatorname{codim}_{\mathscr{X}_0} \mathscr{Z} \cap \mathscr{X}_0 \geq 3$. Then, by taking $\mathscr{X}$ as a sufficiently small neighborhood of x, it follows that $\operatorname{codim}_{\mathscr{X}} \mathscr{Z} \geq 3$. Now we can apply Proposition 9.1.9.

We will now study the rationality of the singularities under a deformation.

Theorem 9.1.12 (Elkik [El1]). *Let $\pi : (\mathscr{X}, x) \to (C, 0)$ be a deformation of a rational singularity (X, x). Then, by taking C and $\mathscr{X}$ sufficiently small, we have that $\mathscr{X}_t$ has rational singularities of every $t \in C$. The singularity $(\mathscr{X}, x)$ is also rational.*

Next we consider deformations of canonical singularities. Kawamata [Ka] proved a deformation of canonical singularities is canonical. We give here the proof according to [KMM, 7-2-4] by using [BCHM].

Theorem 9.1.13. *Let $\pi : (\mathscr{X}, x) \to (C, 0)$ be a 1-parameter deformation of a canonical singularity (X, x). By taking $\mathscr{X}$ and C sufficiently small, we obtain that $\mathscr{X}_t$ has at worst canonical singularities for every $t \in C$. We also have that $(\mathscr{X}, x)$ is a canonical singularity.*

Proof. If $(\mathscr{X}, x)$ is a canonical singularity, then a sufficiently small neighborhood of x has at worst canonical singularities. Then, by Proposition 8.3.1 the fiber $\mathscr{X}_t$ has at worst canonical singularities for t close to 0. Therefore it is sufficient to prove that $(\mathscr{X}, x)$ is a canonical singularity. By [BCHM], there exists a canonical model $f : \mathscr{Y} \to \mathscr{X}$ of $\mathscr{X}$. Let $\mathscr{X}_0'$ be the strict transform of $\mathscr{X}_0$ in $\mathscr{Y}$, let $\mathscr{Y}_0 := f^*\mathscr{X}_0$ and let $\sigma : Z \to \mathscr{X}_0' \subset \mathscr{Y}$ be the normalization. Let r be a common multiple of the indexes of $\mathscr{Y}$ and of $\mathscr{X}_0$. By [Nak, Lemma 1], there is the trace map:

$$\sigma_*\omega_Z^{[r]} \longrightarrow \omega_{\mathscr{Y}}^{[r]} \otimes \mathscr{O}_{\mathscr{Y}_0}.$$

Then, we obtain:

$$\omega_Z^{[r]} \longrightarrow \sigma^*(\omega_{\mathscr{Y}}^{[r]} \otimes \mathscr{O}_{\mathscr{Y}_0}). \tag{9.1}$$

On the other hand, let $\nu := f \circ \sigma$, then as $\mathscr{X}_0$ has canonical singularity, we have:

$$\nu^*\omega_{\mathscr{X}_0}^{[r]} \longrightarrow \omega_Z^{[r]}. \tag{9.2}$$

By composing (9.1) and (9.2), we obtain:

$$\nu^*\omega_{\mathscr{X}_0}^{[r]} \longrightarrow \sigma^*(\omega_{\mathscr{Y}}^{[r]} \otimes \mathscr{O}_{\mathscr{Y}_0}).$$

Therefore, we can write:

$$\sigma^*K_{\mathscr{Y}} = \nu^*K_{\mathscr{X}_0} + (\text{effective exceptional divisor}).$$

Here, as $K_{\mathscr{Y}}$ is f-ample, $\sigma^*K_{\mathscr{Y}}$ is ν-ample. Hence, by Lemma 8.2.11, ν is isomorphic and the equality $K_{\mathscr{Y}} \mid_Z = K_{\mathscr{X}} \mid_{\mathscr{X}_0}$ holds. On the other hand, by Corollary 9.1.11 the total space $\mathscr{X}$ also has $\mathbb{Q}$-Gorenstein singularities, which gives the expression of the $\mathbb{Q}$-Cartier divisor

$$K_{\mathscr{Y}} = f^*K_{\mathscr{X}} + \Delta$$

(where Δ is a $\mathbb{Q}$-Cartier divisor with the support on the exceptional divisor). As $K_{\mathscr{Y}}$ is f-ample, we have $\Delta < 0$ and the support of Δ is the whole exceptional divisor if f is not isomorphic. Then, we have

$$\begin{aligned} &rK_{\mathscr{X}} \mid_{\mathscr{X}_0} = rf^*K_{\mathscr{X}} \mid_Z \\ &> r(f^*K_{\mathscr{X}} + \Delta)|_Z = rK_{\mathscr{Y}} \mid_Z = rK_{\mathscr{X}} \mid_{\mathscr{X}_0}, \end{aligned}$$

which is a contradiction. Therefore f must be isomorphic, i.e., the singularity $(\mathscr{X}, x)$ is canonical.

Next we are going to prove that a deformation of a terminal singularity is a terminal singularity. For the 3-dimensional case, this is proved by Kollár and Shepherd-Barron [KSB] and for an arbitrary dimension by Nakayama [Nak]. The following is the proof using the recent work by [BCHM].

Theorem 9.1.14. *Let $\pi : (\mathscr{X}, x) \to (C, 0)$ be a 1-parameter deformation of a terminal singularity (X, x). If we take $\mathscr{X}$ and C sufficiently small, then $\mathscr{X}_t$ has at worst terminal singularities for every $t \in C$. And $(\mathscr{X}, x)$ is also a terminal singularity.*

Proof. By the result of Theorem 9.1.13, taking a sufficiently small $\mathscr{X}$, we may assume that $\mathscr{X}$ has at worst canonical singularities. It is sufficient to prove that $(\mathscr{X}, x)$ is terminal by Proposition 8.3.1. By [BCHM] there exists a projective minimal model $f : \mathscr{Y} to \mathscr{X}$. By Theorem 8.2.12 (ii), we have $K_{\mathscr{Y}} = f^*(K_{\mathscr{X}})$. Let $\mathscr{Y}_0$ be the strict transform of $\mathscr{X}_0$ in $\mathscr{Y}$ and $f_0 : \mathscr{Y}_0 \to \mathscr{X}_0$ be the restriction of f onto $\mathscr{Y}_0$. Denote the pull-back divisor of $f^*(\mathscr{X}_0)$ by

$$f^*(\mathscr{X}_0) = \mathscr{Y}_0 + \sum_i r_i E_i,$$

where E_i is a prime divisor and $r_i > 0$ if the second part of the right-hand side exists. Here, as $\mathscr{Y}$ has terminal singularities, it follows that $\operatorname{codim}(\mathscr{Y}_{\text{sing}}, \mathscr{Y}) \geq 3$. Then, $\mathscr{Y}_0$ becomes a hypersurface up to codimension 2, therefore it is Gorenstein up to codimension 2. Thus, we obtain the equalities:

$$\begin{aligned} K_{\mathscr{Y}_0} &= (K_{\mathscr{Y}} + \mathscr{Y}_0)|_{\mathscr{Y}_0} = (f^* K_{\mathscr{X}} + \mathscr{Y}_0)|_{\mathscr{Y}_0} \\ &= f_0^* K_{\mathscr{X}_0} + (\mathscr{Y}_0|_{\mathscr{Y}_0}) = f_0^* K_{\mathscr{X}_0} - \sum_i r_i E_i \end{aligned}$$

outside a closed subset of codimension ≥ 2. Let $\nu : \mathscr{Y}_0' \to \mathscr{Y}_0$ be the normalization. Then, outside a closed subset of codimension ≥ 2, it follows that

$$K_{\mathscr{Y}_0'} < \nu^* K_{\mathscr{Y}_0} = \nu^*(f_0^* K_{\mathscr{X}_0} - \sum_i r_i E_i).$$

But $f_0 \circ \nu : \mathscr{Y}_0' \to \mathscr{X}_0$ is a partial resolution of terminal singularities, which yields

$$K_{\mathscr{Y}_0'} - \nu^* f_0^* K_{\mathscr{X}_0} > 0.$$

Therefore, there is no E_i, i.e., $f^* \mathscr{X}_0 = \mathscr{Y}_0$ and the partial resolution $f_0 : \mathscr{Y}_0 \to \mathscr{X}_0$ is a small birational morphism (i.e., there is no divisor contracted to a smaller dimensional subset). By this, replacing $\mathscr{X}$ by a sufficiently small neighborhood of x there is no horizontal exceptional divisor on $\mathscr{Y}$. Here, a horizontal exceptional divisor is an exceptional prime divisor which dominates C and a vertical exceptional divisor is the one not dominating C. As there is no vertical exceptional divisor mapped to $0 \in C$, we can take small neighborhoods $\mathscr{X}$ of x and C of 0 such that $\mathscr{Y} \to \mathscr{X}$ is a small partial resolution. Therefore, $\mathscr{X}$ has terminal singularities.

Remark 9.1.15. For a deformation $\pi : (\mathscr{X}, x) \to (C, 0)$ of a log terminal (resp. log canonical) singularity (X, x), by Examples 9.1.7 and 9.1.8 the singularities $(\mathscr{X}, x)$, $\mathscr{X}_t$ are not necessarily $\mathbb{Q}$-Gorenstein, therefore of course these are not log terminal (resp. log canonical). Then, what if we assume a $\mathbb{Q}$-Gorenstein property on the total space $\mathscr{X}$? The next theorems show that deformations of log terminal singularities (resp. log canonical singularities) are also log terminal (resp. log canonical singularities).

Theorem 9.1.16. *Let $\pi : (\mathscr{X}, x) \to (C, 0)$ be a 1-parameter deformation of a log terminal singularity (X, x). Assume that $(\mathscr{X}, x)$ is a $\mathbb{Q}$-Gorenstein singularity. Then, by replacing $\mathscr{X}$ and C sufficiently small, $\mathscr{X}_t$ $(t \in C)$ has at worst log terminal singularities. The singularity $(\mathscr{X}, x)$ is also log terminal.*

Proof. Let $\rho : \mathscr{Y} \to \mathscr{X}$ be the canonical cover of $(\mathscr{X}, x)$. Then $\pi' := \pi \circ \rho$ becomes a 1-parameter deformation of $\mathscr{Y}_0 := \pi'^{-1}(0)$. As $\mathscr{Y}_0$ does not have a ramification divisor, by Theorem 6.2.9 it has a log terminal singularity of index one, i.e., it is a rational Gorenstein singularity. Then, by Theorems 9.1.6 and 9.1.12, $\mathscr{Y}$ and $\mathscr{Y}_t$ $(t \in C)$ have at worst rational Gorenstein singularities. Therefore the images $\mathscr{X}$, $\mathscr{X}_t$ of these varieties by the finite morphisms without ramification divisors have log terminal singularities by Theorem 6.2.9.

Theorem 9.1.17. *Let $\pi : (\mathscr{X}, x) \to (C, 0)$ be a 1-parameter deformation of a log canonical singularity (X, x). Assume that $(\mathscr{X}, x)$ is a $\mathbb{Q}$-Gorenstein singularity. Then, by replacing $\mathscr{X}$ and C sufficiently small, $\mathscr{X}_t$ $(t \in C)$ has at worst log canonical singularities. The singularity $(\mathscr{X}, x)$ is also log canonical.*

Proof. It is sufficient to prove that $(\mathscr{X}, x)$ is log canonical. This follows immediately from Kawakita's inversion of adjunction of log canonicity [Kaw]. Consider a pair (X, S) consisting of normal variety X and a normal reduced divisor S on X such that $K_X + S$ is a $\mathbb{Q}$-Cartier divisor. He proved that (X, S) is log canonical if and only if S is log canonical. Apply this result to our case. Then, as $\mathscr{X}_0$ is log canonical, it follows that $(\mathscr{X}, \mathscr{X}_0)$ is log canonical. In particular, $\mathscr{X}$ is log canonical since $\mathscr{X}$ is $\mathbb{Q}$-Gorenstein.

Remark 9.1.18. Let $\pi : (\mathscr{X}, x) \to (C, 0)$ be a 1-parameter deformation of a quotient singularity (X, x). We saw in Example 9.1.8 that $(\mathscr{X}, x)$ is not a $\mathbb{Q}$-Gorenstein singularity, therefore it is not a quotient singularity. The question whether $\mathscr{X}_t$ $(t \in C)$ are quotient singularities is answered for the 2-dimensional case as follows:

Theorem 9.1.19 (Esnault–Viehweg [EV]). *Let $\pi : (\mathscr{X}, x) \to (C, 0)$ be a 1-parameter deformation of a 2-dimensional quotient singularity (X, x). Then, by replacing $\mathscr{X}$ and C sufficiently small, we have that $\mathscr{X}_t$ has at worst quotient singularities for every $t \in C$.*

Now we study singularities which have no non-trivial deformations.

Definition 9.1.20. A singularity (X, x) is called a *rigid singularity* if every deformation $\pi : (\mathscr{X}, x) \to (C, 0)$ of (X, x) becomes the trivial deformation $p_2 : X \times C \to C$.

Theorem 9.1.21 (Schlessinger [Sch2]). *A quotient singularity* (X, x) *such that* $\operatorname{codim}_X X_{\mathrm{sing}} \geq 3$ *is rigid.*

Next we will study how the invariants of the singularities vary under deformations of singularities. First we will consider the embedding dimension.

Theorem 9.1.22. *The embedding dimension* $\mathrm{emb}(X, x)$ *of a singularity* (X, x) *is upper semi-continuous under a 1-parameter deformation of* (X, x)*. More precisely, for a 1-parameter deformation* $\pi : \mathscr{X} \to C$ *of* (X, x) *and a section* $\sigma : C \to \mathscr{X}$ *such that* $\sigma(0) = x$*, by replacing* $\mathscr{X}$ *and* C *sufficiently small, we have*

$$\mathrm{emb}(X, x) \geq \mathrm{emb}(\mathscr{X}_t, \sigma(t)), \text{ for every } t \in C.$$

In particular, if (X, x) *is a hypersurface singularity, then* $(\mathscr{X}_t, \sigma(t))$ *are also hypersurface singularities for all* $t \in C$.

Proof. For a 1-parameter deformation $\pi : \mathscr{X} \to C$ of (X, x), by the proof of Theorem 4.1.7, (vi) we obtain

$$\dim \mathfrak{m}_{\mathscr{X},y}/\mathfrak{m}^2_{\mathscr{X},y} = \dim \Omega_{\mathscr{X},y}/\mathfrak{m}_{\mathscr{X},y}\Omega_{\mathscr{X},y}, \quad \forall y \in \mathscr{X}.$$

By Nakayama's lemma, $\Omega_{\mathscr{X},x}$ is generated by $\dim \mathfrak{m}_{\mathscr{X},x}/\mathfrak{m}^2_{\mathscr{X},x}$ elements, therefore by taking $\mathscr{X}$ and C sufficiently small, for every $y \in \mathscr{X}$ we have

$$\dim \mathfrak{m}_{\mathscr{X},x}/\mathfrak{m}^2_{\mathscr{X},x} \geq \dim \mathfrak{m}_{\mathscr{X},y}/\mathfrak{m}^2_{\mathscr{X},y}.$$

On the other hand, for $t \in C$, $y \in \mathscr{X}_t$ it follows that $\mathfrak{m}_{\mathscr{X}_t,y}/\mathfrak{m}^2_{\mathscr{X}_t,y} = \mathfrak{m}_{\mathscr{X},y}/(\mathfrak{m}^2_{\mathscr{X},y} + (\xi_t))$, where ξ_t is a generator of the maximal ideal at $t \in C$. Therefore we have

$$\dim \mathfrak{m}_{\mathscr{X}_t,y}/\mathfrak{m}^2_{\mathscr{X}_t,y} = \dim \mathfrak{m}_{\mathscr{X},y}/\mathfrak{m}^2_{\mathscr{X},y} \text{ or } \dim \mathfrak{m}_{\mathscr{X},y}/\mathfrak{m}^2_{\mathscr{X},y} - 1.$$

Here by taking C and $\mathscr{X}$ sufficiently small, it is sufficient to prove for $0 \neq t \in C$, $y = \sigma(t) \in \mathscr{X}_t$

$$\dim \mathfrak{m}_{\mathscr{X}_t,y}/\mathfrak{m}^2_{\mathscr{X}_t,y} = \dim \mathfrak{m}_{\mathscr{X},y}/\mathfrak{m}^2_{\mathscr{X},y} - 1$$

to show $\dim \mathfrak{m}_{\mathscr{X}_0,x}/\mathfrak{m}^2_{\mathscr{X}_0,x} \geq \dim \mathfrak{m}_{\mathscr{X}_t,y}/\mathfrak{m}^2_{\mathscr{X}_t,y}$. Let $\mathscr{I}$ be the reduced defining ideal of the image $\sigma(C)$. Then for $y = \sigma(t) \in \mathscr{X}_t$ we have

$$(\xi_t) + \mathscr{I}_y = \mathfrak{m}_{\mathscr{X},y}.$$

If $\xi_t \in \mathfrak{m}^2_{\mathscr{X},y}$, then by

$$\mathscr{I} \otimes \mathscr{O}/\mathfrak{m} = \mathscr{I}/\mathfrak{m}\mathscr{I} \twoheadrightarrow \mathscr{I}/\mathfrak{m}^2 \cap \mathscr{I} = \mathfrak{m}/\mathfrak{m}^2 = \mathfrak{m} \otimes \mathscr{O}/\mathfrak{m}$$

we have $\mathfrak{m}/\mathscr{I} \otimes \mathscr{O}/\mathfrak{m} = 0$. Then, by Nakayama's lemma it follows that $\mathscr{I} = \mathfrak{m}$, which is a contradiction. Thus it must be $\xi_t \notin \mathfrak{m}^2_{\mathscr{X},y}$. Therefore we obtain

$$\dim \mathfrak{m}_{\mathscr{X}_t,y}/\mathfrak{m}^2_{\mathscr{X}_t,y} = \dim \mathfrak{m}_{\mathscr{X},y}/\mathfrak{m}^2_{\mathscr{X},y} - 1.$$

Next we will study the geometric genus p_g.

Theorem 9.1.23 (Elkik [El1]). *For an n-dimensional normal isolated singularity (X, x), define*

$$\chi_i(X, x) = \sum_{j=0}^{n-i-1} (-1)^j \dim_{\mathbb{C}} R^{n-j-i} f_* \mathscr{O}_Y,$$

where $f : Y \to X$ is a resolution of the singularities. Then, $\chi_i(X, x)$ $(i = 1, \cdots, n-1)$ is upper semi-continuous under a deformation. I.e., taking C and $\mathscr{X}$ sufficiently small, we obtain:

$$\chi_i(\mathscr{X}_0, x) \geq \sum_{y \in \mathscr{X}_t} \chi_i(\mathscr{X}_t, y), \quad \forall t \in C.$$

In particular, considering $\chi_1 + \chi_2$ we obtain the following formula on the geometric genus p_g:

$$p_g(\mathscr{X}, x) \geq \sum_{y \in \mathscr{X}_t} p_g(\mathscr{X}_t, y), \quad \forall t \in C.$$

For plurigenera δ_m we obtain a similar statement.

Theorem 9.1.24 (Ishii [I5]). *The plurigenera $\delta_m(X, x)$ of a normal isolated singularity (X, x) is upper semi-continuous under deformations. I.e., by taking C and $\mathscr{X}$ sufficiently small, we obtain the following:*

$$\delta_m(\mathscr{X}_0, x) \geq \sum_{y \in \mathscr{X}_t} \delta_m(\mathscr{X}_t, y), \quad \forall t \in C.$$

The second statement of Theorem 9.1.23 is a corollary of this theorem by putting $m = 1$.

The fact that a deformation of a 2-dimensional quotient singularity is again a quotient singularity (Theorem 9.1.19) is also a corollary of this theorem.

Theorem 9.1.24 shows another proof of invariance of log canonicity and log terminally under deformations assuming $\mathbb{Q}$-Gorenstein property on the total space of a deformation.

Another plurigenera γ_m is also proved to hold the upper semi-continuity under deformation.

Theorem 9.1.25. *Let $\pi : (\mathscr{X}, x) \to (C, 0)$ be a 1-parameter deformation of a normal isolated singularity (X, x). Then the plurigenera $\gamma_m(X, x)$ is upper semi-continuous, i.e., taking $\mathscr{X}$ and C sufficiently small, we have the following:*

$$\gamma_m(\mathscr{X}_0, x) \geq \sum_{y \in \mathscr{X}_t} \gamma_m(\mathscr{X}_t, y), \quad \forall t \in C.$$

Here, if the equalities hold then π admits the simultaneous canonical model.

Proof. In [I9] this theorem is proved under the condition that $(\mathscr{X}, x)$ has a minimal model. As in [BCHM] the existence of a minimal model of $(\mathscr{X}, x)$ is proved, we obtain the required theorem.

Next we study the behavior of the type of isolated 1-Gorenstein singularities under deformations.

Theorem 9.1.26 (Ishii [I9]). *Let $\pi : (\overline{\mathscr{X}}, x) \to (C, 0)$ be a 1-parameter deformation of a 1-Gorenstein singularity $(\overline{X}, x)$ with $\kappa_\delta(\overline{X}, x) = 0$ such that:*

(i) $\pi : \overline{\mathscr{X}} \to C$ is projective.
(ii) Each $\overline{\mathscr{X}}_t$ $(t \in C)$ has a unique singularity $(\overline{\mathscr{X}}_t, x_t)$ with $\kappa_\delta(\mathscr{X}_t, x_t) = 0$.

Then by taking C sufficiently small, the type $(0, j)$ of $(\overline{\mathscr{X}}_t, x_t)$ $t \in C^* = C \setminus \{0\}$ is constant. Let the type of $(\overline{\mathscr{X}}_0, x)$ be $(0, k)$, then $k \leq j$. In particular, consider $\pi : (\mathscr{X}, x) \to (C, 0)$, a deformation of 2-dimensional singularities. Then if $(\mathscr{X}_0, x)$ is simple elliptic, then $(\mathscr{X}_t, x_t)$, $t \in C^*$ are all simple elliptic. If $(\mathscr{X}_0, x)$ is a cusp singularity, then $(\mathscr{X}_t, x_t)$, $t \in C^*$ are either cusp or simple elliptic.

9.2 Semi-universal Deformations

Is it possible to grasp all deformations of a singularity? In this section we will see that it is possible.

Definition 9.2.1. A deformation $\pi : (\mathscr{X}, x) \to (S, 0)$ of a singularity (X, x) is called a *semi-universal deformation* if the following conditions hold:

(1) For every deformation $\pi' : (\mathscr{X}', x) \to (S', 0)$ of a singularity (X, x), there is a morphism $\rho : (S', 0) \to (S, 0)$ such that $\mathscr{X}' \cong \mathscr{X} \times_S S'$.
(2) A morphism ρ in (1) induces a $\mathbb{C}$-linear map $d\rho_0 : T_{S',0} := \mathrm{Hom}_{\mathscr{O}_{S',0}}(\mathfrak{m}_{S',0}/\mathfrak{m}_{S',0}^2, \mathbb{C}) \to T_{S,0} := \mathrm{Hom}_{\mathscr{O}_{S,0}}(\mathfrak{m}_{S,0}/\mathfrak{m}_{S,0}^2, \mathbb{C})$. The linear map $d\rho_0$ is unique for π'.

Such an S is denoted by $\mathrm{Def}(X, x)$.

Condition (1) means that the deformation π contains all deformations of (X, x). Such a π is not unique, but condition (2) requires a weak uniqueness.

Theorem 9.2.2 (Tjurina [Tj], Grauert [Gr3], Donin [Do]). *For every isolated singularity (X, x) there exists a semi-universal deformation in the category of analytic spaces.*

Definition 9.2.3. Consider deformations $\pi : (\mathscr{X}, x) \to (S, 0)$ of a singularity (X, x) over S. The set of equivalence classes of those deformations by isomorphism over S is denoted by $\mathrm{Def}_S(X, x)$. In particular, if S is an affine scheme $\mathrm{Spec}\,\mathbb{C}[\varepsilon]$ defined by a $\mathbb{C}$-algebra $\mathbb{C}[\varepsilon]$, $\varepsilon^2 = 0$, $\mathrm{Def}_S(X, x)$ is denoted by $T^1_{X,x}$ and an element of this set is called an *infinitesimal deformation*. Note that the set $T^1_{X,x}$ becomes a $\mathbb{C}$-vector space.

Proposition 9.2.4. *If (X, x) is an isolated singularity, $T^1_{X,x}$ is isomorphic to the tangent space of* $\mathrm{Def}(X, x)$ *at 0.*

Proof. The tangent space $T_{\mathrm{Def}(X,x),0}$ of $\mathrm{Def}(X, x)$ at 0 is isomorphic to $\mathrm{Hom}_0\big(\mathrm{Spec}\,\mathbb{C}[\varepsilon], \mathrm{Def}(X, x)\big)$, the set of morphisms $(\mathrm{Spec}\,\mathbb{C}[\varepsilon], 0) \to (\mathrm{Def}(X, x), 0)$ (see, for example, [Mu, III, §4, Th3]). By the definition of a semi-universal deformation, for an element $\big[\pi : \mathscr{Y} \to \mathrm{Spec}\,\mathbb{C}[\varepsilon]\big] \in T^1_{X,x}$ there exists a morphism $\rho : \big(\mathrm{Spec}\,\mathbb{C}[\varepsilon], 0) \to (\mathrm{Def}(X, x), 0\big)$ such that $d\rho : T_{\mathrm{Spec}\,\mathbb{C}[\varepsilon],0} \to T_{\mathrm{Def}(X,x),0}$ is uniquely determined. As $T_{\mathrm{Spec}\,\mathbb{C}[\varepsilon],0} \cong \mathbb{C}$, we can define a linear map $\Phi : T^1_{X,x} \to T_{\mathrm{Def}(X,x),0}$ by $[\pi] \mapsto d\rho(1)$.

Conversely, for $\zeta \in T_{\mathrm{Def}(X,x),0} = \mathrm{Hom}_0\big(\mathrm{Spec}\,\mathbb{C}[\varepsilon], \mathrm{Def}(X, x)\big)$, by pulling back the semi-universal deformation $\mathscr{X} \to \mathrm{Def}(X, x)$ by the morphism $\zeta : \mathrm{Spec}\,\mathbb{C}[\varepsilon] \to \mathrm{Def}(X, x)$, we obtain $\big[\pi : \mathscr{X} \times_{\mathrm{Def}(X,x)} \mathrm{Spec}\,\mathbb{C}[\varepsilon] \to \mathrm{Spec}\,\mathbb{C}[\varepsilon]\big] \in T^1_{X,x}$. This is the inverse of Φ.

Proposition 9.2.5 ([A3, 6.1]). *For an isolated singularity (X, x) it follows that*

$$T^1_{X,x} \cong \mathrm{Ext}^1_{\mathscr{O}_{X,x}}(\Omega_{X,x}, \mathscr{O}_{X,x}).$$

Theorem 9.2.6 Tjurina[Tj]). *For a normal isolated singularity (X, x) assume $T^2_{X,x} := \mathrm{Ext}^2_{\mathscr{O}_{X,x}}(\Omega_{X,x}, \mathscr{O}_{X,x}) = 0$, then* $\mathrm{Def}(X, x)$ *is non-singular.*

Example 9.2.7. Let $(X, 0) \subset (\mathbb{C}^{n+1}, 0)$ be a hypersurface isolated singularity defined by an equation $f = 0$. Then $T^2_{X,0} = 0$ and $\mathrm{Def}(X, 0)$ is a finite-dimensional $\mathbb{C}$-vector space $\mathbb{C}[x_0, \cdots, x_n]/\big(f, \frac{\partial f}{\partial x_0}, \cdots, \frac{\partial f}{\partial x_n}\big)$.

Indeed by the exact sequence:

$$\begin{array}{ccccccccc} 0 & \longrightarrow & \mathscr{O}_X & \xrightarrow{\alpha} & \Omega_{\mathbb{C}^{n+1}} \otimes \mathscr{O}_X & \longrightarrow & \Omega_X & \longrightarrow & 0 \\ & & \Psi & & \Psi & & & & \\ & & 1 & \longmapsto & df & & & & \end{array}$$

we obtain the following exact sequence:

$$\begin{aligned}&\mathrm{Hom}_{\mathcal{O}_{X,0}}(\Omega_{\mathbb{C}^{n+1},0}\otimes\mathcal{O}_{X,0},\mathcal{O}_{X,0})\xrightarrow{\alpha^*}\mathrm{Hom}_{\mathcal{O}_X}(\mathcal{O}_{X,0},\mathcal{O}_{X,0})\\&\xrightarrow{\delta}\mathrm{Ext}^1_{\mathcal{O}_{X,0}}(\Omega_{X,0},\mathcal{O}_{X,0})\longrightarrow\mathrm{Ext}^1_{\mathcal{O}_{X,0}}(\Omega_{\mathbb{C}^{n+1},0}\otimes\mathcal{O}_{X,0},\mathcal{O}_{X,0})\\&\longrightarrow\mathrm{Ext}^1_{\mathcal{O}_{X,0}}(\mathcal{O}_{X,0},\mathcal{O}_{X,0})\longrightarrow\mathrm{Ext}^2_{\mathcal{O}_{X,0}}(\Omega_{X,0},\mathcal{O}_{X,0})\\&\longrightarrow\mathrm{Ext}^2_{\mathcal{O}_{X,0}}(\Omega_{\mathbb{C}^{n+1},0}\otimes\mathcal{O}_{X,0},\mathcal{O}_{X,0}).\end{aligned}$$

Here, by Proposition 3.3.7, we have

$$\mathrm{Ext}^i_{\mathcal{O}_{X,0}}(\Omega_{\mathbb{C}^{n+1},0}\otimes\mathcal{O}_{X,0},\mathcal{O}_{X,0})=\mathrm{Ext}^i_{\mathcal{O}_{X,0}}(\mathcal{O}_{X,0},\mathcal{O}_{X,0})=0\qquad(i\geq 1),$$

therefore $\mathrm{Ext}^2_{\mathcal{O}_{X,0}}(\Omega_{X,0},\mathcal{O}_{X,0})=0$ and δ is surjective. Hence

$$T^1_{X,0}=\mathrm{Ext}^1_{\mathcal{O}_{X,0}}(\Omega_{X,0},\mathcal{O}_{X,0})=\mathrm{Hom}_{\mathcal{O}_{X,0}}(\mathcal{O}_{X,0},\mathcal{O}_{X,0})/\mathrm{Im}\,\alpha^*.$$

Here a basis of $\mathrm{Hom}(\Omega_{\mathbb{C}^{n+1},0}\otimes\mathcal{O}_{X,0},\mathcal{O}_{X,0})$ is written as $\langle\varphi_0,\cdots,\varphi_n\rangle$ such that $\varphi_i(dx_j)=\delta_{ij}\in\mathcal{O}_{X,0}$. As $\alpha^*(\varphi_i)(1)=\varphi_i\circ\alpha(1)=\varphi_i(df)=\frac{\partial f}{\partial x_i}$, by using the isomorphism

$$\mathrm{Hom}_{\mathcal{O}_{X,0}}(\mathcal{O}_{X,0},\mathcal{O}_{X,0})\cong\mathcal{O}_{X,0},\ \varphi\mapsto\varphi(1)$$

we obtain $\mathrm{Im}\,\alpha^*\cong\left(\frac{\partial f}{\partial x_0},\cdots,\frac{\partial f}{\partial x_n}\right)$. Therefore

$$T^1_{X,0}\cong\mathbb{C}[x_0,\cdots,x_n]\big/\left(f,\frac{\partial f}{\partial x_0},\cdots,\frac{\partial f}{\partial x_n}\right).$$

Here, let $p_i(x)$ $(i=1,\cdots,\ell)$ be a basis of $\mathbb{C}[x_0,\cdots,x_n]\big/\left(f,\frac{\partial f}{\partial x_0},\cdots,\frac{\partial f}{\partial x_n}\right)$ as a $\mathbb{C}$-vector space, then a semi-universal deformation $\pi:(\mathscr{X},x)\to(\mathrm{Def}(X,x),0)=(\mathbb{C}^\ell,0)$ is given by

$$\begin{array}{ccc}\mathscr{X}=\left\{(x_0,\cdots,x_n,t_1,\cdots,t_\ell)\in\mathbb{C}^{n+1+\ell}\mid f+\sum_{i=1}^{\ell}t_ip_i=0\right\}&\longrightarrow&\mathbb{C}^\ell\\ \Psi&&\Psi\\(x_0,\cdots,x_n,t_1,\cdots,t_\ell)&\longmapsto&(t_1,\cdots,t_\ell).\end{array}$$

This construction is given in (Kas–Schlessinger [KS]).

Example 9.2.8 (*Pinkham [Pin, §8], Artin [A3, §13]*). Pinkham obtained T^1 and a semi-universal deformation of a 2-dimensional quotient singularity $X_{n,1}$ $(n\geq 4)$. According to his result, $\dim T^1=2n-4$ and a semi-universal deformation for $n\geq 5$ is as follows. The base space has an embedding component at the origin 0 (as a scheme) and its reduced structure is of $(n-1)$-dimensional non-singular space and the total space of the deformation is in $\mathbb{C}^{n+1} \times \mathbb{C}^{n-1}$ defined by the 2×2 minors of the following matrix = 0:

$$\begin{pmatrix} x_0 & x_1 & x_2 & \cdots & x_{n-2} & x_{n-1} \\ x_1 - t_1 & x_2 - t_2 & \cdots & \cdots & x_{n-1} - t_{n-1} & x_n \end{pmatrix}.$$

For $n = 4$, the base space of a semi-universal deformation is reduced and there are two non-singular irreducible components $D_1 \cong \mathbb{C}^3$, $D_2 \cong \mathbb{C}^1$. The total space on D_1 is in $\mathbb{C}^5 \times \mathbb{C}^3$ defined by 2×2 minors of the following matrix = 0:

$$\begin{pmatrix} x_0 & x_1 & x_2 & x_3 \\ x_1 - t_1 & x_2 - t_2 & x_3 - t_3 & x_4 \end{pmatrix}.$$

The total space on D_2 is in $\mathbb{C}^5 \times \mathbb{C}$ defined by 2×2 minors in the following matrix = 0:

$$\begin{pmatrix} x_0 & x_1 & x_2 - t \\ x_1 & x_2 & x_3 \\ x_2 - t & x_3 & x_4 \end{pmatrix}.$$

In general, it is difficult to calculate $T^i_{X,x}$, in particular $T^2_{X,x}$. We have some results on 2-dimensional rational singularities, in particular quotient singularities **(author?)** [Ri2, BeK, BKR].

Chapter 10
Quick Overview of Recent Developments on Singularities

You never catch on until after the test.
– Natalie's Law of Algebra (Young [Mur])

In this chapter, we will present a quick overview of recent developments on singularities after the publication of the first Japanese version (1997) of this book. The goal of the chapter is just to show the directions of the research which is now going on. The reader who would like to step further in these directions is recommended to read some of the references cited in this chapter.

10.1 Thinking of Pairs and the Log Discrepancies

In this section a variety X is assumed to be an affine variety, since our interest is local. In the classical work on singularities in the previous chapters we work only on the "space" in which the singular point is located. However, working on singularities under the birational viewpoint, it turns out to be more convenient to think of the singularity of a "pair".

One can see the origin of this idea in the adjunction formula in Proposition 5.3.11.

Proposition 10.1.1 (Proposition *5.3.11***).** *Let X be a quasi-projective integral variety and $D \subset X$ a Cartier divisor. If X is a Cohen–Macaulay variety,*

$$\omega_D \cong \omega_X(D) \otimes_{\mathscr{O}_X} \mathscr{O}_D,$$

where $\omega_X(D) = \omega_X \otimes_{\mathscr{O}_X} \mathscr{O}_X(D)$. This formula is called the adjunction formula.

The definitions of canonical / log canonical / terminal / log terminal singularities of a variety X are based on the canonical sheaf ω_X. Therefore the adjunction formula

S. Ishii, *Introduction to Singularities*,
https://doi.org/10.1007/978-4-431-56837-7_10

shows that the singularities on D are related to the singularities of the "pair" (X, I_D), where I_D is the ideal of the divisor D in $\mathcal{O}_X$.

We study a more general pair $(X, \mathfrak{a}^n)$, where $\mathfrak{a}$ is a non-zero coherent ideal sheaf on X and n is a non-negative real number. We will give a precise definition of the discrepancy of the singularity of a pair $(X, \mathfrak{a}^n)$ and define canonical / log canonical / terminal / log terminal singularities which coincide with the original definitions when $\mathfrak{a} = \mathcal{O}_X$.

Definition 10.1.2. Let X be a variety. A prime divisor $E \subset Y$ on a partial resolution $f : Y \to X$ is called a *prime divisor over* X and the image $f(E)$ is called the *center* of E on X. The center of E on X is sometimes denoted by $c_X(E)$.

A prime divisor E over X is called an *exceptional prime divisor* if the morphism $f : Y \to X$ is not isomorphic at general points of E.

Definition 10.1.3. Let X be a $\mathbb{Q}$-Gorenstein variety with the index r, $\mathfrak{a}$ a non-zero coherent ideal sheaf on X and n a non-negative real number. Let $\omega^r_{Y/X}$ be the unique invertible subsheaf of constant sheaf $K(Y_{\mathrm{reg}})$ on Y_{reg}, such that it is isomorphic to $\omega^r_Y \otimes_{\mathcal{O}_Y} (f^*\omega^{[r]}_X)^{-1}$ on Y_{reg} and trivial on the open subset $U \subset Y_{\mathrm{reg}}$ on which f is isomorphic. Then the *log discrepancy* of the pair $(X, \mathfrak{a}^n)$ at the prime divisor E is defined as follows:

$$a(E; X, \mathfrak{a}^n) = \frac{1}{r}\nu_E(\omega^r_{Y/X}) - n \cdot \nu_E(\mathfrak{a}) + 1,$$

where ν_E is the valuation on $K(Y)$ corresponding to E and $\nu_E(\mathfrak{a})$ is defined as:

$$\nu_E(\mathfrak{a}) = \min\{\nu_E(f) \mid f \in \mathfrak{a}\}.$$

Up to this moment, we think only of closed point x of a variety X. But it is sometimes convenient for us to think also of non-closed point of X, which corresponds to a prime ideal in the affine coordinate ring $A(X)$, while a closed point of X corresponds to a maximal ideal. For details refer to Chap. 2 of [Ha2].

Definition 10.1.4. Let X be $\mathbb{Q}$-Gorenstein at a (not necessarily closed) point $\eta \in X$, $\mathfrak{a}$ a non-zero coherent ideal sheaf on X and n a non-negative real number. We say that the pair $(X, \mathfrak{a}^n)$ is *canonical / log canonical / terminal / log terminal* at η, if

$$a(E; X, \mathfrak{a}^n) \geq 1 / \geq 0 / > 1 / > 0 \quad \textit{(respectively)}$$

for every exceptional prime divisor E over X such that $\eta \in c_X(E)$.

Remark 10.1.5. The definitions in 10.1.4 do not need resolutions of the singularities. If resolutions of the singularities exist, then the definition of these singularities of the pair $(X, \mathcal{O}_X)$ coincides with the definition in 6.2.4. Therefore we can take this as the definition of those named singularities for a pair $(X, \mathfrak{a}^n)$ even for a variety X over the base field of positive characteristic, on which the existence of the resolutions is not yet proved.

Definition 10.1.6. Let X be a variety of dimension d. Here, we do not assume that X is normal or $\mathbb{Q}$-Gorenstein. Note that the projection

$$\pi : \mathbb{P}_X\left(\bigwedge^d \Omega_X\right) \longrightarrow X$$

is an isomorphism over the smooth locus $X_{\text{reg}} \subseteq X$. In particular, we have a section $\sigma : X_{\text{reg}} \to \mathbb{P}_X(\bigwedge^n \Omega_X)$.

The closure of the image of the section σ is called the *Nash blow-up* of X, and is denoted by $\widehat{X}$.

It is known that the Nash blow-up $\phi : \widehat{X} \to X$ has the following universality:

(1) the quotient of the pull-back $\phi^*(\bigwedge^d \Omega_X)$ by the torsion part is invertible and
(2) for a birational morphism $\varphi : Y \to X$ such that $\varphi^*(\bigwedge^d \Omega_X)$ has the same property as in (1), then φ factors through ϕ.

Definition 10.1.7. Let E be a prime divisor over X, then the *Mather discrepancy* $\widehat{k}_E \in \mathbb{Z}_{\geq 0}$ and the *Jacobian discrepancy* $j_E \in \mathbb{Z}_{\geq 0}$ at E are defined as follows:

Let $\varphi : Y \to X$ be a partial resolution of X such that E appears on Y, let $\widehat{X} \to X$ be the Nash blow-up and let $Y \times_X \widehat{X}$ be the fiber product. In the fiber product, the irreducible component dominating X is called the main part.

Replacing Y by the normalization of the main part of the fiber product, we may assume that $\varphi : Y \to X$ factors through the Nash blow-up $\widehat{X} \to X$.

By the universality of the Nash blow-up, the image *Im* of the following homomorphism is invertible:

$$\varphi^*\left(\bigwedge^d \Omega_X\right) \longrightarrow \omega_Y,$$

where $d = \dim X$. Restricting the above homomorphism to the smooth locus Y_{reg}, we can describe

$$Im|_{Y_{\text{reg}}} = \mathcal{I}\omega_{Y_{\text{reg}}}$$

with some invertible ideal sheaf $\mathcal{I}$, as $\omega_{Y_{\text{reg}}}$ is invertible. As Y is normal, general points of E are in Y_{reg}. Let us define

$$\widehat{k}_E = \nu_E(\mathcal{I})$$

and call it the Mather discrepancy of X at E. We note that if $\mathcal{I} = \mathcal{O}_Y(-\widehat{K}_{Y/X})$ is expressed with a Cartier divisor $\widehat{K}_{Y/X}$ on Y_{reg}, then $\widehat{k}_E$ is the coefficient of $\widehat{K}_{Y/X}$ at E.

Note that if $\varphi : Y \to X$ factors through the blow-up of X by the ideal $\mathfrak{a}$, then $\nu_E(\mathfrak{a})$ is the coefficient at E of the divisor Z defined by $\mathcal{O}_Y(-Z) = \mathfrak{a}\mathcal{O}_Y$. In particular, when $\mathfrak{a}$ is the Jacobian ideal $\mathcal{J}_X$ of X, we define

$$j_E = \nu_E(\mathcal{J}_X)$$

and call it the Jacobian discrepancy of X at E.

Then, the difference $\widehat{k}_E - j_E$ is called the *Mather–Jacobian discrepancy* (MJ discrepancy, for short) of X at E.

Definition 10.1.8. Let X be a variety over k, let $\mathfrak{a} \subset \mathcal{O}_X$ be a non-zero coherent ideal sheaf, and let n be a non-negative real number. A real number

$$a_{MJ}(E; X, \mathfrak{a}^n) = \widehat{k}_E - j_E - n \cdot \nu_E(\mathfrak{a}) + 1$$

is called the *MJ-log discrepancy* of the pair $(X, \mathfrak{a}^n)$ at E.

Definition 10.1.9. Let X be a variety, $\eta \in X$ a (not necessarily closed) point, $\mathfrak{a}$ a coherent ideal sheaf on X and n a non-negative real number. We say that the pair $(X, \mathfrak{a}^n)$ is *MJ-canonical / MJ-log canonical / MJ-terminal / MJ-log terminal* at η if

$$a_{MJ}(E; X, \mathfrak{a}^n) \geq 1 / \geq 0 / > 1 / > 0 \quad (\textit{respectively})$$

for every exceptional prime divisor E over X such that $\eta \in c_X(E)$.

Proposition 10.1.10. *Let $\mathfrak{a}$ be a non-zero coherent ideal sheaf on a variety X and n a non-negative real number. If X is normal locally a complete intersection at a point $\eta \in X$, then for every exceptional prime divisor E with $\eta \in c_X(E)$, the following equality holds:*

$$a(E; X, \mathfrak{a}^n) = a_{MJ}(E; X, \mathfrak{a}^n).$$

Therefore, at the point η the notions MJ-canonical / MJ-log canonical / MJ-terminal / MJ-log terminal coincide with canonical / log canonical / terminal / log terminal, respectively.

Proof. As X is locally a complete intersection at η, we have

$$\bigwedge^d \Omega_X = \mathcal{J}_X \cdot \omega_X$$

around the point η. (See for example, Proposition 9.1 in [EM].) Therefore, by pulling back this equality onto the partial resolution $\varphi : Y \to X$ where the exceptional prime divisor E appears, we obtain

$$k_E = \widehat{k}_E - j_E,$$

which yields the required equality.

According to the two kinds of log discrepancies, we define "*minimal log discrepancy*" for each.

Definition 10.1.11. Let $\eta \in X$ be a (not necessarily closed) point. We define the following:

Assume X is $\mathbb{Q}$-Gorenstein around η,

$$\mathrm{mld}(\eta; X, \mathfrak{a}^n) = \inf \left\{ a(E; X, \mathfrak{a}^n) \,\middle|\, \begin{array}{l} E \text{ is a prime exceptional divisor} \\ \text{over } X \text{ with the center } c_X(E) = \overline{\{\eta\}} \end{array} \right\},$$

when $\dim X \geq 2$.

For a point η which is not necessarily $\mathbb{Q}$-Gorenstein or even normal,

$$\mathrm{mld}_{\mathrm{MJ}}(\eta; X, \mathfrak{a}^n) = \inf \left\{ a_{MJ}(E; X, \mathfrak{a}^n) \,\middle|\, \begin{array}{l} E \text{ is a prime exceptional divisor} \\ \text{over } X \text{ with the center } c_X(E) = \overline{\{\eta\}} \end{array} \right\},$$

when $\dim X \geq 2$.

When $\dim X = 1$ and the right-hand side of each definition is ≥ 0, then we define mld and $\mathrm{mld}_{\mathrm{MJ}}$ by the right-hand side of each. Otherwise, we define $\mathrm{mld} = -\infty$ or $\mathrm{mld}_{\mathrm{MJ}} = -\infty$.

Remark 10.1.12. For every $\mathbb{Q}$-Gorenstein point $\eta \in X$ either of the following holds:

$$\mathrm{mld}(\eta, X, \mathfrak{a}^n) \geq 0, \quad \text{or} \quad \mathrm{mld}(\eta, X, \mathfrak{a}^n) = -\infty.$$

For every point $\eta \in X$ (not necessarily $\mathbb{Q}$-Gorenstein or normal) either of the following holds:

$$\mathrm{mld}_{\mathrm{MJ}}(\eta, X, \mathfrak{a}^n) \geq 0, \quad \text{or} \quad \mathrm{mld}_{\mathrm{MJ}}(\eta, X, \mathfrak{a}^n) = -\infty.$$

For the proof, we have only to prove that if $\mathrm{mld}(\eta, X, \mathfrak{a}^n) < 0 \,/\, \mathrm{mld}_{\mathrm{MJ}}(\eta, X, \mathfrak{a}^n) < 0$, then there exists a sequence $\{E_i\}_{i \geq 1}$ of exceptional prime divisors over X such that the centers $c_X(E_i) = \overline{\{\eta\}}$ and $a(E_i; X, \mathfrak{a}^n) \to -\infty / a_{MJ}(E_i; X, \mathfrak{a}^n) \to -\infty$ when $i \to \infty$. Such a sequence is obtained by appropriate blow-ups of a partial resolution Y on which an exceptional prime divisor E with $a(E; X, \mathfrak{a}^n) < 0 / a_{MJ}(E; X, \mathfrak{a}^n) < 0$ appears.

Proposition 10.1.13. *(i) If X is canonical / MJ-canonical at η, then*

$$\mathrm{mld}(\eta, X, \mathfrak{a}^n) \geq 1 / \mathrm{mld}_{\mathrm{MJ}}(\eta, X, \mathfrak{a}^n) \geq 1$$

respectively.

(ii) X is log canonical / MJ-log canonical at η, if and only if

$$\mathrm{mld}(\eta, X, \mathfrak{a}^n) \geq 0 / \mathrm{mld}_{\mathrm{MJ}}(\eta, X, \mathfrak{a}^n) \geq 0$$

respectively, where we assume that X is $\mathbb{Q}$-Gorenstein for the statement on canonicity and log canonicity.

(Sketch of proof) The statements of (i) and the "only if" part of (ii) follow immediately from the definitions. The proof of "if" part of (ii) is as follows: If X is not log canonical / MJ-log canonical, there exists an exceptional prime divisor E such that $a(E; X, \mathfrak{a}^n) < 0$ / $a_{MJ}(E; X, \mathfrak{a}^n) < 0$ and $\eta \in c_X(E)$. Then by successive blow-ups at appropriate centers on a partial resolution on which E appears, we obtain an exceptional prime divisor F over X such that $a(F; X, \mathfrak{a}^n) < 0$ / $a_{MJ}(F; X, \mathfrak{a}^n) < 0$ and $\overline{\{\eta\}} = c_X(F)$; this yields

$$\mathrm{mld}(\eta, X, \mathfrak{a}^n) = -\infty \,/\, \mathrm{mld}_{\mathrm{MJ}}(\eta, X, \mathfrak{a}^n) = -\infty.$$

For the details about MJ-singularities the reader can refer to [DD, EI, EIM, I15, IR] et al.

10.2 The Spaces of Jets and Arcs

Roughly speaking, an arc of a variety X is a very small portion of a curve on X and an m-jet is an approximation of order m of a small portion of a curve on X. The space of arcs is the set of all arcs on X and the space of m-jets is the set of all m-jets on X. These spaces have the natural scheme structures and reflect the properties of the singularities of the variety X. The first step of relating the space of arcs to birational geometry was made by John F. Nash in the paper [Nash]. He posed a so-called "Nash problem" in the preprint of 1968, which was published much later [Nash]. The problem has been solved recently by the contributions of many people. The reader can refer to [IK, BP, DF]. There are many other intermediate results for this problem. One can find these references in the above papers. The Nash problem plays a role to bridge of the space of arcs and birational geometry. In this section, we introduce the space of m-jets and the space of arcs of a variety X and associate them to the birational invariants of singularities of X introduced in the previous section.

Now we start with the definition of a jet and an arc.

Definition 10.2.1. Let X be a variety over k and $K \supset k$ a field extension. For $m \in \mathbb{Z}_{\geq 0}$ a k-morphism

$$\mathrm{Spec}\ K[t]/(t^{m+1}) \longrightarrow X$$

is called an *m-jet* of X and k-morphism

$$\mathrm{Spec}\ K[[t]] \longrightarrow X$$

is called an *arc* (or ∞-jet) of X. Here, we denote the closed points of Spec $K[t]/(t^{m+1})$ and Spec $K[[t]]$ corresponding to the canonical projections $K[t]/(t^{m+1}) \to K$ and $K[[t]] \to K$, respectively by the same symbol, 0.

Proposition 10.2.2. *Let X be a variety over k, (Sch/k) the category of k-schemes and (Set) the category of sets. A contravariant functor $F_m^X : (Sch/k) \to (Set)$ is defined as follows:*

$$F_m^X(Z) = \mathrm{Hom}_k(Z \times_{\mathrm{Spec}\, k} \mathrm{Spec}\, k[t]/(t^{m+1}), X).$$

Then F_m^X is representable by a scheme X_m of finite type over k, in particular, there is a bijection as follows:

$$\mathrm{Hom}_k(Z, X_m) \simeq \mathrm{Hom}_k(Z \times_{\mathrm{Spec}\, k} \mathrm{Spec}\, k[t]/(t^{m+1}), X).$$

This scheme X_m is called the space of m-jets of X.

This proposition is proved in [BLR], p. 276 but more concretely, we can construct X_m for an affine variety X and patch them together for a general variety.

We show the construction of the affine case, as it is useful to have a good insight on the space of m-jets.

Note 10.2.3 (Construction of X_m). Let $X = \mathrm{Spec}\ R$ and $R = k[x_1, \ldots, x_n]/(f_1, \ldots, f_r)$. It is sufficient to find X_m with the bijectivity in the theorem for an arbitrary k-scheme Z. Let $A = \Gamma(Z, \mathcal{O}_Z)$, then we have

$$\begin{aligned} &\mathrm{Hom}_k(Z \times \mathrm{Spec}\, k[t]/(t^{m+1}), X) \simeq \mathrm{Hom}_k(R, A[t]/(t^{m+1})) \\ \simeq &\{\varphi \in \mathrm{Hom}_k(k[x_1, \ldots, x_n],\ A[t]/(t^{m+1})) \mid \varphi(f_i) = 0,\ i = 1, \ldots, r\}. \end{aligned} \tag{10.1}$$

Here, we represent $\varphi(x_j) = a_j^{(0)} + a_j^{(1)}t + a_j^{(2)}t^2 + \cdots + a_j^{(m)}t^m$ ($a_j^{(l)} \in A$), then there exist polynomials $F_i^{(s)}$ on variables $a_j^{(l)}$ satisfying the equations:

$$\varphi(f_i) = F_i^{(0)}(a_j^{(l)}) + F_i^{(1)}(a_j^{(l)})t + \cdots + F_i^{(m)}(a_j^{(l)})t^m.$$

Then the set (10.1) is described as

$$\begin{aligned} &\{\varphi \in \mathrm{Hom}_k(k[x_j^{(0)}, x_j^{(1)}, \ldots, x_j^{(m)} \mid j = 1, \ldots, n],\ A) \mid \varphi(x_j^{(l)}) = a_j^{(l)},\ F_i^{(s)}(a_j^{(l)}) = 0\} \\ = &\mathrm{Hom}_k\left(k[x_j^{(0)}, x_j^{(1)}, \ldots, x_j^{(m)}]/(F_i^{(s)}(x_j^{(l)})),\ A\right). \end{aligned}$$

If we put $X_m = \mathrm{Spec}\ k[x_j^{(0)}, x_j^{(1)}, \ldots, x_j^{(m)}]/(F_i^{(s)}(x_j^{(l)}))$, then the last set is bijective to

$$\mathrm{Hom}_k(Z, X_m).$$

Note 10.2.4. Let X be a scheme of finite type over k. Then, by the construction, the space of m-jets is also a scheme of finite type over k.

We also observe the following:
For a pair of non-negative integers $m < m'$, the natural projection $k[t]/(t^{m'+1}) \to k[t]/(t^{m+1})$ induces a morphism

$$\psi_{m',m} : X_{m'} \to X_m.$$

This morphism $\psi_{m',m}$ is called a *truncation morphism*. In particular, for $m = 0$, the truncation morphism $\psi_{m',0} : X_{m'} \to X$ is denoted by π_m.

As we already did in the discussion above, m-jet α : Spec $K[t]/(t^{m+1}) \to X$ and the corresponding point in X_m are denoted by the same symbol, α. Here, we should note that $\pi_m(\alpha) = \alpha(0)$, where α in the left-hand side is a point on X_m and α in the right-hand side is a k-morphism Spec $K[t]/(t^{m+1}) \to X$.

Definition 10.2.5. As the system $\{\psi_{m',m} : X_{m'} \to X_m\}_{m<m'}$ is a projective system, let $X_\infty = \varprojlim_m X_m$ be its projective limit and call it the *space of arcs* of X. Here, we note that X_∞ is not of finite type over k if $\dim X > 0$.

Definition 10.2.6. Let $\eta \in X$ be a point. Then the fiber $\pi_m{}^{-1}(\eta)$ is denoted by $X_m(\eta)$. The dimension $\dim X_m(\eta)$ is defined as the dimension $\dim \overline{X_m(\eta)}$ of the closure $\overline{X_m(\eta)}$. Here, we note that it is different from $\dim X_m(\overline{\eta})$ in general. We also denote $\pi_m{}^{-1}(W)$ by $X_m(W)$ for a closed subset $W \subset X$.

One of the most important contributions of the space of arcs and the space of jets to the singularity theory is to describe the invariants of singularities. By making use of the space of arcs, we obtain the following "Inversion of Adjunction". It is first proved for locally complete intersection variety X by [EM] for the "usual mld" and generalized to any varieties for the "MJ-mld" by [DD, I15] independently. Here, these are for varieties defined over a field of characteristic 0. Then, it is also proved for positive characteristic case by [IR2].

Theorem 10.2.7 (Inversion of Adjunction). *Let X be a variety over an algebraically closed field k of arbitrary characteristic and A a smooth variety containing X as a closed subscheme of codimension c. Let $\widetilde{\mathfrak{a}} \subset \mathcal{O}_A$ be a coherent ideal sheaf such that its image $\mathfrak{a} := \widetilde{\mathfrak{a}}\mathcal{O}_X \subset \mathcal{O}_X$ is non-zero. Denote the defining ideal of X in A by I_X. Then, for a proper closed subset W of X, we have*

$$\mathrm{mld}_{\mathrm{MJ}}(W; X, \mathfrak{a}^n) = \mathrm{mld}(W; A, \widetilde{\mathfrak{a}}^n I_X^c).$$

For a point $\eta \in X$, we have

$$\mathrm{mld}_{\mathrm{MJ}}(\eta; X, \mathfrak{a}^n) = \mathrm{mld}(\eta; A, \widetilde{\mathfrak{a}}^n I_X^c).$$

Corollary 10.2.8. *Let X be a variety of dimension d defined over an algebraically closed field of arbitrary characteristic. Let $\eta \in X$ be a point and $W \subset X$ a closed subset. Then, we have the equalities:*

$$\mathrm{mld_{MJ}}(W; X) := \mathrm{mld_{MJ}}(W; X, \mathcal{O}_X) = \inf_m \{(m+1)d - \dim X_m(W)\},$$

$$\mathrm{mld_{MJ}}(\eta; X) := \mathrm{mld_{MJ}}(\eta; X, \mathcal{O}_X) = \inf_m \{(m+1)d - \dim X_m(\eta)\}.$$

The following is essentially proved in [Mus, EMY] for characteristic 0. Here, by making use of the above corollary we give its proof for an arbitrary characteristic.

Corollary 10.2.9. *Let X be a variety of dimension d over an algebraically closed base field of arbitrary characteristic. Assume that X is locally a complete intersection, then the following hold:*

(i) X has MJ-log canonical singularities if and only if X_m is locally a complete intersection for every $m \in \mathbb{N}$. If moreover X is normal, then it is also equivalent to the fact that X has log canonical singularities.
(ii) X has (MJ-) canonical singularities if and only if X_m is irreducible for every $m \in \mathbb{N}$.
(iii) X has (MJ-) terminal singularities if and only if X_m is normal for every $m \in \mathbb{N}$.

Proof. As X is locally a complete intersection, X is locally defined by $c := N - d$ equations in a non-singular variety A of dimension N. Then, X_m is locally defined by $(m+1)c$ equations in a non-singular variety A_m of dimension $(m+1)N$ (cf. the construction of X_m). Therefore, we have

$$\dim X_m \geq (m+1)N - (m+1)c = (m+1)d, \tag{10.2}$$

where the equality holds if and only if X_m is locally a complete intersection.

First we show the equivalence in (i). We know that the restriction ${\pi_m}^{-1}(X_{\mathrm{reg}}) \to X_{\mathrm{reg}}$ of π_m is a smooth morphism of relative dimension md. Therefore, by the formula in Corollary 10.2.8, X has MJ-log canonical singularities if and only if for every $m \in \mathbb{N}$, the following inequality holds:

$$(m+1)d - \dim X_m(W) \geq 0,$$

where W is the singular locus X_{sing} of X. This is equivalent to the fact that the equality holds in (10.2).

For the second statement of (i), we have only to note that for a normal locally complete intersection variety, "MJ-log canonical" and "log canonical" are the same.

For the proof of (ii), we may assume that X_m is locally a complete intersection of dimension $(m+1)d$. Now, again by the formula in Corollary 10.2.8, X has MJ-canonical singularities if and only if for every $m \in \mathbb{N}$ and the singular locus $W \subset X$, the following inequality holds:

$$(m+1)d - \dim X_m(W) \geq 1,$$

which yields $\dim X_m(W) < (m+1)d$. This is equivalent to the fact that none of the irreducible components of $X_m(W)$ can be an irreducible component of X_m, since

X_m is of pure dimension $(m+1)d$. This holds if and only if X_m is irreducible for every $m \in \mathbb{N}$.

As for the statement of "canonicity" in (ii), we have only to note that for a locally complete intersection variety, "MJ-canonical" and "canonical" are equivalent. This follows from the fact that an MJ-canonical locally complete intersection variety is normal, as it is non-singular in codimension one ([IR], Prop. 4.3).

For the proof of (iii), we may assume that X_m is irreducible and locally a complete intersection of dimension $(m+1)d$. Then X_m has the property S_2. Now, X has MJ-terminal singularities if and only if for every $m \in \mathbb{N}$ and for the singular locus $W \subset X$, the following inequality holds:

$$(m+1)d - \dim X_m(W) \geq 2,$$

which yields $\dim X_m(W) \leq (m+1)d - 2$. Here we note that the singular locus of X_m is just $X_m(W)$. Indeed, it is obvious that the singular locus of X_m is contained in $X_m(W)$, as the complement ${\pi_m}^{-1}(X_{\text{reg}})$ of $X_m(W)$ is non-singular. To show the opposite inclusion, denote the local Jacobian matrix of the embedding $X_m \subset A_m$ by J and the Jacobian matrix of the embedding $X \subset A$ by J_0. Then J has the following form:

$$J = \begin{pmatrix} J_0 & O \\ * & * \end{pmatrix}.$$

As we may assume that $X_m \subset A_m$ is a complete intersection, X_m is non-singular at a point p if and only if the Jacobian matrix J has full rank at p. Here, if $p \in X_m(W)$, then J_0 does not have full rank, therefore J cannot have full rank.

Hence, the inequality $\dim X_m(W) \leq (m+1)d - 2$ is equivalent to the fact that X_m is normal by the Serre's criteria for normality.

As for the terminality, we can apply the last discussion in the proof of (ii).

For further studies in this direction, the reader can refer to [EM, I14] as well as the references in the previous section.

There is a lot of room for developing singularity theory in a positive characteristic by using jet schemes. At this moment the existence of resolutions of singularities is not yet confirmed in general. Therefore, many facts which hold in characteristic 0 are not yet proved for the positive characteristic case. For example, openness of canonicity or log canonicity is not yet proved in the positive characteristic case, which is very basic for working on singularities in this case. But there is a possibility for openness of MJ-canonicity or MJ-log canonicity to be proved by making use of the spaces of jets (see [I16]).

10.3 F-Singularities in Positive Characteristic

In this section, we study singularities appearing in varieties defined over a field k of positive characteristic p. For those varieties, at this moment we do not have resolutions of singularities in general. Therefore, we cannot expect discussions similar to the case of characteristic 0. On the other hand, singularity theory for the positive characteristic case has developed in its own way, using a Frobenius map. Around the end of the 20th century, it was unveiled that these singularities and the singularities of characteristic 0 are closely related by the "reduction modulo p" (see also, for example, [HH07]). Concerning the relation of F-singularities and the singularities in a birational viewpoint, it was first observed by N. Hara [Hara] and K. E. Smith [Sm] and then many discoveries in this direction appeared in [HW, HWY, MeSr, MS, Scw] and the research in this direction is still active.

In this section, we introduce the study in this direction very briefly to help the reader to get a rough idea about this. The reader who would like to study further is encouraged to read the expository papers [HH07, TaW].

First we start with a very general setting that R is a Noetherian integral domain of positive characteristic p. But actually we think of the local ring of a singular point as R.

Definition 10.3.1. Let R be a Noetherian integral domain of positive characteristic p. Then the *Frobenius map* $F : R \to R$ is defined as $F(a) = a^p$ for every element $a \in R$.

As the characteristic of R is p, we have the relation

$$(a+b)^p = a^p + b^p$$

for every $a, b \in R$. This guarantees that F is a ring homomorphism. Based on this fact, $F : R \to R$ gives another R-algebra structure on R. In order to distinguish these algebras, we introduce the following convention:

Let $q = p^e$ ($e \in \mathbb{N}$) and define an R-algebra

$$R^{1/q} := \{x \in \overline{Q(R)} \mid x^q \in R\},$$

where $\overline{Q(R)}$ is the algebraic closure of the quotient field $Q(R)$ of R. Then, we have the inclusion $R \hookrightarrow R^{1/q}$ and the isomorphism $R^{1/q} \simeq R; a \mapsto a^q$. As we have the following commutative diagram

$$\begin{array}{ccc} R & \hookrightarrow & R^{1/q} \\ \| & & \|\wr \\ R & \xrightarrow{F^e} & R, \end{array}$$

we can regard the inclusion $R \hookrightarrow R^{1/q}$ as the e times composition F^e of the Frobenius map F.

Definition 10.3.2. We say that a ring R is *F-finite* if $R^{1/q}$ is a finitely generated R-module.

Example 10.3.3. (1) A perfect field k is F-finite, because the map $k \hookrightarrow k^{1/p} = k$ is surjective.
(2) A ring of finite type or essentially of finite type over an F-finite field is F-finite.
(3) A complete local ring with the F-finite residue field is F-finite.

As in our geometric situation the local ring of the singular point of a variety is essentially of finite type over an algebraically closed field k, in this section we always assume that a ring R is F-finite.

Theorem 10.3.4 (Kunz [KZ]). *For a Noetherian local integral domain R of characteristic p, the following are equivalent:*

(i) R is regular;
(ii) $R^{1/p}$ is a free R-module;
(iii) $R^{1/q}$ is free for all $q = p^e$.

Example 10.3.5. Let $R = k[[x_1, \dots, x_n]]$ for a perfect field k. Then $R^{1/p} = k^{1/p}[[x_1^{1/p}, \dots, x_n^{1/p}]] = k[[x_1^{1/p}, \dots, x_n^{1/p}]]$

$$= \bigoplus_{0 \le i_j \le p-1} R \cdot (x_1^{i_1} \cdots x_n^{i_n})^{1/p},$$

which shows that $R^{1/p}$ is a free R-module with R a direct summand.

We will think of this situation under more general settings.

Definition 10.3.6. A Noetherian integral domain R of characteristic p is called an *F-pure ring* if the map $R \hookrightarrow R^{1/p}$ splits as R-modules homomorphism. It means that there exists $\varphi \in \operatorname{Hom}_R(R^{1/p}, R)$ such that the composite $R \hookrightarrow R^{1/p} \xrightarrow{\varphi} R$ is the identity map of R.

Proposition 10.3.7. *For a Noetherian integral domain R of characteristic $p > 0$, the following are equivalent:*

(i) R is F-pure;
(ii) there exists $q = p^e$ such that the canonical map $R \hookrightarrow R^{1/p}$ splits;
(iii) for every $q = p^e$, the canonical map $R \hookrightarrow R^{1/p}$ splits;
(iv) there exists $q = p^e$ such that $\operatorname{Hom}_R(R^{1/q}, R) \to R\,;\ \varphi \mapsto \varphi(1)$ is surjective;
(v) for every $q = p^e$, the canonical map $\operatorname{Hom}_R(R^{1/q}, R) \to R\,;\ \varphi \mapsto \varphi(1)$ is surjective;
(vi) there exists $q = p^e$ such that R is a direct summand of $R^{1/q}$;
(vii) for every $q = p^e$, R is a direct summand of $R^{1/q}$.

Example 10.3.8. A regular local ring is F-pure.

The following is a useful criterion for F-purity.

Theorem 10.3.9 (Fedder [Fe]). *Let S be an F-finite regular local ring with the maximal ideal $\mathfrak{n}$ of characteristic $p > 0$. Let $I \subset S$ be an ideal and define the quotient ring $R := S/I$. Then R is F-pure if and only if $(I^{[p]} : I) \not\subset \mathfrak{n}^{[p]}$.*

In particular, if I is generated by a regular sequence $f_1, \ldots, f_r$ in S, then R is F-pure if and only if $(\prod_i f_i)^{p-1} \notin \mathfrak{n}^{[p]}$.

Definition 10.3.10. A Noetherian integral domain R of characteristic p is called a *strong F-regular* ring if for every non-zero element $c \in R$ there exists $e \in \mathbb{N}$ such that

$$cF^e : R \longrightarrow R^{1/p^e}, \quad x \longmapsto c^{1/p^e} x$$

splits as an R-module homomorphism.

Definition 10.3.11. We say that a variety X defined over a field of characteristic $p > 0$ is strong F-regular or F-pure at a point $x \in X$, if the local ring $\mathscr{O}_{X,x}$ is strong F-regular or F-pure, respectively. A variety X is called strong F-regular or F-pure if X is strong F-regular or F-pure at every point of X, respectively.

It is clear that strong F-regular implies F-pure by taking $c = 1$. Besides F-purity and strong F-regularity, there are other notions of F-singularities; e.g., F-rational, F-regular, weakly F-regular, F-injective, and so on. But here we focus only on F-purity and strong F-regularity. The reader who is interested in the other notions can refer to [TaW].

Modulo p Reduction

Let X_0 be an affine variety in $\mathbb{C}^n$ defined by an ideal $I \subset \mathbb{C}[x_1, \ldots, x_n]$. We write $I = (f_1, ..., f_m)$ and $R = \mathbb{C}[x_1, \ldots, x_n]/I$. For simplicity, we first assume that all of the coefficients of the f_i are integers. We set $R_{\mathbb{Z}}$ to be the ring $\mathbb{Z}[x_1, \ldots, x_n]/(f_1, ..., f_m)$. For each prime integer p, consider the ring

$$R_p := R_{\mathbb{Z}}/pR_{\mathbb{Z}} \simeq (\mathbb{Z}/p\mathbb{Z})[x_1, \ldots, x_n]/(f_1(\text{mod } p), \ldots, f_m(\text{mod } p))$$

and the "variety" X_p defined by $(f_1(\text{mod } p), \ldots, f_m(\text{mod } p))$ in $(\mathbb{Z}/p\mathbb{Z})^n$. Here, we should note that X_p may not be irreducible or reduced, even if X_0 is irreducible and reduced. The variety X_p is called the characteristic p model for X_0 or *the modulo p reduction of X_0*.

If the coefficients of f_i are not all in $\mathbb{Z}$, the modulo p reduction of X_0 is defined as follows:

Instead of working with $R_{\mathbb{Z}}$, we work with $R_A = A[x_1, \ldots, x_n]/(f_1, \ldots, f_m)$, where A is the $\mathbb{Z}$-algebra generated by the coefficients of the f_i's. In this case a modulo p reduction X_p is a variety in $(A/\mathfrak{m})^n$ defined by the ideal

$$(f_1(\text{mod } \mathfrak{m}), \ldots, f_m(\text{mod } \mathfrak{m})) \subset (A/\mathfrak{m})[x_1, \ldots, x_n],$$

where $\mathfrak{m}$ is a maximal ideal of A such that $\mathfrak{m} \cap \mathbb{Z} = (p)$.

Remark 10.3.12. In the history of commutative algebra, close relations between X_0 and X_p for a prime number p were observed by many people. Actually, X_0 and X_p share many properties for large $p \gg 0$. For example, R_p is regular for large $p \gg 0$ if and only if R_0 is regular, [HH07].

Theorem 10.3.13. ([HW, Tak, Tak2]). *Let (X, x) be a singularity on a variety X over $\mathbb{C}$. Assume that X is $\mathbb{Q}$-Gorenstein around x. Then the following hold:*

(i) (X, x) is log terminal if and only if there exists $p_0 \in \mathbb{N}$ such that X_p is strong F-regular for all prime numbers $p \geq p_0$ at the point corresponding to x;
(ii) If there are infinitely many prime numbers p such that X_p is F-pure at the point corresponding to x, then (X, x) is log canonical.

Example 10.3.14. Let X_0 be a surface in $\mathbb{C}^3$ defined by the equation $x^3 + y^3 + z^3 = 0$. Then $(X_0, 0)$ is log canonical singularity. On the other hand, we have the following modulo p reductions: The singularity $(X_p, 0)$ is F-pure if and only if $p \equiv 1 (\mathrm{mod}\ 3)$.

By this example, it turns out that we cannot expect the equivalence of "X_0 has log canonical singularities" and "X_p is F-pure for all prime numbers such that $p \gg 0$".

Now we can pose the following conjecture:

Conjecture 10.3.15. X_0 has log canonical singularities, if and only if X_p is $\mathbb{Q}$-Gorenstein and F-pure for infinitely many prime numbers p.

Remark 10.3.16. In a similar way to Sect. 10.1, there are notions of F-singularities of the pairs $(X, \mathfrak{a}^n)$ consisting of a variety X and non-negative real power n of an ideal $\mathfrak{a}$. Theorem 10.3.13 is generalized for pairs. For this direction the reader can refer to [HW, Tak, Tak1, Tak2, Tak3].

Definition 10.3.17. We define that a variety X defined over a field of positive characteristic is *adjunctive strong F-regular / adjunctive F-pure* at a point $x \in X$ if there is a closed immersion $X \subset A$ into a non-singular variety A locally around the point x and it satisfies that the pair (A, I_X^c) is strongly F-regular / F-pure at the point x, respectively. Here, I_X is the defining ideal of X in A and c is the codimension of X in A. Note that the definitions are independent of the choice of local ambient space A.

By virtue of the generalized theorem of Theorem 10.3.13 for pairs (see [TaW, Tak1, Tak3]) we obtain the MJ-version of the theorem:

Corollary 10.3.18. *Let (X, x) be a singularity on a variety X over $\mathbb{C}$. Then the following hold:*

(i) (X, x) is MJ-canonical if and only if there exists $p_0 \in \mathbb{N}$ such that X_p is adjunctive strong F-regular at the point corresponding to x for all prime numbers $p \geq p_0$;
(ii) If there are infinitely many prime numbers p such that X_p is adjunctive F-pure at the point corresponding to x, then (X, x) is MJ-log canonical.

Note that if Conjecture 10.3.15 for pairs holds true, then the converse of the statement (ii) in the above corollary also holds.

References

[Ab] Abhyankar, S.S.: Local Analytic Geometry. Academic Press, New York, London (1964)

[AK] Altman, K., Kleiman, S.: Introduction to Grothendieck Duality Theory. Lecture Notes in Mathematics, vol. 146. Springer, Berlin (1970)

[A1] Artin, M.: Some numerical criteria for contractability of curves on algebraic surfaces. Am. J. Math. **84**, 485–496 (1962)

[A2] Artin, M.: On isolated rational singularities of surfaces. Am. J. Math. **88**, 129–136 (1966)

[A4] Artin, M.: On the solutions of analytic equations. Invent. Math. **5**, 277–291 (1968)

[A5] Artin, M.: Algebraic approximation of structures over complete local rings. Publ. Math. IHES **36**, 23–58 (1969)

[A6] Artin, M.: Algebraization of formal moduli I, collection of mathematical papers in Honor of K. Kodaira, pp. 21–71. University of Tokyo Press, Tokyo (1970)

[A7] Artin, M.: Algebraization of formal moduli II. Ann. Math. **91**(1), 88–135 (1970)

[A3] Artin, M.: Lecture on deformations of singularites. Tata Institute of Fund Research. Springer, Bombay (1976)

[AM] Atiyah, M.F., Macdonald, I.G.: Introduction to Commutative Algebra, Addison-Wesley Series in Mathematics (1969)

[AHV] Aroca, J.M., Hironaka, H., Vicente, J.L.: Desingularization Theorems. Memorias de Mathematica del Instituto "Jorge Juan" 30, Madrid (1977)

[BKR] Behnke, K., Kahn, C., Riemenschneider, O.: Infinitesimal deformations of quotient surface singularities, in "Singularities", vol. 20, pp. 31–66. Banach Center Publication, Warsaw (1988)

[BeK] Behnke, K., Knörrer, H.: On infinitesimal deformations of rational surface singularities. Compositio Math. **61**, 103–127 (1987)

[BeC] Behnke, K., Christophersen, J.A.: Hypersurface sections and obstructions. rational surface singularities. Compositio Math. **77**, 233–258 (1991). (with appendix by J. Stevens)

[BCHM] Birkar, C., Cascini, P., Hacon, C., McKernan, J.: Existence of minimal models for varieties of log general type. J. AMS **23**, 405–468 (2010)

[Mur] Bloch, A.: The Complete Murphy's Law (revised version). Price Stern Sloan, Inc. New York (1991)

[B] Borelli, M.: Divisorial varieties. Pac. J. Math. **13**, 375–388 (1963)

S. Ishii, *Introduction to Singularities*,
https://doi.org/10.1007/978-4-431-56837-7

[BP] Bobadilla, J.F., Pereira, M.P.: The nash problem for surfaces. Ann. of Math. **176**, 2003–2029 (2012)

[BLR] Bosch, S., Lütkebohmert, W., Raynaud, M.: Néron Models. Ergebnisse der Mathematik und ihrer Grenzgebiete. Springer, Berlin (1990)

[Br] Brieskorn, E.: Rationale Singularitäten komplexer Flächen. Invent. Math. **4**, 336–358 (1968)

[CL] Call, F., Lyubeznik, G.: A simple proof of Grothendieck's theorem on the parafactoriality of local rings. Contemp. Math. **159**, 15–18 (1994)

[Ch] Chevalley, C.: Invariants of finite groups generated by reflections. Am. J. Math. **77**, 778–782 (1955)

[De] Deligne, P.: Theórie de Hodge II, III, Publ. Math. IHES, vol. 40, pp. 5–58 (1971); vol. 44, pp. 50–77 (1975)

[DF] De Fernex, T.: Three-dimensional counter-examples to the nash problem. Compositio Math. **149**, 1519–1534 (2013)

[DD] De Fernex, T., Docampo, R.: Jacobian discrepancies and rational singularities. J. Eur. Math. Soc. **16**, 165–199 (2014)

[Do] Donin, I.F.: Complete families of deformations of germs of complex spaces. Math. USSR Sbornik **18**, 397–406 (1972)

[DB] Du Bois, P.: Complex de de Rham fitré d'une variété singulière. Bull. Soc. Math. France **118**, 75–114 (1981)

[Du] Durfee, A.: Fifteen characterizations of rational double points and simple critical points. L'Enseignement XXV **1–2**, 131–163 (1979)

[EI] Ein, L., Ishii, S.: Singularities with respect to Mather-Jacobian discrepancies, vol. 67, pp. 125–168. MSRI Publications (2015)

[EIM] Ein, L., Ishii, S., Mustaţă, M.: Multiplier ideals via Mather discrepancy. Adv. Stud. Pure Math. **70**, 1–20 (2016)

[EMY] Ein, L., Mustaţă, M., Yasuda, T.: Jet schemes, log-discrepancies and inversion of adjunction. Invent. Math. **153**, 519–535 (2003)

[EM] Ein, L., Mustaţă, M.: Jet schemes and singularities. Proc. Symp. Pure Math. **80**(2), 505–546 (2009)

[El1] Elkik, R.: Singularités rationnelles et déformations. Invent. Math. **47**, 139–147 (1978)

[El2] Elkik, R.: Rationalité des singularités canoniques. Invent. Math. **64**, 1–6 (1981)

[EV] Esnault, H., Viehweg, E.: Two-dimensional quotient singularities deform to quotient singularities. Math. Ann. **271**, 439–449 (1985)

[Fe] Fedder, R.: F-purity and rational singularity. Trans. Am. Math. Soc. **278**, 461–480 (1983)

[Fis] Fischer, G.: Complex Analytic Geometry. Lecture Notes in Mathematics, vol. 538. Springer, Berlin (1976)

[F] Fletcher, A.R.: Plurigenera of 3-folds and weighted hypersurfaces. Ph.D thesis submitted to the Unversity of Warwick, London Mathematical Society, Lecture Series, vol. 281 (1988)

[Fo] Fossum, R.: The Divisor Class Groups of Krull Domain, Ergebnisse der Math, u. Ihrer Grenz., vol. 74, Springer, Berlin (1973)

[Fjo] Fujino, O.: On isolated log canonical singularities with index one. J. Math. Sci. Univ. Tokyo **18**, 299–323 (2011)

[Fu] Fujita, T.: A relative version of Kawamata-Viehweg's vanishing theorem, Proceedings of Alg. Geom. in Hiroshima

[Gr1] Grauert, H.: Ein Theorem der analytischen Garbentheorie und die Modulräume komplexer Sturkturen, Inst. Publ. Math. IHES. 5 (1960)

[Gr2] Grauert, H.: Über Modifikationen und exzeptionelle analytischen Mengen. Math. Ann. **146**, 331–368 (1962)

[Gr3] Grauert, H.: Über die Deformation isolierter Singularitäten analytischen Mengen. Invent. Math. **15**, 171–198 (1972)

[GR] Grauert, H., Riemenschneider, O.: Verschwindungssatze für analytische Kohomologiegruppen auf komplexen Räumen. Invent. Math. **11**, 263–292 (1970)

[Gr] Grothendieck, A.: Local Cohomology. Lecture Notes in Mathematics, vol. 41. Springer, Berlin (1967)

[SGA] Grothendieck, A.: Cohomologie locale des faisceaux cohérents et théorèmes de Lefschetz locaux et globaux (SGA2). North-Holland, Amsterdam (1968)

[EGA II, III, IV] Grothendieck, A., Dieudonné, J.: Eléments de Géometrie Algébrique II, III, IV, Publ. Math. IHES, vol. 8 (1961); Ibid, vol 11 (1961); vol 17 (1963)

[Ha1] Hartshorne, R.: Residues and Duality. Lecture Notes in Mathematics, vol. 20. Springer, Berlin (1966)

[Ha2] Hartshorne, R.: Algebraic Geometry. Graduate Texts in Mathematics, vol. 52. Springer, Berlin (1977)

[Hara] Hara, N.: A characterization of rational singularities in term of injectivity of Frobenius map. Am. J. Math. **120**, 981–996 (1998)

[HW] Hara, N., Watanabe, K-i: F-regular and F-pure rings vs. log terminal and log canonical singularities. J. Algebr. Geom. **11**, 363–392 (2002)

[HWY] Hara, N.: Watanabe, K-i: F-rationality of Rees algebras. J. Algebra **247**, 153–190 (2002)

[HH07] Hochster, M., Huneke, C.: Tight Closure in Equal Characteristic Zero: With an Introduction to the Characteristic p Theory. Springer Monographs in Mathematics, Berlin (2007)

[H] Hironaka, H.: Resolution of singularities of an algebraic variety over a field of characteristic zero: I. II. Ann. Math. **79**, 109–326 (1964)

[HR] Hironaka, H., Rossi, H.: On the equivalence of embeddings of exceptional complex spaces. Math. Ann. **156**, 313–368 (1964)

[Hol] Holmann, H.: Quotienten komplexer Räume. Math. Ann. **142**, 407–440 (1961)

[I1] Ishii, S.: On the pluri-genera and mixed Hodge structures of isolated singularities. Proc. Jpn. Acad. **59**, 355–357 (1983)

[I2] Ishii, S.: On isolated Gorenstein singularities. Math. Ann. **270**, 541–554 (1985)

[I3] Ishii, S.: Du Bois singularities on a normal surface. Advanced Study in Pure Mathematics, Complex Analytic Singularities **8**, 153–163 (1986)

[I4] Ishii, S.: Isolated Q-Gorenstein singularities of dimension three. Advanced Study in Pure Mathematics, Complex Analytic Singularities **8**, 165–198 (1986)

[I5] Ishii, S.: Small deformations of normal singularities. Math. Ann. **275**, 139–148 (1986)

[I6] Ishii, S.: Two dimensional singularities with bounded pluri-genera δ_m are Q-Gorenstein singularities. In: Proceedings of Symposium of Singularities, Iowa 1986, Contemporary Mathematics, vol. 90, pp. 135–145 (1989)

[I7] Ishii, S.: The asymptotic behavior of pluri-genera for a normal isolated singularity. Math. Ann. **286**, 803–812 (1990)

[I8] Ishii, S.: Quasi-Gorenstein Fano 3-folds with the isolated non-rational loci. Compositio Math. **77**, 335–341 (1991)

[I9] Ishii, S.: Simultaneous canonical modifications of deformations of isolated singularities. In: Algebraic Geometry and Analytic Geometry, Proceedings of the Satellite Conference of ICM 90, pp. 81–100. Springer, Berlin (1991)

[I10] Ishii, S.: The canonical modifications by weighted blowups. J. Alg. Geom. **5**, 783–799 (1996)

[I11] Ishii, S.: Minimal, canonical and log-canonical models of hypersurface singularities. Contemp. Math. **207**, 63–77 (1997)

[I12] Ishii, S.: The minimal model theorem for divisors of toric varieties. Tohoku Math. J. **51**, 213–226 (1999)

[I14] Ishii, S.: Jet schemes, arc spaces and the nash problem. C.R. Math. Rep. Acad. Can. **29**, 1–21 (2007)

[I13] Ishii, S.: A supplement to Fujino's paper: on isolated log canonical singularities with index one. J. Math. Sci. Univ. Tokyo. **19**, 135–138 (2012)

[I15] Ishii, S.: Mather discrepancy and the arc spaces. Ann. de l'Institut Fourier **63**, 89–111 (2013)

[I16] Ishii, S.: Finite determination conjecture for Mather-Jacobian minimal log discrepancies and its applications, preprint, arXiv:1703.08500, to appear in Eur. J. Math

[IK] Ishii, S., Kollár, J.: The nash problem on arc families of singularities. Duke Math. J. **120**(3), 601–620 (2003)

[IW] Ishii, S., Watanabe, K.: A geometric characterization of a simple K3-singularity. Tohoku Math. J. **44**, 19–24 (1992)

[IR] Ishii, S., Reguera, A.: Singularities with the highest Mather minimal log discrepancy. Math. Zeitschrift. **275**(3–4), 1255–1274 (2013)

[IR2] Ishii, S., Reguera, A.: Singularities in arbitrary characteristic via jet schemes, arXiv:1510.05210, to appear in "Hodge Theory and L^2 Analysis"

[KS] Kas, A., Schlessinger, M.: On the versal deformation of a complex space with an isolated singularity. Math. Ann. **196**, 23–29 (1972)

[Kaw] Kawakita, M.: Inversion of adjunction on log canonicity. Invent. Math. **167**, 129–133 (2007)

[Ka] Kawamata, Y.: Deformations of canonical singularities. J. AMS **12**(1), 85–92 (1999)

[KMM] Kawamata, Y., Matsuda, K., Matsuki, K.: Introduction to the minimal model problem. In: Oda, Advanced Studies, in Pure Mathematics, vol. 10, (eds.) Algebraic Geometry in Sendai 1985, pp. 283–360. Kinokuniya, Tokyo and North-Holland, Amsterdam, New York, Oxford (1987)

[Kl] Klein, F.: Vorlesungen über das Ikosaeder und die Auflösung der Gleihungen vom fünften Grade. Teubner, Leibzig (1884)

[Klm] Kleiman, S.: Toward a numerical theory of ampleness. Ann. Math. **84**, 293–344 (1966)

[Kn] Knöller, F.W.: 2-dimensionale Singularitäten und Differentialformen. Math. Ann. **206**, 205–213 (1973)

[Kol] Kollár, J.: Shafarevich Maps and Automorphic Forms. Princeton, UP (1995)

[KZ] Kunz, E.: On Noetherian rings of characteristic p. Am. J. Math. **98**, 999–1013 (1976)

[KM] Kollár, J., Mori, S.: Birational Geometry of Algebraic Varieties. Cambridge University Press, Cambridge (2008)

[KSB] Kollár, J., Shepherd-Barron, N.: Threefolds and deformations of surface singularities. Invent. Math. **91**, 299–338 (1988)

[KK] Kollár, J., Kovács, S.: Log canonical singularities are Du Bois. J. Am. Math. Soc. **23**, 791–813 (2010)

[Kov] Kovács, S.: Rational, log canonical, Du Bois singularities: II Kodaira vanishing and small deformations. Compositio Math. **121**, 123–133 (2000)

[KS] Kovács, S., Schwede, K.: Du Bois singularities deform. Adv. Stud. Pure Math. (to appear)

[La1] Laufer, H.B.: Taut two dimensional singularities. Math. Ann. **205**, 131–164 (1973)

[La2] Laufer, H.B.: On minimally elliptic singularities. Am. J. Math. **99**, 1257–1295 (1977)

[Li] Lichtenbaum, S.: Curves over discrete valuation rings. Doctoral dissertation, Harvard (1964)

[Lip] Lipman, J.: Rings with discrete divisor class group: theorem of Danilov-Samuel. Am. J. Math. **101**, 203–211 (1979)

[Ma1] Matsumura, H.: Commutative Algebra. W.A. Benjamin Co., New York (1970)

[Ma2] Matsumura, H.: Commutative Ring Theory. Cambridge Studies in Advanced Mathematics, vol. 8. Cambridge University Press, Cambridge (1986)

[MeSr] Mehta, V.B., Srinivas, V.: A characterization of rational singularities. Asian J. Math. **1**(2), 249–271 (1997)

[Mi] Milnor, J.: Singular Points of Complex Hypersurfaces. Annals of mathematics studies, vol. 61. Princeton, UP (1974)

[MS] Miller, L.E., Schwede, K.: Semi-log canonical vs F-pure singularities. J. Algebra **349**, 150–164 (2012)

[Mor] Morales, M.: Calcul de quelques invariants des singularités de surface normale, Comptes rendes de séminaire tenu aux Plans- sur-Bex, pp. 191–203 (1982)

[Mo2] Mori, S.: On 3-dimensional terminal singularities. Nagoya Math. J. **98**, 43–66 (1985)

[Mo1] Mori, S.: Flip theorem and the existence of minimal models for 3-folds. J. Am. Math. Soc. **1**, 117–253 (1988)

[MK] Morrow, J., Kodaira, K.: Complex Manifolds. Rinehart and Winston Inc., New York, Holt (1971)

[Mu] Mumford, D.: The Red Book of Varieties and Schemes. Lecture Note in Mathematics, vol. 1358. Springer, Berlin (1988)

[GIT] Mumford, D., Forgaty, J., Kirwan, F.: Geometric Invariant Theory, Ergebnisse der Math. u. ihrer Grenz., vol. 34. Springer, Berlin (1994)

[Mus] Mustaţă, M.: Jet schemes of locally complete intersection canonical singularities, with an appendix by David Eisenbud and Edward Frenkel. Invent. Math. **145**, 397–424 (2001)

[Nak] Nakayama, N.: Invariance of the plurigenera of algebraic varieties under minimal model conjecture. Toplogy **25**(2), 237–251 (1986)

[Nar] Narashimham, R.: Introduction to the Theory of Analytic Spaces. Lecture Note Mathematics, vol. 25. Springer, Berlin (1966)

[Nash] Nash, J.F.: Arc structure of singularities. Duke Math. J. **81**, 31–38 (1995)

[Od1] Oda, T.: Lectures on torus embeddings and applications (based on joint work with Katsuya Miyake), Tata Institute of Fundamental Research, vol. 58. Springer, Berlin (1978)

[Od2] Oda, T.: Convex Bodies and Algebraic Geometry – An introduction to the Theory of Toric Varieties, Ergeb. Math. Grenzgeb, vol. 15 (1988)

[Ok] Okuma, T.: The pluri-genera of surface singularities. Tohoku Math. J. **50**(1), 119–132 (1998)

[Ok2] Okuma, T.: The plurigenera of Gorenstein surface singularities. Manuscripta Math. **94**, 187–194 (1997)

[Pin] Pinkham, H.C.: Deformation of algebraic varieties with G_m action. Astérisque **20** (1974)

[Re1] Reid, M.: Canonical 3-folds. In: Proceedings Algebraic Geometry Anger (Sijthoff and Nordhoff), pp. 273–310 (1979)

[Re2] Reid, M.: Young person's guide to canonical singularities. Proceedings of Symposia in Pure Mathematics. **46**, 345–414 (1987)

[Ri1] Riemenschneider, O.: Deformationen von Quotientensingularitäten (nach zyklischen Gruppen). Math. Ann. **209**, 211–248 (1974)

[Ri2] Riemenschneider, O.: Dihedral singularities: invariants, equations and infinitesimal deformations. Bull. Am. Math. Soc. **82**, 725–747 (1976)

[S] Šafarevič, I.R.: Algebraic Surfaces. In: Proceedings of the Steklov Institute of Mathematics, vol. 75. AMS (1967)

[Sa] Saito, K.: Einfach elliptische Singularitäten. Univ. Math. **23**, 289–325 (1974)

[Sch1] Schlessinger, M.: Functors of Artin rings. Trans. Am. Math. Soc. **130**, 208–222 (1968)

[Sch2] Schlessinger, M.: Rigidity of quotient singularities. Invent. Math. **14**, 17–26 (1971)

[Scw] Schwede, K.: F-injective singularities are Du Bois. Am. J. Math. **131**, 445–473 (2009)

[Se1] Serre, J.-P.: Faiseaux algébriques cohérents. Ann. Math. **61**, 197–278 (1955)

[Se2] Serre, J.-P.: Sur la cohomologie des variétés algēbriques. J. de Math. Pures et Appl. **36**, 1–16 (1957)

[Se3] Serre, J.-P.: Géométrie Algébrique et Géométrie Analytique. Ann. Inst. Fourier **6**, 1–42 (1956)

[Sm] Smith, K.E.: F-rational rings have rational singularities. Am. J. Math. **119**, 159–180 (1997)

[Sn] Snapper, E.: Polynomials associated with divisors. J. Math. Mech. **9**, 123–129 (1960)

[Spa] Spanier, E.: Algebraic Toplogy. McGraw-Hill Book Co., New York (1966)

[Sp] Springer, T.A.: Invariant Theory. Lecture Notes in Mathematics, vol. 585. Springer, Berlin (1977)

[St] Stevens, J.: On canonical singulrities as total spaces of deformations. Abh. Math. Sem. Univ. Hambg. **58**, 275–283 (1988)

[ST] Schwede, K., Tucker, K.: A survey of test ideals, Progress in Commutative Algebra 2, Closures, Finiteness and Factorization, pp. 39–99. Walter de Gruyter GmbH & Co. KG, Berlin (2012)

[Tak] Takagi, S.: An interpretation of multiplier ideals via tight closure. J. Algebr. Geom. **13**, 393–415 (2004)

[Tak1] Takagi, S.: F-singularities of pairs and inversion of adjunction of arbitrary codimension. Invent. Math. **157**, 123–146 (2004)

[Tak2] Takagi, S.: A characteristic p analogue of plt singularities and adjoint ideals. Math. Z. **259**, 321–341 (2008)

[Tak3] Takagi, S.: Adjoint ideals and a correspondence between log canonicity and F-purity. Algebra Number Theory **7**(4), 917–942 (2013)

[TaW] Takagi, S., Watanabe, K-i.: F-singularities, to appear in Sugaku-expository. AMS

[Tj] Tjurina, G.N.: Locally semi-universal flat deformations of isolated singularities of complex spaces. Math. USSR. Izv. **3**, 976–1000 (1969)

[Ts1] Tsuchihashi, H.: Higher dimensional analogues of periodic continued fractions and cusp singularities. Tôhoku Math. J. **35**, 607–639 (1983)

[Ts2] Tsuchihashi, H.: Three-dimensional cusp singularities. Advanced Studies in Pure Mathematics, vol. 8, Complex Analytic, Singularities, pp. 649–679 (1986)

[Ts3] Tsuchihashi, H.: Simple K3 singularities which are hypersurface sections of toric singularities. Publ. RIMS, Kyoto University **27**, 783–799 (1991)

[ToW] Tomari, M.: On L^2-plurigenera of not-log-canonical Gorenstein isolated singularities. Am. Math. Soc. **109**, 931–935 (1990)

[V] Varchenco, A.N.: Zeta-function of monodromy and Newton's diagram. Invent. Math. **37**, 253–262

[W1] Watanabe, K.: On plurigenera of normal isolated singularities I. Math. Ann. **250**, 65–94 (1980)

[W2] Watanabe, K.: On plurigenera of normal isolated singularities II. Advanced Studies in Pure Mathematics, Complex Analytic Singularities **8**, 671–685 (1986)

[WH] Watanabe, K., Higuchi, T.: On certain class of purely elliptic singularities in dimension > 2. Sci. Rep. Yokohama Nat. Univ. Sect. **I**(30), 31–35 (1983)

[WKi] Watanabe, K-i.: Certain invariant subrings are Gorenstein, I, II. Osaka J. Math. **11**(1–8), 379–388 (1974)

[Wh] Whitney, H.: Differential manifolds. Ann. Math. **37**, 645–680 (1936)

[Y] Yonemura, T.: Hypersurface simple K3 singularities. Tohoku Math. J. Second Ser. **42**, 351–380 (1990)

[Z] Zariski, O.: The reduction of the singularities of an algebraic surface. Ann. Math. **40**, 639–689 (1939)

[ZS] Zariski, O., Samuel, P.: Commutative Algebra I, II. Van Nostrand Co., Inc., Princeton (1958, 1960)

Index

S. Ishii, *Introduction to Singularities*,
https://doi.org/10.1007/978-4-431-56837-7

Printed in the United States
By Bookmasters